Lignocellulosic Biomass Refining for Second Generation Biofuel Production

This book compiles research aspects of second-generation (2G) biofuel production derived specifically from lignocellulose biomass using biorefinery methods. It focuses on the valorization of different sources of 2G biofuels and their relative importance. The constituents of lignocelluloses and their potential characteristics different methods of treating lignocellulose, various means of lignocellulose bioconversion, and biofuel production strategies are discussed.

Features:

- Describes technological advancements for bioethanol production from lignocellulosic waste.
- Provides the roadmap for the production and utilization of 2G biofuels.
- Introduces the strategic role of metabolic engineering in the development of 2G biofuels.
- Discusses technological advancements, life cycle assessment, and prospects.
- Explores the novel potential lignocellulosic biomass for 2G biofuels.

This book is aimed at researchers and professionals in renewable energy, biofuel, bioethanol, lignocellulose conversion, fermentation, and chemical engineering.

Novel Biotechnological Applications for Waste to Value Conversion

Series Editors: Neha Srivastava, *IIT BHU Varanasi, Uttar Pradesh, India*
Manish Srivastava, *IIT BHU Varanasi, Uttar Pradesh, India*

Solid waste and its sustainable management are considered a major global issue due to industrialization and economic growth. Effective solid waste management (SWM) is a major challenge in areas with high population density, and despite significant development in social, economic and environmental areas, SWM systems are still increasing environmental pollution day by day. Thus, there is an urgent need to address this issue for a green and sustainable environment. Therefore, the proposed book series is a sustainable attempt to cover waste management and its conversion into value-added products.

Utilization of Waste Biomass in Energy, Environment and Catalysis
Dan Bahadur Pal and Pardeep Singh

Nanobiotechnology for Safe Bioactive Nanobiomaterials
Poushpi Dwivedi, Shahid S. Narvi, Ravi Prakash Tewari and Dhanesh Tiwary

Sustainable Microbial Technologies for Valorization of Agro-Industrial Wastes
Jitendra Kumar Saini, Surender Singh and Lata Nain

Enzymes in Valorization of Waste
Enzymatic Pre-treatment of Waste for Development of Enzyme-based Biorefinery (Vol I)
Pradeep Verma

Enzymes in Valorization of Waste
Enzymatic Hydrolysis of Waste for Development of Value-added Products (Vol II)
Pradeep Verma

Enzymes in Valorization of Waste
Next-Gen Technological Advances for Sustainable Development of Enzyme-based Biorefinery (Vol III)
Pradeep Verma

Biotechnological Approaches in Waste Management
Rangabhashiyam S, Ponnusami V and Pardeep Singh

Agricultural and Kitchen Waste
Energy and Environmental Aspects
Edited by Dan Bahadur Pal and Amit Kumar Tiwari

Lignocellulosic Biomass Refining for Second Generation Biofuel Production
Edited by Ponnusami V., Kiran Babu Uppuluri, Rangabhashiyam S and Pardeep Singh

For more information about this series, please visit: www.routledge.com/Novel-Biotechnological-Applications-for-Waste-to-Value-Conversion/book-series/NVAWVC

Lignocellulosic Biomass Refining for Second Generation Biofuel Production

Edited by

Ponnusami V., Kiran Babu Uppuluri,
Rangabhashiyam S and Pardeep Singh

CRC Press
Taylor & Francis Group
Boca Raton London New York

CRC Press is an imprint of the
Taylor & Francis Group, an **informa** business

First edition published 2023
by CRC Press
6000 Broken Sound Parkway NW, Suite 300, Boca Raton, FL 33487-2742

and by CRC Press
4 Park Square, Milton Park, Abingdon, Oxon, OX14 4RN

CRC Press is an imprint of Taylor & Francis Group, LLC

ISBN: 9781032067001 (hbk)
ISBN: 9781032067018 (pbk)
ISBN: 9781003203452 (ebk)

DOI: 10.1201/9781003203452

Typeset in Times
by codeMantra

Contents

Preface

For the past four decades, constant overexploitation of fossil fuel reserves has been observed to meet the growing energy demand. On the other hand, the rapid growth of industries in various sectors, population growth, and increased vehicle usage are predominant factors contributing to the uncertainty of the existing energy source of fossil fuels. In addition, the use of fossil fuels leads to a higher level of air-pollutant contamination, which causes more environmental damage through global warming and toxicity to living systems. Hence, the supply of fossil fuels has been declining very rapidly from its reserves and is one of the significant environmental threats that have made biofuels an option for an alternative energy source for the future. The development of second-generation (2G) biofuel from lignocellulosic biomass (LB) is the most promising, commercially viable process and offers many benefits from both an energy and environmental perspective.

Lignocellulosic biomass (LB) is the most abundant, economical, renewable, and sustainable feedstock to produce 2G biofuels. Nevertheless, the 'crystallinity, complexity, and insolubility' due to the cellulose, hemicellulose, and lignin restrict the commercial usage of LB. The LB is predominantly composed of cellulose 40%–50%, hemicellulose 25%–35%, and lignin 15%–20%. The LB composition depends on various factors, including biomass types, soil quality, climatic conditions, and locality. The LB is expected to provide approximately 45%–55% fermentable sugars. Therefore, the development of 2G biofuel from LB would be the most promising, commercially viable process, and offer many benefits from both an energy and environmental perspective. But the major bottleneck in refining LB for 2G biofuel production is the release of sugars from LB due to its structural complexity. In LB biorefinery, sugar polymers, cellulose, and hemicellulose are separated by pretreatment, hydrolyzed to sugars, fermented to fuels, and the residual biomass used for other applications.

The various biorefinery strategies, technologies, and policies have been invented, validated, and applied successfully at an industrial scale that utilizes various LB comprehensively and efficiently. At the same time, the LB biorefinery concept emphasizes the sustainable processing of biomass into value-added products and energy and offers an excellent bioeconomy. Realizing the importance of LB for biofuels from both an energy and environmental perspective, the present book emphasizes refining LB for 2G biofuel production. The book's three unique sections on the potential of LB for biofuel production, lignocellulose conversion strategies for biofuel production, and trends in 2G biofuel production form a complete framework pertinent to narrate the 2G biofuel production sourced from LBs.

The physical and physicochemical treatments for lignocellulose conversion have been reviewed in detail in Chapter 1. Chapter 2 gives an overview of the various biorefining approaches to valorize lignocellulosic resources. Various inhibitors generated during the pretreatment and hydrolysis affect the microbial growth and fermentation of biomass into biofuels. Various solutions available to tackle the same are presented in Chapter 3. The microbial strains for 2G biofuels must survive at elevated conditions, including higher concentrations of inhibitors and liquid biofuels, and grow on various complex carbon sources. Metabolic engineering is a reliable and efficient technology to construct the required microbes, a key for circular bioeconomy. Therefore, Chapter 4 emphasizes the role of metabolic engineering in converting LB to develop 2G biofuels. Various fermentation techniques to convert the LB hydrolysate into effective biofuels are discussed in Chapter 5. Furthermore, for the large-scale conversion of LB to biofuels, the choice and design of a suitable bioreactor are critical. Various factors that need to be considered in designing various bioreactors for the production of biofuels from LB are presented in Chapter 6.

The next part of the book highlights the advanced topics in the LB biorefinery. Chapter 7 gives an overview of various advanced pretreatments of LB in the context of more efficient, reliable, economic, and environmentally benign LB biorefineries. The green biorefinery is the most sought-after solution to overcome the problems associated with the separation of liquid biofuels in the LB biorefinery. In Chapter 8, the potential of various ionic liquids as alternate solvents for the separation of biobutanol in the LB biorefinery is described in detail. Chapters 9 and 10 provide an extensive overview of various approaches, merits, demerits, and challenges in intensifying the production and separation of biobutanol towards producing biofuels at a larger scale. Chapter 11 presents the production of aviation fuels from the LB biorefinery. The last chapter gives a comprehensive review of up-to-date research on the role of thermophilic microorganisms and thermostable enzymes in biofuel production through the conversion of LCB through fermentation.

Editors

Ponnusami V. is presently working as Associate Dean in the School of Chemical & Biotechnology, SASTRA Deemed University, India. He graduated from A. C. Tech., Anna University and did his doctoral research at SASTRA Deemed University, India. He has rich industrial and academic experience. His research interests include domestic and industrial wastewater treatment, bioconversion of lignocellulose, and bioprocessing. His group is working on photocatalysis, production of renewable fuels, and microbial polysaccharides. He had published over 50 research papers in international journals.

Kiran Babu Uppuluri was awarded his PhD from the Faculty of Chemical Engineering, Andhra University, Visakhapatnam, in 2011. He is currently working as an Associate Professor in the Department of Biotechnology, School of Chemical and Biotechnology, SASTRA Deemed University, Thanjavur. His group is working on the development, modeling, optimization, and intensification of bioprocesses for various biochemicals such as biofuels, therapeutic enzymes, and biopolymers.

Rangabhashiyam S is currently working as Assistant Professor in the School of Chemical & Biotechnology, SASTRA Deemed University, India. He received his Doctor of Philosophy degree from National Institute of Technology Calicut, India. He has received Post-Doctoral Fellowship from Max Planck Institute for Dynamics of Complex Technical Systems, Germany, received National Post-Doctoral Fellowship from SERB-DST, India, selected as Young Scientist by DST, India for the BRICS Conclave held in Durban, South Africa, and received Hiyoshi Young Leaf Award from Hiyoshi Ecological Services, Hiyoshi Corporation, Japan. His major research interests are bioremediation and wastewater treatment. He is Editorial Board Member in *Separation & Purification Reviews; Scientific Reports; Biomass Conversion and Biorefinery; Environmental Management*, Associate Editor in *International Journal of Environmental Science and Technology, IET Nanobiotechnology, Frontiers in Environmental Chemistry*, and Academic Editor of *Adsorption Science and Technology*. He has published more than 100 papers and also contributed to several book chapters. He is listed in the World's Top 2% Scientists published by Elsevier, Stanford.

Pardeep Singh is presently an Assistant Professor in the Department of Environmental Study, PGDAV College, University of Delhi, New Delhi, India. He obtained his doctorate degree from the Indian Institute of Technology (Banaras Hindu University) Varanasi. He was also selected as Young Scientist by DST, India for the BRICS Conclave held in Durban, South Africa. His research interests include waste management, wastewater treatment, water scarcity, and global climate change. He has published more than 75 research/review papers in international journals in the fields of waste and wastewater treatment/management. He has also edited more than 40 books with various international publishers.

Contributors

Adeolu Adesoji Adediran
Department of Mechanical Engineering
Landmark University
Omu-Aran, Kwara State, Nigeria

Arumugam A.
Bioprocess Intensification Laboratory, Centre for Bioenergy, School of Chemical & Biotechnology
SASTRA Deemed University
Thirumalaisamudram, Thanjavur, Tamil Nadu, India

Arumukam Ramasubramanian
Department of Energy and Environment
National Institute of Technology
Tiruchirappalli
Tamil Nadu, India

Kalirajan Arunachalam
Department of Science and Mathematics, School of Science, Engineering and Technology
Mulungushi University
Kabwe, Zambia

Muhammad Bilal
School of Life Science and Food Engineering
Huaiyin Institute of Technology
Huaian, China

Ayesha Butt
Institute of Biochemistry and Biotechnology
University of Veterinary and Animal Sciences
Lahore, Pakistan

Nhamo Chaukura
Department of Physical and Earth Sciences
Sol Plaatje University
Kimberley, South Africa

Ngonidzashe Chimwani
Institute for the Development of Energy for African Sustainability
University of South Africa
Science Campus-Rooderport, Johannesburg, South Africa

Ziaul Hasan
Centre for Interdisciplinary Research in Basic Sciences
Jamia Millia Islamia
Jamia Nagar, New Delhi, India

Asimul Islam
Centre for Interdisciplinary Research in Basic Sciences
Jamia Millia Islamia
Jamia Nagar, New Delhi, India

Diana Jose
Department of Biotechnology and Microbiology, School of Life Sciences
Kannur University
Thalassery Campus, Kannur, Kerala, India
and
King Mongkut's University of Technology North Bangkok (KMUTNB)
Bangkok, Thailand

Itha Sai Kireeti
Bioprocess Intensification Laboratory, Centre for Bioenergy, School of Chemical & Biotechnology
SASTRA Deemed University
Thirumalaisamudram, Thanjavur, Tamil Nadu, India

Nichapat Kittiborwornkul
Biorefinery and Process Automation Engineering Center (BPAEC), Department of Chemical and Process Engineering, The Sirindhorn International Thai-German Graduate School of Engineering (TGGS)
King Mongkut's University of Technology North Bangkok (KMUTNB)
Bangkok, Thailand

Muneera Lateef
Faculty of Agricultural Sciences and Technology, Department of Agricultural Genetic Engineering
Nigde Omer Halisdemir University
Nigde, Turkey

Dayavathi Madhavan
Biomass, Bioenergy and Bioproducts Laboratory, School of Chemical and Biotechnology
SASTRA Deemed University
Thirumalaisamudram, Thanjavur, Tamil Nadu, India

Bisma Meer
Department of Biotechnology
Quaid-i-Azam University
Islamabad, Pakistan

Kushif Meer
Institute of Chemistry
University of the Punjab
Lahore, Pakistan

Tahir Mehmood
Institute of Biochemistry and Biotechnology
University of Veterinary and Animal Sciences
Lahore, Pakistan

Kalyani A. Motghare
Department of Chemical Engineering
Visvesvaraya National Institute of Technology (VNIT)
Nagpur, Maharashtra, India

Fareeha Nadeem
Institute of Biochemistry and Biotechnology
University of Veterinary and Animal Sciences
Lahore, Pakistan

Vinod Kumar Nathan
School of Chemical & Biotechnology
SASTRA Deemed University
Thirumalaisamudram, Thanjavur, Tamil Nadu, India

Ponnusami V.
Biomass, Bioenergy and Bioproducts Laboratory, School of Chemical & Biotechnology
SASTRA Deemed University
Thirumalaisamudram, Thanjavur, Tamil Nadu, India
and
Bioprocess Intensification Laboratory, Centre for Bioenergy, School of Chemical & Biotechnology
SASTRA Deemed University
Thirumalaisamudram, Thanjavur, Tamil Nadu, India

Govindarajan Ramadoss
School of Chemical & Biotechnology
SASTRA Deemed University
Thirumalaisamudram, Thanjavur, Tamil Nadu, India

Charles Rashama
Institute for the Development of Energy for African Sustainability
University of South Africa
Science Campus-Rooderport, Johannesburg, South Africa

Kittipong Rattanaporn
Faculty of Agro-Industry, Department of Biotechnology
Kasetsart University
Bangkok, Thailand

Ruben Sudhakar D.
Department of Energy and Environment
National Institute of Technology
Tiruchirappalli, Tamil Nadu, India

Saravanan Ramiah Shanmugam
Biomass, Bioenergy and Bioproducts Laboratory, School of Chemical & Biotechnology
SASTRA Deemed University
Tirumalaisamudram, Thanjavur, Tamil Nadu, India

Diwakar Shende
Department of Chemical Engineering
Visvesvaraya National Institute of Technology (VNIT)
Nagpur, Maharashtra, India

Ramachandran Sivaramakrishnan
Faculty of Science, Laboratory of Cyanobacterial Biotechnology, Department of Biochemistry
Chulalongkorn University
Bangkok, Thailand

Subramaniyasharma Sivaraman
Biomass, Bioenergy and Bioproducts Laboratory, School of Chemical & Biotechnology
SASTRA Deemed University
Thirumalaisamudram, Thanjavur, Tamil Nadu, India

Aparna Ganapathy Vilasam Sreekala
School of Chemical and Biotechnology
SASTRA Deemed University
Thirumalaisamudram, Thanjavur, Tamil Nadu, India

Malinee Sriariyanun
Biorefinery and Process Automation Engineering Center (BPAEC), Department of Chemical and Process Engineering, The Sirindhorn International Thai-German Graduate School of Engineering (TGGS)
King Mongkut's University of Technology North Bangkok (KMUTNB)
Bangkok, Thailand

Bhuvaneshwari Veerapandian
Biomass, Bioenergy and Bioproducts Laboratory, School of Chemical and Biotechnology
SASTRA Deemed University
Thirumalaisamudram, Thanjavur, Tamil Nadu, India

Kailas L. Wasewar
Advance Separation and Analytical Laboratory (ASAL), Department of Chemical Engineering
Visvesvaraya National Institute of Technology (VNIT)
Nagpur, Maharashtra, India

1 Physical and Physicochemical Pretreatment Methods for Lignocellulosic Biomass Conversion

Ruben Sudhakar D. and
Arumukam Ramasubramanian
National Institute of Technology Tiruchirappalli

CONTENTS

DOI: 10.1201/9781003203452-1

1.1 INTRODUCTION

With the steep decline in availability of conventional fuels, raising fuel prices and climatic changes due to global warming are major concerns for human beings in recent times. Humans are forced to move toward energy technologies that are efficient and eco-friendly. This paves the way for the effective use of alternative energy sources. Hence, the world is shifting toward the use of alternative fuels and biomass. This chapter gives a broad overview of how lignocellulosic biomass is pretreated and the different types of pretreatment methods available. Pretreatment contributes to approximately 20% of the total production, requiring a good share of the total expenses of the production process of biofuels. The conventional methods of pretreatment are not cost-effective, as they require high energy input and involve tackling toxic inhibitors. In this chapter, different methods of pretreatment are discussed, ranging from conventional methods to recent advances in pretreatment of lignocellulosic biomass.

1.1.1 LIGNOCELLULOSE

Lignocellulose, the remains of plant waste, is widely used for producing biofuels. Lignocellulose is one of the easily and widely available resources, which forms a major source of energy if utilized properly. Hence, a lot of research interest on better conversion of lignocellulosic biomass to useful products is observed in the literature, recently. Lignocellulose is used in different industries such as biological industry, pharmaceutical industry, sugar industry, and chemical industry.

Table 1.1 gives the different components of various lignocellulosic biomass. Its main components are as follows:

1. Carbohydrate polymers (cellulose and hemicellulose)
2. Noncarbohydrate phenolic polymer (lignin)

Lignin keeps cellulose attached to the plant cell wall, thus strengthening it. Cellulose accounts for more than 30%–50% of the dry weight of lignocellulose [1]. It has a linear chain of beta (1→4) linked to d-glucose [2]. The second polysaccharide part of lignocellulose is hemicellulose, which occupies 15%–30% of the plant cell wall.

TABLE 1.1
Different Compositions of Cellulose, Hemicellulose, and Lignin in Different Biomass

Lignocellulosic Material	Cellulose (%)	Hemicellulose (%)	Lignin (%)
Softwood	45–50	25–35	25–35
Hardwood	40–55	24–40	18–25
Shell of nuts	25–30	25–30	30–40
Corn	45	35	15
Grass	25–40	35–50	10–30
Paper	85–99	0	0–15
Straw after wheat removal	30	50	15
Refuse	60	20	20
Plant leaves	15–20	80–85	0
Cotton seed hairs	80–95	5–20	0
Paper (newspaper)	40–55	25–40	18–30
Waste papers got from chemical pulps	60–70	10–20	5–10
Primary solid water waste	8–15		
Solid cattle manure	1.6–4.7	1.4–3.3	2.7–5.7
Coastal Bermuda grass	25	35.7	6.4
Switchgrass	45	31.4	12
Swine waste	6.0	28	na

Source: Reproduced from Ref. [4].

Lignin is found in all vascular plants. It is the second most available carbon source, followed by cellulose. Figure 1.1 shows the chemical composition of various lignocellulosic biomasses.

1.1.2 Need for Pretreatment of Lignocellulosic Biomass

Pretreatment is one of the major steps in lignocellulose transformation processes and is important to expose the molecules of cellulose to make them easier to transform into fermentable sugars. Figures 1.2 and 1.3 depict the structure of plant cells and fibers, giving an idea of how the cellulose molecules are internally intertwined in plants. When a pretreatment is done, the cellulose is exposed by breaking the complex structure of lignocellulose, and the available cellulose is processed further according to the purpose of application. Recalcitrance of lignocellulose is broken by combined effect of physical and chemical changes to the carbohydrates and lignin. Figure 1.4 depicts diagrammatically the need for pretreatment of lignocellulosic biomass.

There are different types of pretreatments of lignocellulose, namely, chemical, biological, physical, and physiochemical processes, and their combinations, to speed up hydrolysis by improving the accessibility of cellulose.

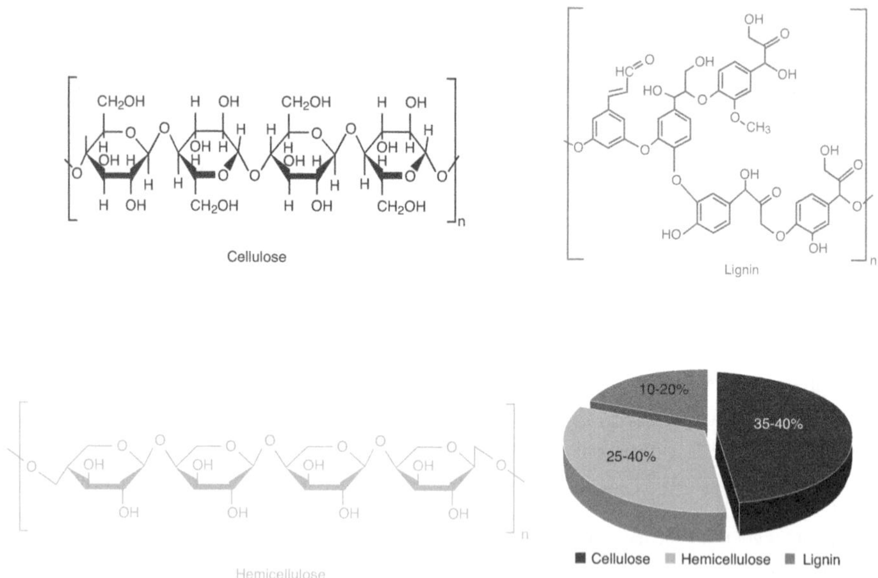

FIGURE 1.1 Lignocellulose chemical composition [3].

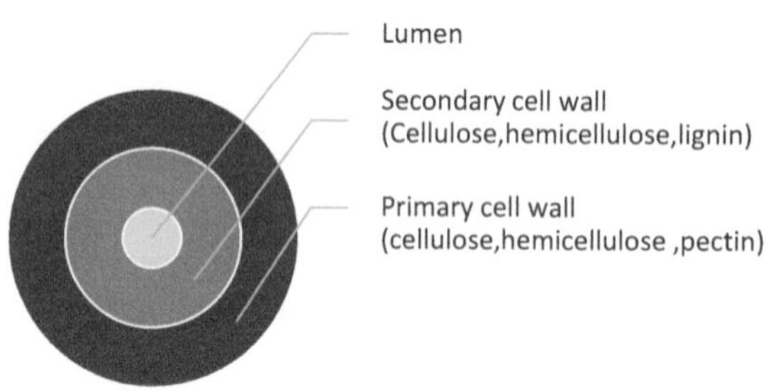

FIGURE 1.2 Structure of a plant cell.

1.2 TYPES OF PRETREATMENTS FOR LIGNOCELLULOSIC BIOMASS

The pretreatment methods for better utilization of lignocellulosic biomass can be broadly classified into physical, chemical, physiochemical, and biological methods. Each of these methods includes varieties of techniques for pretreatment, which have been shown in Figures 1.5–1.7.

1.3 PRETREATMENT METHODS

An efficient pretreatment process must have the following traits: (1) reduction of cost and (2) minimum energy utilization, unaltering fractions of hemicellulose, especially pentose, without the degradation of glucose, minimal inhibitor formation and ease of

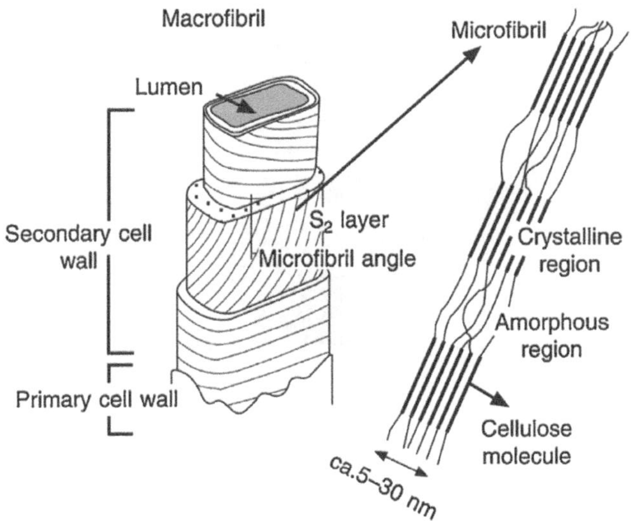

FIGURE 1.3 Structure of a plant fiber (cross-sectional view) [5].

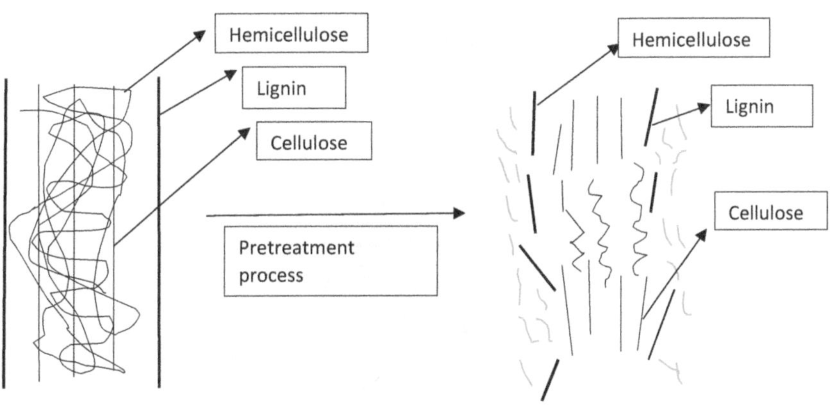

FIGURE 1.4 Diagrammatic representation showing the role of pretreatment in exposing the cellulose for enzymic/chemical attack in further process.

the fermentation process, and recovery of lignin without damage for the production of valuable coproducts [6].

1.3.1 Physical Methods

Pretreatment of lignocellulose includes a reduction in particle size. The physical methods are an attempt to minimize the size of the particle to improve heat and mass transfer in the further pretreatment process. Apart from pretreatment of lignocellulose, microfibrillated cellulose has a wide range of applications, starting from the paper industry to biological applications. The different techniques included in the

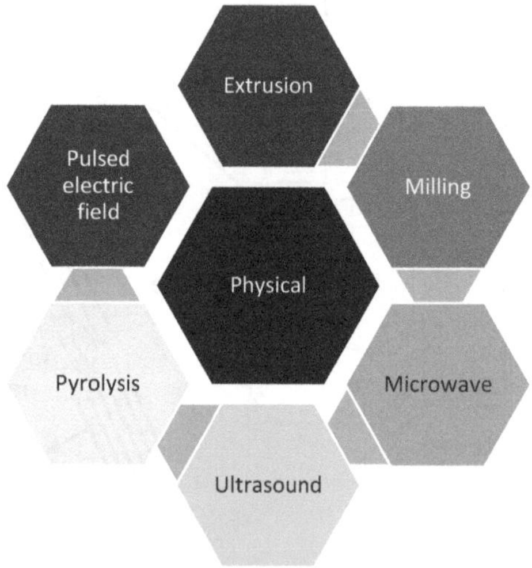

FIGURE 1.5 Various physical treatment techniques.

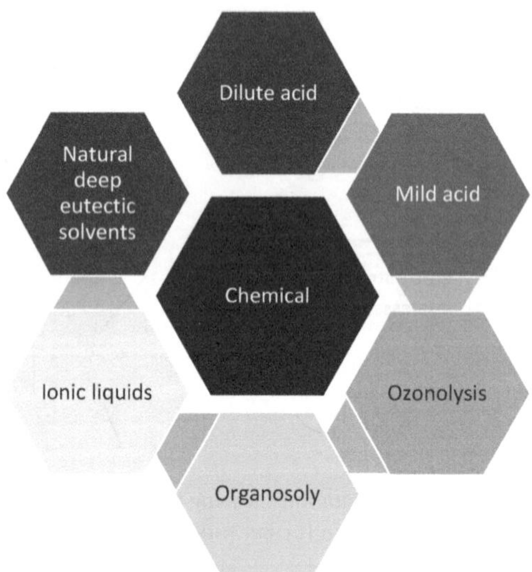

FIGURE 1.6 Various chemical treatment techniques.

physical treatment are given in detail below. Figure 1.8 shows the typical requirement of sizes to be achieved for different further processing of lignocellulosic fibers into fuels, chemicals, and value-added products.

1.3.1.1 Extrusion

Extrusion is defined as the process of obtaining an object with a desired cross section by forcing it through a die. In the lignocellulose process, extrusion is done to reduce

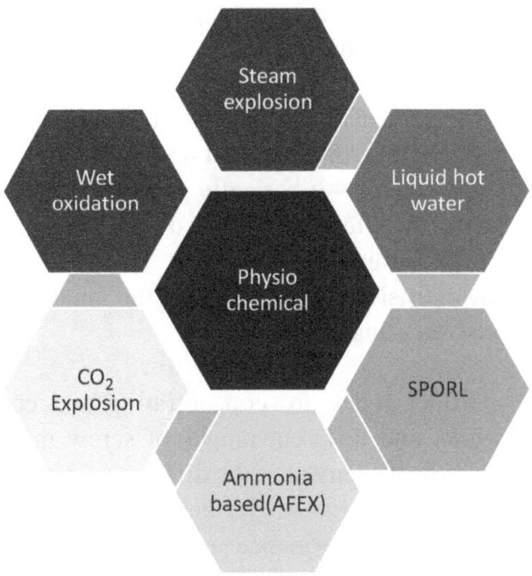

FIGURE 1.7 Various physicochemical treatment techniques.

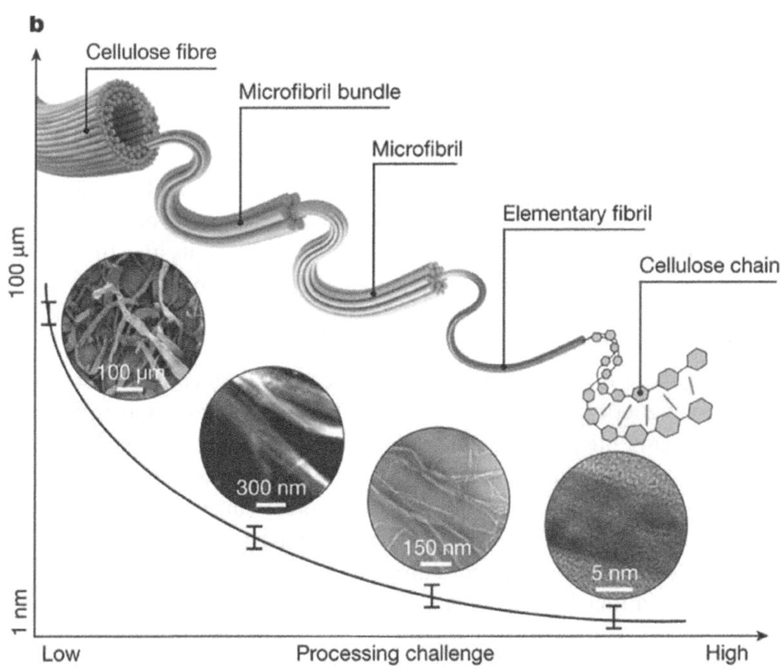

FIGURE 1.8 Reduction in size of fibers (microfibrillated cellulose). (The microscopy images were taken from Refs. [7–10].)

the fiber size. A screw extruder is a screw element, and the biomass is squeezed through the screws, including the following:

1. The forward screw elements – It primarily transports large material with varying width and length with the lowest degree of mixing and shearing.
2. The mixing screw elements – It basically performs through mixing of biomasses, and the shearing effect is based on different tool angles in combination with a low front conveying effect.
3. The reverse screw extrusion is made to push the material backward, which creates mixing and shearing effects [11].

The most common extruder types are copenetrating and corotating twin screw extruders (Dziezak, 1989), and different ranges of screw materials are used. The screw dimensions explain the placement of the different screw elements and the characteristics of the screw, such as stagger angle, length, and pitch, in different positions. The main factors influencing performance are product transformation, residence time distribution, and mechanical energy input.

1.3.1.2 Milling

Milling utilizes different mechanical stresses such as effect, pressure, contact, and shear to make chemical and mechanical impacts on the lignocellulosic biomass. It is powerful in reducing size, level of polymerization, and cellulose crystallinity, and upsets the inflexible construction of lignocellulosic biomass. These impacts ultimately lead to a further improved rate and yield in subsequent processes. The most widely recognized kinds of milling processing strategies utilized for lignocellulosic biomass pretreatment are two-roll processing, bar processing, ball processing, wet circle processing, hammer processing, diffusive processing, and colloid processing [12,13]. The choice of one of these processing strategies depends on a few constraints, such as the physical and synthetic properties of the biomass, its moisture content, the final molecule size and molecule dispersion, and the last expected application [14].

1.3.1.3 Microwave

Microwave radiation is another way of pretreating lignocellulose. A microwave is an electromagnetic wave with a wavelength between radio and infrared waves. Pretreatment using microwaves was first introduced by. Conventional acid or alkali pretreatment of lignocellulose in an elevated temperature and pressure environment is an energy-intensive process. Microwave treatment can further develop cellulose availability and compound reactivity. This innovation has a simple process, is energy-productive, and can increase temperature in a short time under ordinary temperature and pressure. Additionally, it has been effectively applied in chemical reactions, and a few investigations on the utilization of microwaves in fiber pretreatment have been completed [15–21]. There are different methods for microwave treatments mentioned below which were discussed in detail by Binod et al. [16] (Figure 1.9).

1. Microwave pretreatment
2. Microwave–alkali pretreatment

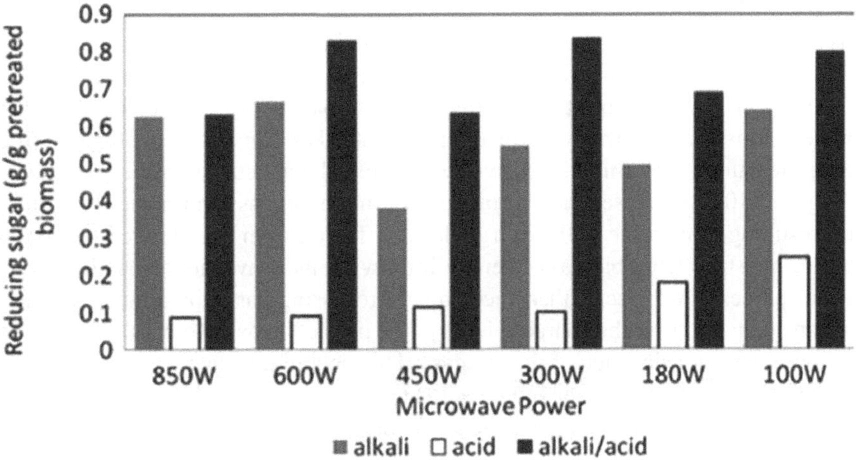

FIGURE 1.9 Maximum reducing sugar yields various microwave power using different techniques [16].

3. Microwave–acid pretreatment
4. Microwave combined with alkali followed by acid pretreatment

1.3.1.4 Ultrasound

Ultrasound-assisted approaches reduce the time required for pretreatment and also decrease the enzyme required in the subsequent process [22]. The ultrasound technique is a green, recent technique that is currently in the research stage. Ultrasound generates pressure differences to improve physical and chemical processes. The process takes place at a frequency beyond the human hearing range. Ultrasound is generally created by piezoelectric and magnetostrictive transducers.

The piezoelectric material will convert alternating current pressure into ultrasound in a specific frequency range. Ultrasound technology is one of the most eco-friendly techniques for use in the chemical processing step of pretreatment because it reduces reaction times and chemical loading. The ultrasound technique enhances the separation, improving the subsequent processing of lignocellulose. Ultrasound enhances both physical and chemical processes. Ultrasound improves physical mechanisms by shearing and eroding the surface, and chemical mechanisms are improved by producing oxidizing radicals. The above-mentioned physical and chemical mechanisms facilitate the splitting of the lignin macromolecule to enhance the splitting of lignin and hemicellulose and degrade the lignin compound by phenolic ring and hydroxyl reactions. Ultrasound also helps in the splitting and depolymerization of polysaccharides. Low frequency is chosen for its lower contribution to physical effects, and vice versa. Ultrasonic treatment can be used along with chemical treatment to enhance the process.

1.3.1.5 Torrefaction

It is the process by which the usability of biomass is enhanced. Torrefaction subjects the biomass to a low-temperature treatment of about 150°C–300°C, in the absence of oxygen or in a semi-inert environment with a partial presence of oxygen environment. It is kind of pyrolysis treatment, but it takes place at a low temperature, hence the name "low-temperature pyrolysis [23]." The torrefaction process is shown in Figure 1.10. It increases the carbon content in the biomass and reduces moisture, thus resulting in increase in the energy density. Torrefaction is mainly carried out to remove moisture from biomass, thereby improving its conversion capability in the further process. At present, thermochemical processing has gained much attention with an aim to improve the characteristics of the lignocellulosic biomass. Torrefaction improves the calorific value of the biomass. The yield of torrefaction depends on the type of reactor used to heat the biomass, the temperature to which it is subjected, and the environment in which the biomass torrefaction is done.

1.3.1.6 Pulsed Electric Field

A pulsed electric field creates pores in the cell membrane, and the cellulose inside gets exposed for chemical/enzymic attack. A very high voltage of about 5–20 kV/cm is applied suddenly from nanoseconds to milliseconds and continued in a pulsed format. Lignocellulosic biomass is placed in this pulsed electric field environment. $E = V/d$, where 'V' is the voltage and 'd' is the distance between the plates and electrodes. The equation gives the relation between the electric field strength and voltage. The electric field strength is directly proportional to the voltage and inversely proportional to the distance between the plate and the electrode. When the electric field is applied, there is mass permeability and tissue rupture. The pulsed electric field treatment setup consists of a control system, pulse generator, material handling equipment, and data acquisition system [24]. The benefits of the pulsed electric treatment include its simple design and short duration, which saves energy and time [25]. Function generator pulses are used to apply very high voltage. Formation of pores in this process depends mainly on two factors: (1) pulse duration and (2) electric field strength.

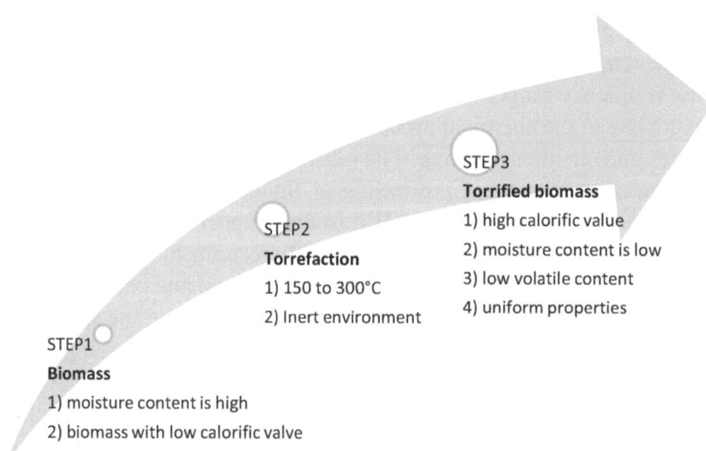

FIGURE 1.10 Torrefaction process.

1.3.1.7 Chipping

Lignocellulose often requires reduction in size through a mechanical process. Large woody biomass needs significant mechanical size reduction or comminution to access the cellulose. Woody and bulky biomass must undergo size reduction when compared to agricultural biomasses. Chipping is the size reduction process of lignocellulosic pretreatment.

1.3.1.8 Briquetting

Briquetting is the process of converting lignocellulose into small briquettes. This process involves several steps: the first is to collect the biomass feedstock, then physically process it, and finally make it denser by adding binders, resulting in densified briquettes. The briquettes are compressed lignocellulosic material, which is further subjected to conversion by either a chemical, physiochemical, or physical combustion process. The details of each step involved in the briquetting process are given below.

Step 1 (Cleaning of lignocellulosic feedstock):

The biomass is initially subjected to cleaning and sorting to eliminate unwanted materials; this is done by a sieve and the magnetic conveyors, which eliminate the plastics, dust, soil, and so on, accumulated during the storage and transportation. The other means of cleaning is by using water to remove agricultural pesticides, chemicals, and fertilizers. The cleaning of biomass improves the further processes in which it is used.

Step 2 (Drying):

The cleaned biomass is subjected to drying to remove the moisture. Drying the lignocellulosic biomass improves its efficiency, but care must be taken not to overdry the biomass. There are several methods such as solar drying and heating of biomass using waste heat available from other processes.

Step 3 (Size reduction):

This step is important in the briquetting process since it partially breaks down the lignin surface. There are different types of size reduction processes as follows:

1. Chopping
2. Shredding
3. Milling
4. Crushing
5. Chipping
6. Grinding

Generally, size reduction is broadly classified into two major steps as follows:

1. Breaking biomass into larger-sized material makes it easier to handle on the conveyors.
2. The second step involves further breaking biomass into smaller pieces to make them ready for thermal and thermochemical processes. Hammer and cutting mills are the most preferred forms to further reduce the size of the particle.

Step 4 (Binder addition):

Binder is added to the biomass to keep the particles of the biomass in their fixed shape and to improve the integrity and quality of the final briquettes formed.

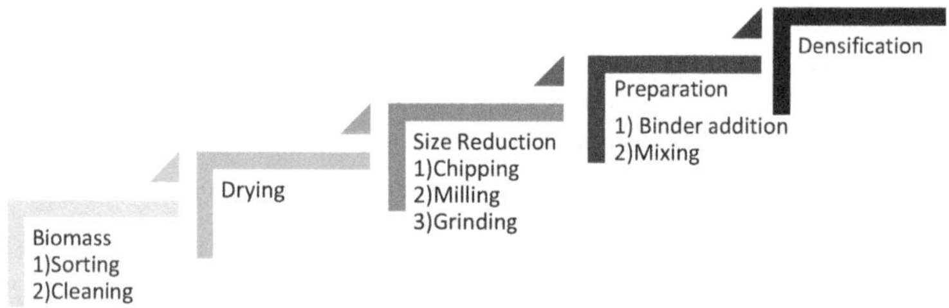

FIGURE 1.11 Step-by-step process in briquetting.

Figure 1.11 shows the different steps involved in the briquetting of lignocellulosic biomass. Binders are added before the densification process. There are different types of binders used: (1) organic binders, (2) inorganic binders, and (3) compound binders.

Some of the most commonly used inorganic binders are (1) clay, (2) lime, (3) cement, (4) plaster, and (5) sodium silicate. Some of the most commonly used organic binders are (1) wastepaper, (2) pulp, (3) starch, (4) molasses, (5) tar, (6) pitch, (7) petroleum bitumen binder, (8) lignosulphonate binder, and (9) polymer binder [26–28]. A combination of both the binders is the compound binders.

Binders influence the thermal stability, strength, combustion efficiency, and finally the cost of briquette. Inorganic binders provide good compaction and strength in comparison with organic binders but also have the disadvantage of high ash formation and reduced calorific value [26,29–31].

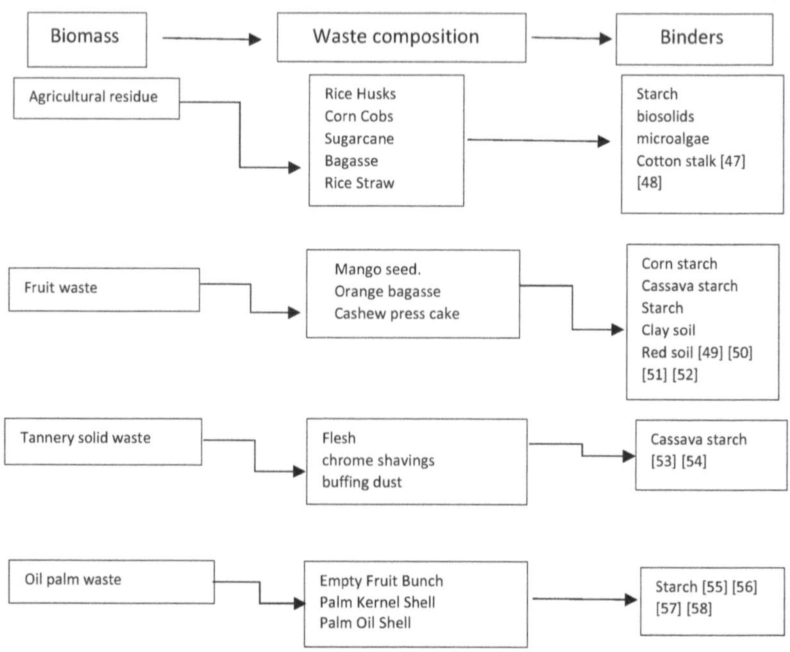

Step 5 (Biomass densification):

Densification is the process of increasing the density of the biomass by shaping it into briquettes. This process is carried out under pressure. Figures 1.12 and 1.13 show the densification of lignocellulosic biomass using a hydraulic press and a roller press, respectively. There are many advantages of densification as follows:

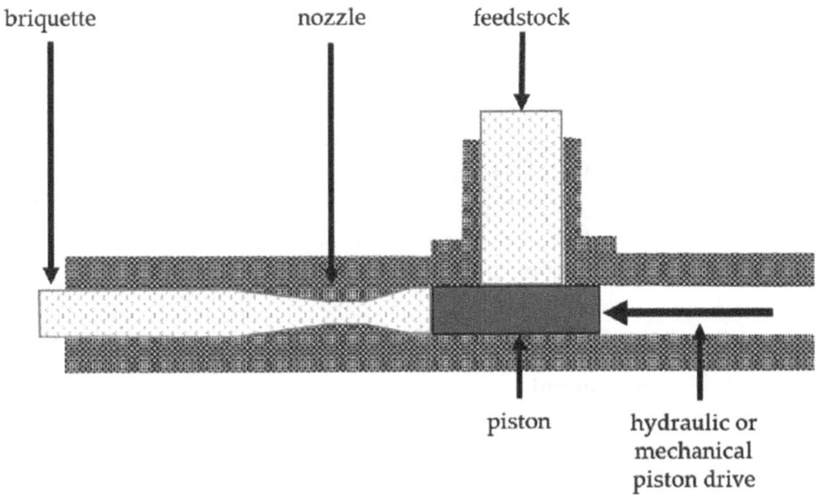

FIGURE 1.12 Hydraulic press for briquette manufacturing. (Adapted from Ref. [32].)

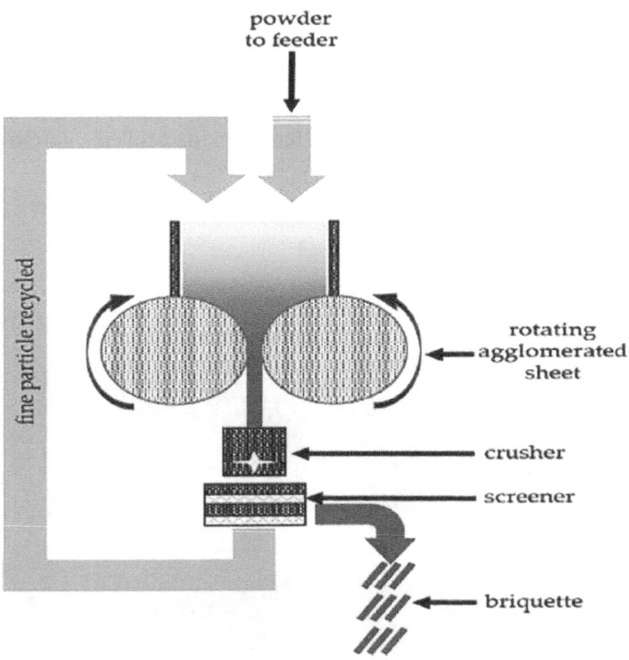

FIGURE 1.13 Roller press. (Adapted from Ref. [33].)

1. They improve the handling of biomass.
2. Uniform density and particle distribution.

Different Biomass Densification Techniques

Briquetting is an agglomeration process that converts biomass into high-density particles of fixed shape [34]. Densification is achieved by applying force and bringing the particles together.

1.3.1.9 Pelletization

Pellets and briquettes have the main difference in their diameter. Pellets are smaller than the briquettes. The process of pelletization is similar to that of briquettes. First, the biomass is subjected to cleaning, followed by a reduction in size, and then the densification process, by which the pellets are made along with binder [35–45]. The choice between pellets and briquettes is made mainly based on application. Figure 1.14 shows the size comparison of the pellets and briquettes.

1.3.2 The Chemical Pretreatment Process

1.3.2.1 Acid Pretreatment

In the acid pretreatment process, acids are used to treat the lignocellulosic biomass. The acid pretreatment is a less attractive option due to the inhibitory products generated during the process. There are two types of acid pretreatment based on duration: one is short duration (1–7 min), and the other is long duration (30–90 min). A separate hydrolysis process is not required since the hydrolysis is already carried out by acid. For acid pretreatment, resistive reactors are to be used because of the corrosive, toxic, and hazardous environment. Therefore, acid pretreatment is very expensive. Concentrated acid is retrieved at the end of the pretreatment, so that it can be reused to save cost and make the process more cost-effective [46].

Acid pretreatments are further classified based on the acid concentration as follows:

1. Dilute acid
2. Concentrated acid

FIGURE 1.14 Size comparison between pellets and briquettes.

The most commonly used acids are dilute and concentrated sulfuric and hydrochloric acids.

Dilute acid pretreatments are divided into three types, namely:

1. High-temperature acid treatment (>160°C, solid loadings [5%–10%]) [47]
2. Continuous flow process at low temperature for low solid loadings (<160°C, solid loadings [10%–40%]) [48]
3. Low temperature and subjected to a batch process and with heavy loading of lignocellulose.

The two most commonly used acids are sulfuric acid and hydrochloric acid. There are other acids, such as maleic acid and oxalic acid, used in acid pretreatment.

1.3.2.2 Ozonolysis

In the ozonolysis technique, ozone acts as the oxidizing agent to break down lignin. Water dissolves ozone, and the resulting mixture is used as an oxidizing agent to break down lignin. This process does not release any toxic or hazardous waste or any inhibitors, and also releases less molecular weight and soluble compounds. Figure 1.15 shows a typical setup used for ozonolysis of lignocellulosic biomass. Ozone is used to degrade lignin and hemicellulose. The ozonolysis experimental setup has an iodine trap used for testing the efficiency of spectrophotometer, automatic gas flow control valve, pressure regulation valve, ozone generator, catalyst, ozone UV oxygen cylinder, process gas humidifier, ozone catalytic destroyer, vent, and three-way valve [24,49–51].

1.3.2.3 Organosoly

Organosoly uses the organic solvents such as acetone, methanol, ethylene glycol, and ethanol as solvents to pretreat the lignocellulosic biomass. This process is carried out at a predefined temperature and pressure. A salt catalyst, acid, and base are used in

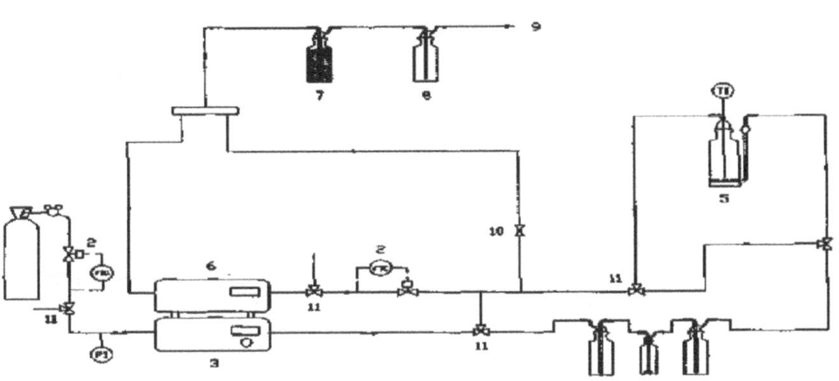

FIGURE 1.15 Experimental ozonolysis setup: (1) oxygen cylinder, (2) automatic gas flow control valve, (3) ozone generator, (4) process gas humidifier, (5) reactor, (6) ozone UV spectrophotometer, (7) ozone catalytic destroyer, (8) iodide trap to test catalyst efficiency, (9) vent, (10) pressure regulation valve, and (11) three-way valve. (Reproduced from Ref. [50].)

the organosoly technique. The temperature in the organosoly reaches about 200°C. The temperature used depends on the type of biomass used and also on the catalyst used in the process. The useful by-product lignin is removed in this process. The variables affecting the process output are as follows [52]:

1. Temperature
2. Concentration of solvent
3. Reaction time

1.3.2.4 Ionic Liquids

Ionic liquid has cations and anions. These ionic liquids are used as solvents in the pretreatment of biomass because of their high temperature resistance, polarity, low melting point, and low vapur pressure. The major drawback of ionic liquid is that it reacts with cellulose, rendering it inactive at low viscosity and temperature. So when using ionic liquids, the viscosity plays a major role. If viscosity is high, energy consumption is also high. Hence, there is a trade-off between energy consumption and process efficiency. High temperatures have disadvantages such as triggering side reactions and decreasing ionic stability [53,54].

1.3.2.5 Natural Deep Eutectic Solvents

Deep eutectic solvents have nonsymmetric ions. It also has very low lattice energy and a low melting point (Smith et al. 2014).

The charge delocalization occurs between the hydrogen bond donor and the hydrogen bond acceptor. The freezing point of a deep eutectic solvent is very low when compared with the individual components by which it is made of.

For example:

ChCl: Urea → 12°C
ChCl → 302°C
Urea → 133°C

Types of deep eutectic solvents are as follows:

1. Metal salt + organic salt
2. Metal salt hydrate + organic salt
3. Hydrogen bond donor + organic salt
4. Zinc/Aluminum chloride + hydrogen bond donor

Deep Eutectic Solvents (DESs) are capable of donating and accepting protons, and this characteristic enables the formation of hydrogen bonds with other compounds, which enhances their solvation properties (Pandey et al. 2017).

1.4 PHYSIOCHEMICAL METHODS

1.4.1 Steam Explosion

In this approach, high-pressure saturated steam is used to treat lignocellulosic biomass, and then the pressure is quickly lowered, resulting in explosive decompression of the biomass. The instigation temperature for the steam explosion is 160°C–260°C with a pressure of 0.69–4.83 MPa for a few seconds to minutes. The lignocellulosic biomass is exposed and kept at atmospheric pressure for a certain period of time. This causes hemicellulose hydrolysis, which is followed by explosive decompression, which ends the whole process. The acids are produced during the steam explosion pretreatment and play a role in the hydrolysis of hemicellulose. The lignocellulosic biomass material is broken due to turbulent material flow and rapid flashing of material to atmospheric pressure. The use of sulfuric acid or carbon dioxide in steam explosion pretreatment reduces time and temperature, inhibits product generation, and increases hydrolysis efficiency, resulting in total hemicellulose removal. Pretreating softwoods with a steam explosion is an ineffective method; nevertheless, adding an acid catalyst during the process is required to make the substrate available to hydrolytic enzymes. One of the advantages of steam is that it quickly achieves the targeted temperature during the process without any excessive dilution. At the end of the process, a sudden release of pressure quenches the entire process and decreases the temperature. The steam explosion is mainly affected by the four parameters: moisture content, residence time, chip size, and temperature. The optimal hemicellulose hydrolysis and solubilization can be done in one of two ways: high temperature and short residence time, or low temperature and long residence time. Steam explosion pretreatment has a low energy need. However, mechanical pretreatment requires 70% more energy than steam explosion pretreatment to achieve the same reduced particle size. Steam explosion pretreatment is a very effective pretreatment technique for hardwood and agricultural wastes.

1.4.2 Ammonia Fiber Expansion

The principle of ammonia fiber expansion (AFEX) is very similar to steam explosion. This method is a low-temperature technique; concentrated ammonia (0.3–2 kg ammonia/kg of dry weight) is used as a catalyst. In a high-pressure reactor, ammonia is added to biomass; after 5–45 min of cooking, pressure is released rapidly. This operation is usually carried out at a temperature of roughly 90°C. This ammonia catalyst can be retrieved and used again because of its volatile nature. The fermentation rate is improved by using the AFEX technique on several types of grass and herbaceous crops. The AFEX pretreatment method is mainly used to treat alfalfa, wheat chaff, and wheat straw. The AFEX pretreatment technique cannot remove the hemicellulose and lignin content. As the lignin content of biomass increases, the effectiveness of the AFEX technique will decrease. Hence, AFEX pretreatment is not recommended for the treatment of biomass with high lignin content.

Another ammonia-based approach is ammonia recycle percolation (ARP). In this procedure, aqueous ammonia (10–15 wt%) is employed. Aqueous ammonia travels through biomass in this pretreatment at a fluid velocity of 1 cm/min at a temperature of 150°C–170°C and a residence duration of 14 min, and ammonia is collected afterward. Ammonia reacts with lignin under these conditions, causing lignin links to break down. The AFEX process uses liquid ammonia, whereas the ARP method/technique uses aqueous ammonia.

1.4.3 CO_2 Explosion

In this method, the supercritical carbon dioxide behaves like a solvent. The carbon dioxide is dissolved in water and forms carbonic acid, which is less corrosive due to its unique properties. Due to their small size, the carbon dioxide molecules enter tiny pores in lignocellulosic biomass throughout the process. The carbon dioxide pretreatment was carried out at a low temperature, which helped to prevent acid degradation of sugar. When carbon dioxide pressure is released, the cellulosic structure is disturbed, making the substrate more accessible to cellulolytic enzymes for hydrolysis.

1.4.4 SPORL Method

SPORL means sulfite pretreatment to overcome the recalcitrance of lignocellulose.

The following are the two steps of SPORL:

1. Step 1: Lignin and hemicellulose are removed by treatment with calcium or magnesium sulfite.
2. Step 2: Biomass is reduced in size.

Yin et al. [55] and Kumar et al. [24] studied the SPORL method under certain conditions. The result was that more than 90% of the substrate was converted to cellulose.

1.4.5 WET Oxidation

Wet oxidation method uses oxygen and water or hydrogen peroxide at a high temperature above 120°C for about 30 min at 0.5–2 MPa pressure [56,57]. The factors affecting the wet oxidation process are as follows:

1. Time taken for reaction
2. Pressure of oxygen supplied
3. Temperature at which the process is carried out

At high temperature, water behaves like acid because the hydrogen ion concentration increases when the temperature increases, which in turn decreases the pH value and causes hydrolysis to take place. Hemicellulose breaks down, pentose is formed, and the lignin oxidizes, but the cellulose remains unaffected. The lesser the temperature, the lesser the inhibitory compounds formed as the result of reaction. The number of inhibitors formed is lesser.

The two main reasons that make this process a major disadvantage in terms of industrial applications are as follows:

1. Oxygen aids combustion, and extreme carefulness in handling is required.
2. High cost of hydrogen peroxide.

1.4.6 Advantages and Disadvantages of Physicochemical Pretreatment Methods

Table 1.2 lists the advantages and disadvantages of each physicochemical pretreatment method used for lignocellulosic biomass.

1.4.7 Recent Advancements in Physicochemical Pretreatment of Lignocellulosic Biomass

An effective fractionation method is required to maximize the efficiency of lignocellulosic biomass conversion. This is achieved by developing an eco-friendly, economical, and sustainable pretreatment method. Pretreatment is essential for several reasons, including altering the lignocellulosic structure to acquire individual constituents, producing highly reactive lignocellulosic material for future use, and preserving the original forms of lignocellulosic constituents. Several problems should be considered while developing lignocellulosic biomass pretreatment, including cost cutting, environmental concerns, and long-term viability. The biomass pretreatment efficiency is mainly affected by the physical and chemical properties of cellulose crystallinity, lignin content, and the specific surface area of the biomass. Using single pretreatment methods such as physical, chemical, physicochemical, and biological does not provide the required biomass degradability. Hence, it is necessary to develop hybrid pretreatment methods that combine any two processes and can improve the efficiency of lignocellulosic pretreatment.

TABLE 1.2
Advantages and Disadvantages of Physicochemical Pretreatment Methods

Pretreatment Method	Advantages	Disadvantages
Steam	High yield of glucose and hemicellulose in a two-step process. Lignin transformation and hemicellulose solubilization. Cost-effective.	Toxic compound generation. Acid catalyst needed to make the process efficient with high lignin content material. Partial hemicellulose degradation.
Ammonia fiber expansion	High effectiveness for herbaceous material and low lignin content biomass. Cellulose becomes more accessible. Causes inactivity between lignin and enzymes. Low formation of inhibitors.	Recycling of ammonia is needed. Less effective process with increasing lignin content. Alters lignin structure.
Ammonia recycle percolation	Removes most of the lignin. High cellulose content after pretreatment. Herbaceous materials are most affected.	High energy costs and liquid loading.
Liquid Hot Water (LHW)	Separation of nearly pure hemicellulose from rest of feedstock. No need for catalyst. Hydrolysis of hemicellulose.	High energy/water input. Solid mass left over will need to be dealt with (cellulose/lignin).
Super critical fluid	Low degradation of sugars. Cost-effective. Increases cellulose accessible area.	High pressure requirements. Lignin and hemicelluloses unaffected.

1.5 CONCLUSION

This chapter has given an overview of the various physical, chemical, and physico-chemical pretreatment methods that can be used to pretreat lignocellulosic biomass for the production of biofuels.

REFERENCES

1. Foyle, T.; Jennings, L.; Mulcahy, P. Compositional analysis of lignocellulosic materials: Evaluation of methods used for sugar analysis of waste paper and straw. *Bioresour. Technol.* 2007, 98, 3026–3036. Doi: 10.1016/j.biortech.2006.10.013.
2. Updegraff, D.M. Semimicro determination of cellulose in biological materials. *Anal. Biochem.* 1969, 32(3), 420–424, ISSN 0003-2697.
3. Amin, F.R.; Khalid, H.; Zhang, H. et al. Pre-treatment methods of lignocellulosic biomass for anaerobic digestion. *AMB Expr.* 2017, 7, 72.
4. Jorgensen, H.; Kristensen, J.B.; Felby, C. Enzymatic conversion of lignocellulose into fermentable sugars: Challenges and opportunities. *Biofuels, Bioprod. Bioref.* 2007, 1, 119–134.
5. Baillie, C. Chapter 4: Natural Fibre Sources, in: *Green Composites – Polymer Composites and the Environment*; 49–80, Woodhead Publishing, New Delhi, 2004.
6. National Research Council. Committee on Biobased Industrial Products. *Biobased Industrial Products: Priorities for Research and Commercialization*; National Academy Press: Washington, DC, 2000.
7. Moon, R.J.; Martini, A.; Nairn, J.; Simonsen, J.; Youngblood, J. Cellulose nanomaterials review: Structure, properties and nanocomposites. *Chem. Soc. Rev.* 2011, 40, 3941–3994.
8. Zhu, H. et al. Wood-derived materials for green electronics, biological devices, and energy applications. *Chem. Rev.* 2016, 116, 9305–9374.
9. Wang, Q.Q. et al. Morphological development of cellulose fibrils of a bleached eucalyptus pulp by mechanical fibrillation. *Cellulose* 2012, 19, 1631–1643.
10. Zhu, H. et al. Anomalous scaling law of strength and toughness of cellulose nanopaper. *Proc. Natl. Acad. Sci. USA* 2015, 112, 8971–8976.
11. Rigal, L. Twin-screw extrusion technology and the fractionation of vegetable matter; *Proceedings of the CLEXTRAL Conference*; Firminy, France. 8–10 October 1996
12. Baruah, J; Nath, B.K.; Sharma, R. et al. Recent trends in the pretreatment of lignocellulosic biomass for value-added products. *Front. Energy Res.* 2018, 6, 1–19.
13. Karimi, K.; Taherzadeh, M.J. A critical review of analytical methods in pretreatment of lignocelluloses: Composition, imaging, and crystallinity. *Bioresour. Technol.* 2016, 200, 1008–1018.
14. Barakat, A.; Mayer-laigle, C.; Solhy, A. et al. Mechanical pretreatments of lignocellulosic biomass: Towards facile and environmentally sound technologies for biofuels production. *RSC Adv.* 2014, 4, 48109–48127.
15. Lo, K.V.; Srinivasan, A.; P.H. Liao, Bailey, S. Microwave oxidation treatment of sewage sludge *J. Environ. Sci. Health A* 2015, 50, 882–889.
16. Binod, P.; Satyanagalakshmi, K.; Sindhu, R.; Janu, K.U.; Sukumaran, R.K.; Pandey, A. Short duration microwave assisted pretreatment enhances the enzymatic saccharification and fermentable sugar yield from sugarcane bagasse Renew. *Energy* 2012, 37, 109–116.
17. Mafuleka, S., Kana, E.G. Modelling and optimization of xylose and glucose production from napier grass using hybrid pre-treatment techniques. *Biomass Bioenergy* 2015, 77, 200–208.

18. Issa, A.A.; Al-Degs, Y.S.; Mashal, K.; Al Bakain, R.Z. Fast activation of natural biomasses by microwave heating *J. Ind. Eng. Chem.* 2015, 21, 230–238.
19. Omar, R.; Idris, A.; Yunus, R.; Khalid, K.; Isma, M.A. Characterization of empty fruit bunch for microwave-assisted pyrolysis *Fuel* 2011, 90, 1536–1544.
20. Wang, W.; Dalal, R.C.; Moody, P.W. Evaluation of the microwave irradiation method for measuring soil microbial biomass *Soil Sci. Soc. Am. J.* 2001, 65, 1696–1703.
21. Anis, S.; Zainal, Z. Study on kinetic model of microwave thermocatalytic treatment of biomass tar model compound. *Bioresour. Technol.*, 2014, 151, 183–190.
22. Zou, S.; Wang, X.; Chen, Y.; Wan, H.; Feng, Y. Enhancement of biogas production in anaerobic co-digestion by ultrasonic pretreatment. *Energy Convers. Manag.* 2016, 112, 226–235.
23. Kilzer, F.J.; Broido, A. Speculations on the nature of cellulose pyrolysis. *Pyrodynamics* 1965, 2, 151–163.
24. Kumar, P.; Barrett, D.M.; Delwiche, M.J.; Stroeve, P. Methods for pretreatment of lignocellulosic biomass for efficient hydrolysis and biofuel production. *Ind. Eng. Chem. Res.* 2009, 48, 3713–3729.
25. Kumar, P.; Barrett, D.M.; Delwiche, M.J.; Stroeve, P. Pulsed electric field pretreatment of switchgrass and wood chip species for biofuel production. *Ind. Eng. Chem. Res.* 2011, 50(19), 10996–11001.
26. Zhang, X.; Xu, D.; Xu, Z.; Cheng, Q. The effect of different treatment conditions on biomass binder preparation for lignite briquette. *Fuel Process. Technol.* 2001, 73, 185–196.
27. Lumadue, M.R.; Cannon, F.S.; Brown, N.R. Lignin as both fuel and fusing binder in briquetted anthracite fines for foundry coke substitute. *Fuel* 2012, 97, 869–875.
28. Massaro, M.M.; Son, S.F.; Groven, L.J. Mechanical, pyrolysis, and combustion characterization of briquetted coal fines with municipal solid waste plastic (MSW) binders. *Fuel* 2014, 115, 62–69.
29. Ugwu, K.; Agbo, K. Evaluation of binders in the production of briquettes from empty fruit bunches of Elais Guinensis. *Int. J. Renew. Sustain. Energy* 2013, 2, 176–179.
30. Onchieku, J.M.; Chikamai, B.N.; Rao, M.S. Optimum parameters for the formulation of charcoal briquettes using bagasse and clay as binder. *Eur. J. Sustain. Dev.* 2012, 1, 477–492.
31. Hu, Q.; Shao, J.; Yang, H.; Yao, D.; Wang, X.; Chen, H. Effects of binders on the properties of bio-char pellets. *Appl. Energy* 2015, 157, 508–516.
32. Christoforou, E.; Fokaides, P.A. *Advances in Solid Biofuels. Green Energy and Technology*; Springer: Cham, 2019; pp. 1–130.
33. Tumuluru, S.J.; Wright, C.T.; Hess, J.R.; Kenney, K.L. A review of biomass densifi cation systems to develop uniform feedstock commodities for bioenergy application. *Biofuels Bioprod. Bioref.* 2011, 5, 683–707.
34. Kaliyan, N.; Morey, R.V. Natural binders and solid bridge type binding mechanisms in briquettes and pellets made from corn stover and switchgrass. *Bioresour. Technol.* 2010, 101, 1082–1090.
35. Muazu, R.I.; Stegemann, J.A. Biosolids and microalgae as alternative binders for biomass fuel briquetting. *Fuel* 2017, 194, 339–347.
36. Gill, N.; Dogra, R.; Dogra, B. Influence of moisture content, particle size, and binder ratio on quality and economics of rice straw briquettes. *Bioenergy Res.* 2018, 11, 54–68.
37. Katimbo, A.; Kiggundu, N.; Kizito, S.; Kivumbi, H.B.; Tumutegyereize, P. Potential of densification of mango waste and effect of binders on produced briquettes. *Agric. Eng. Int Cigr J.* 2014, 16, 146–155.
38. Zanella, K.; Gonçalves, J.L.; Taranto, O.P. Charcoal briquette production using orange bagasse and corn starch. *Chem Eng. Trans.* 2016, 49, 313–318.

39. Brunerová, A.; Roubík, H.; Brožek, M.; Herák, D.; Šleger, V.; Mazancová, J. Potential of tropical fruit waste biomass for production of bio-briquette fuel: Using Indonesia as an example. *Energies* 2017, 10, 2119.
40. Sawadogo, M.; Kpai, N.; Tankoano, I.; Tanoh, S.T.; Sidib, S. Cleaner production in Burkina Faso: Case study of fuel briquettes made from cashew industry waste. *J. Clean Prod.* 2018, 195, 1047–1056.
41. Onukak, I.; Mohammed-Dabo, I.; Ameh, A.; Okoduwa, S.; Fasanya, O. Production and characterization of biomass briquettes from tannery solid waste. *Recycling* 2017, 2, 17.
42. Oyelaran, O.A.; Sani, F.M.; Sanusi, O.M.; Balogun, O.; Fagbemigun, A.O. Energy potentials of briquette produced from tannery solid waste. *Makara J. Technol.* 2017, 21, 122–128.
43. Sing, C.Y.; Aris, M.S. An experimental investigation on the handling and storage properties of biomass fuel briquettes made from oil palm mill residues. *J. Appl. Sci.* 2012, 12, 2621–2625.
44. Bazargan, A.; Rough, S.L.; McKay, G. Compaction of palm kernel shell biochars for application as solid fuel. *Biomass Bioenergy* 2014, 70, 489–497.
45. Hamid, M.F.; Idroas, M.Y.; Ishak, M.Z.; Zainal Alauddin, Z.A.; Miskam, M.A.; Abdullah, M.K. An experimental study of briquetting process of torrefied rubber seed kernel and palm oil shell. *Biomed. Res. Int.* 2016, 2016, 1–11.
46. Behera, S.; Arora, R.; Nandhagopal, N.; Kumar, S. Importance of chemical pretreatment for bioconversion of lignocellulosic biomass. *Renew. Sust. Energ. Rev.* 2014, 36, 91–106.
47. Esteghlalian, A.; Hashimoto, A.G.; Fenske, J.J.; Penner, M.H. Modeling and optimization of the dilute-sulfuric-acid pretreatment of corn stover, poplar and switchgrass. *Bioresour. Technol.* 1997, 59(2–3), 129–136.
48. Brennan, W.; Hoagland, W.; Schell, D.J.; Scott, C.D. High temperature acid hydrolysis of biomass using an engineering-scale plug flow reactor: Results of low solids testing. *Biotechnol. Bioeng. Symp.* 1986, 17, 53–70.
49. Quesada, J.; Rubio, M.; Gómez, D. Ozonation of lignin rich solid fractions from corn stalks. *J. Wood Chem. Technol.* 1999, 19(1), 115–137.
50. Vidal, P.F., Molinier, J. Ozonolysis of lignin—Improvement of in vitro digestibility of poplar sawdust. *Biomass* 1988, 16(1), 1–17.
51. Ben-Ghedalia, D, Miron, J. The effect of combined chemical and enzyme treatments on the saccharification and in vitro digestion rate of wheat straw. *Biotechnol. Bioeng.* 1981, 23(4), 823–831.
52. Aftab, N.; Iqbal, I.; Riaz, F.; Karadag, A.; Tabatabaei, M. Different pretreatment methods of lignocellulosic biomass for use in biofuel production. 2019. Doi: 10.5772/intechopen.84995.
53. Zavrel, M.; Bross, D.; Funke, M.; Büchs, J.; Spiess, A.C. High-throughput screening for ionic liquids dissolving (ligno-) cellulose. *Bioresour. Technol.* 2009, 100(9), 2580–2587.
54. Mäki-Arvela, P.; Anugwom, I.; Virtanen, P.; Sjöholm, R.; Mikkola, J.P. Dissolution of lignocellulosic materials and its constituents using ionic liquids—A review. *Ind. Crops Prod.* 2010, 32, 175–201.
55. Yin, C.; Kær, S.K.; Rosendahl, L.; Hvid, S.L. Co-firing straw with coal in a swirl-stabilized dual-feed burner: Modelling and experimental validation. *Bioresour. Technol.* 2010, 101(11), 4169–4178.
56. Haghighi Mood, S.; Hossein Golfeshan, A.; Tabatabaei, M.; Salehi Jouzani, G.; Najafi, G.H.; Gholami, M., et al. Lignocellulosic biomass to bioethanol, a comprehensive review with a focus on pretreatment. *Renew. Sust. Energ. Rev.* 2013, 27, 77–93.
57. Varga, E.; Schmidt, A.S.; Réczey, K.; Thomsen, A.B. Pretreatment of corn stover using wet oxidation to enhance enzymatic digestibility. *Appl. Biochem. Biotechnol.: Part A Enzyme Eng Biotechnol.* 2003, 104(1), 37–50.

2 Biorefining Processes for Valorization of Lignocellulosic Biomass for Sustainable Production of Value-Added Products

Diana Jose
Kannur University
King Mongkut's University of Technology
North Bangkok (KMUTNB)

Kittipong Rattanaporn
Kasetsart University

Nichapat Kittiborwornkul
King Mongkut's University of Technology
North Bangkok (KMUTNB)

Adeolu Adesoji Adediran
Landmark University

Malinee Sriariyanun
King Mongkut's University of Technology
North Bangkok (KMUTNB)

CONTENTS

DOI: 10.1201/9781003203452-2

2.1 INTRODUCTION

The growth of population around the world results in a scarcity of fossil-derived energy and an increased amount of emissions of carbon-based compounds, leading to an alteration in climate and ozone depletion (Nyika et al. 2020). Bioeconomy plays a significant role in fulfilling the energy needs of society and reducing pollution that risks the ecosystem. The activities in the bioeconomy are based on the use of renewable resources to produce value-added products (Erik et al. 2019). Until 2018, renewable energy sources, such as solar, wind, hydro, biomass, geothermal, and so on, had a share of 13.8%, with an impressive annual growth rate of 2.4% since 2000 (GBS 2020). To meet the sustainable development goals (SDG), it is important for renewable energy technologies to grow at a much faster rate, and at the same time, efforts have to be taken to ensure an exit strategy for fossil fuels. These led researchers to develop effective energy derived from renewable resources that will sustain society for the long term. The utilization of life cycle assessment (LCA) tools permits one to more readily design these frameworks. Biorefinery comes under this platform, which provides sustainable energy as well as a green economy and allows the concept of a circular economy (Maria & Gerfried 2019).

The term 'Biorefinery' is defined by the International Energy Agency (IEA) as the sustainable processing of biomass into a high value-added product related to food, feed, chemicals, materials, biofuels, power, and heat (bioenergy) (Antonella et al. 2016). Previously, biorefineries were categorized based on different criteria, for example, use of technology, raw materials, intermediate products, and conversion processes (Kamm & Kamm 2004). Later, in 2008, IEA Bioenergy Task 42 classified types of biorefineries based on four important factors, including chemical platforms, products to differentiate, feedstocks, and conversion processes, to mention different biorefinery systems. The platforms, which are the primary factor, act as links to various biorefinery processes that determine the complexity of the system. It can be an end product, for instance, C5/C6 sugars, syngas, or biogas. Next, the main feature is the product, which includes energy products such as bioethanol, biodiesel, and synthetic biofuels, as well as products related to chemicals, materials, food, and feed. The energy obtained from crops and biomass residues is the difference between the two categories of feedstock. Conversion processes consist of biochemical (e.g., fermentation and enzymatic conversion), thermochemical (e.g., gasification and pyrolysis), chemical (e.g., acid hydrolysis, synthesis, and esterification), and mechanical processes (e.g., fractionation, pressing, and size reduction) (De Jong & Jungmeier 2015).

In order to reduce the adverse effects of global warming, countries such as Canada, Finland, France, and the USA started depending on renewable sources of energy produced from biomass. Biofuel from biomass has the ability to generate high energy output and can also supplant fossil-based energy (Keller et al. 2018). According to the statistics report of U.S. Biomass Energy, almost 5.13 quadrillion British thermal units (Btu) of energy were produced from biomass and consumed in the United States in 2018. This usage will be expected to increase by 5.54 quadrillion Btu in 2050. In 2018, biofuel production peaked at 95.4 million tons of oil equivalent globally (Madhumitha 2021). According to Global Bioenergy Statistics (GBS 2020), bioethanol and biodiesel hold the largest biofuel produced globally with a share of 62% and 26%, respectively. Other biofuels, such as hydrogenated vegetable oil, renewable diesel, and cellulosic ethanol, had a share of 12%. From the year 2000 to 2018, the biofuel sector experienced an annual growth rate of 9%.

Among various types of biomass, lignocellulose is considered to be the most promising resource for biorefinery because it has immense characteristics that include being renewable, sustainable, eco-friendly, noncompetitive with food, having a low cost of feedstock, and having a high availability of raw material (Cherubini 2010, Goldy et al. 2018, Javier et al. 2019). Furthermore, it can serve as a building block chemical in the production of a large numbers of value-added bioproducts through biorefinery processes (Isikgor & Becer 2015). Potential sources of lignocellulosic biomass (LB) include agricultural residues, energy crops, forest residues, and human waste (Merklein et al. 2016, Anuj et al. 2018). Among these sources, LB from agricultural wastes provides the major supply of bioenergy (10%) globally. The rise in the production of crops enables the increase in bioenergy production worldwide (Rahul et al. 2020). During the last decade, Asia was the main producer of agricultural residues, and around 0.2 billion tons of this biomass was produced by India in 2015 (FAOSTAT 2019).

The conversion of LB to various value-added biochemical products (e.g., bioethanol, biobutanol, biohydrogen, biogas, organic acids, alcohol, polysaccharides, and single-cell proteins) is sustainable and helps to relieve environmental pollution.

However, the major obstacle to the commercialization of biorefineries is their economic viability (Kondusamy et al. 2020). Biorefining process is composed of multiprocessing steps that increase the complexity of the operation depending on the characteristics of the raw material used. Complexity indicates the technological and economic risk of the product (De Jong & Jungmeier 2015). Based on the availability and location of feedstock, many countries use different feedstock in biorefineries. For instance, Brazil utilizes corn residue for the production of ethanol, whereas the United States explores LB from miscanthus, switchgrass, and so on for the biorefinery process. Mixed feedstocks were used in India and China (Somerville et al. 2010). Other technical factors that face challenges in the commercialization of biorefineries are the logistics, procuring, and transportation of biomass, equipment, and operational and processing costs (pretreatment, saccharification, fermentation, and valorization) (Thorsell et al. 2004, Anuj et al. 2018). Not only technical problems but also marketing issues also need to be assessed for the successful commercialization of value-added products developed through the biorefinery process. Bioproducts are expected to grow at an 8.1% compound annual growth rate (CAGR) from $586.8 billion to $867.7 billion between 2020 and 2025 (BBC Research 2021). The expected economic value of lignocellulose will reach its peak of up to 300 billion USD with a CAGR of 4.2% from 2019 to 2026 (Faruq et al. 2020). Despite the fact that lignocellulose-based biorefinery faces environmental and technoeconomical challenges, value-added products obtained through a combinatorial model can lower the cost of production and make it economically feasible.

This chapter aims to provide an overview of biorefining processes for the valorization of LB for the sustainable production of value-added products. Detailed research and progress of bioconversion approaches in the biorefinery process, including pretreatment, hydrolysis, and fermentation for the valorization of bioproducts from LB, are described. Moreover, the economic status and cost reduction strategies using consolidated and combinatorial methods have been briefly discussed. In addition, value-added products, techno-economic challenges, and issues present in the commercialization of biorefinery products were assessed at the end of this chapter.

2.2 LIGNOCELLULOSIC BIOMASS

LB is one of the major resources used in biorefinery due to its renewability, sustainability, availability, productivity, and economic feasibility. Naturally, 170 billion metric tons of LB were produced through photosynthesis (Somerville et al. 2010). Using LB obtained from agricultural residues could also reduce the improper management of wastes, especially on-field combustion, which subsequently leads to the release of PM10 and PM2.5. However, the recalcitrant character of LB requires different pretreatment steps to promote bioconversion of LB to value-added products. Therefore, it is necessary to understand the chemical composition and physical properties of LB before designing the pretreatment process. The main sources of LB are agro-forestry residues and industrial–municipal solid wastes, where the composition of LB varies in each plant species based on origin, genetic variants of the same species, and environmental factors (Antonella et al. 2016). The sources of LB are listed in Table 2.1 along with the composition (cellulose, hemicellulose, and lignin) present in each feedstock.

Lignocellulose is produced abundantly in the secondary cell walls of perennial, herbaceous, and woody crops, and it is composed of polysaccharides and aromatic polymers. Cellulose (40%–60% of the biomass weight) and hemicelluloses (20%–35% of the biomass weight) are the polysaccharides, and lignin (15%–40% of the dried matter) (Cheng et al. 2020) is the aromatic polymer. The other remaining fraction contains pectin, proteins, oils, ash, extractives, and inorganic compounds (Goldy et al. 2018). These compounds are highly interconnected to form a complex structure, even though their compositions and properties differ based on the types of LB. Cellulose is the major biopolymer that presents 40%–60% of the dry weight of the plant cell wall and provides mechanical strength. It is structurally linear, unbranched, and parallel polymer chains of 500-400 d-glucose units bonded by β-(1,4)-glycosidic linkages (Figure 2.1). The polymer chains are strongly held together by hydrogen bonds and van der Waals interactions to form microfibrils. These microfibrils are packed together to form cellulose fibrils that are embedded in a lignocellulosic matrix (hemicelluloses, lignin, and pectin). Therefore, cellulose appears as crystalline in its natural form. The crystalline structure of cellulose is resistant to degradation by the cellulase enzyme. The numbers of glucose units in the polymer chains denote the degree of polymerization that plays a significant role in the recalcitrant property of LB. It is proven that hydrolysis occurs easily when the numbers of hydrogen bonds between the polymer chains decrease, whereas it is hard to hydrolyze long polymer chains with strong hydrogen bonds and high numbers of glucose units (Amit et al. 2016, Robak & Balcerek 2018).

Hemicelluloses are heteropolysaccharides that represent 20%–50% of biomass weight. It contains β-(1,4)-linked glycans (Figure 2.1), which show a complex combination of pentose and hexose sugars, such as d-xylose, d-mannose, d-galactose, d-glucose, l-arabinose, 4-O-methyl-d-glucuronic acid, d-galacturonic acid, and d-glucuronic acid, or, rarely, l-rhamnose and l-fucose. Among these sugars, xylan is the major hemicellulose and the second most common biopolymer in biomass after cellulose. The structural unit of xylan is the xylose linked by β-1,4-glycosidic linkage alongside l-arabinose residues. Cellulose and hemicellulose bind tightly with noncovalent bonds to the surface of each cellulose microfibril and they bind with lignin with covalent linkages. Comparing to cellulose, hemicellulose contains a lower degree of polymerization with only 100–200 units. The physical state of hemicellulose is amorphous in nature; therefore, it is easily hydrolyzed using dilute acids, bases, and enzymes (Isikgor & Becer 2015). Hence, during pretreatment, the removal of hemicellulose will increase the enzyme's access to cellulose (Santos et al. 2018). The role of hemicelluloses on LB recalcitrance is not clear as some lignin is often removed with hemicelluloses during pretreatment. Hemicellulose removal was demonstrated to be more efficient than lignin removal from biomass (Leu & Zhu 2013, Lv et al. 2013), while on the contrary, lignin removal was much more important to increase the hydrolysis rate (Kruyeniski et al. 2019). It is found that acetylation of hemicelluloses increases the hydrophobicity of biomass and prevents the binding between cellulose and active site of an enzyme, and it may limit cellulose availability by interacting with enzyme recognition (Pan et al. 2006, Zhao et al. 2012).

TABLE 2.1
Composition of LB in Different Residues

Lignocellulosic Feedstock	Source of Residue	Composition (%)			References
		Cellulose	Hemicellulose	Lignin	
Arundo donax	Fibers	33.7	30.1	10.1	Danping et al. (2021)
Aspen hardwood	Wood	40–55	24–40	18–25	Haiyan et al. (2015)
Bamboo	Trunk	37	16.6	39.2	Jing et al. (2020)
Banana waste	Straw	53	29	15	Silveira et al. (2008)
Barley straw	Straw	37.6	34.9	15.8	Belhadj et al. (2020)
Bast fiber jute	Fibers	45–53	18–21	21–26	Jahirul et al. (2012)
Bast fiber kenaf	Fibers	61.2	18.5	12.9	Samaneh et al. (2013)
Black gram residue	Residue	26.8	32.48	23.14	Kumar et al. (2016)
Cassava pulp	Pulp	22	16	16	Patthra et al. (2020)
Coastal Bermuda grass	Fibers	25	35.7	9–18	Shruti & Kalburgi (2016)
Coconut coir	Fibers	32.69	22.56	42.10	Fredina et al. (2021)
Coffee ground waste	Ground	12.4	39.1	23.9	Lenka et al. (2017)
Corn cob	Cob	45	35	15	Amit et al. (2016)
Corn leaf	Leaf	32.1	18.1	11.9	Mohammad et al. (2021)
Corn stalks	Stalks	43	24	17	Amit et al. (2016)
Corn stover	Stover	40	22	18	Amit et al. (2016)
Cotton gin waste	Gin	78	16	0	Amit et al. (2016)
Cotton liner	Liner	83.9	3.6	5.9	Elieber et al. (2016)
Durian peel	Peel	47.2	9.63	9.89	Saowalak et al. (2020)
Elephant grass	Fiber	22	24	24	Amit et al. (2016)
Esparto grass	Fiber	33–38	27–32	17–19	Amit et al. (2016)
Flax straw	Straw	29	27	22	Amit et al. (2016)
Garlic skin	Skin	41–50	16–26	26–39	Amit et al. (2016)
Grasses (average)	Fiber	25–40	25–50	10–30	Amit et al. (2016)
Hardwood stem	Stem	40–55	24–40	18–25	Amit et al. (2016)
Hemp	Fiber	44.5	32.78	21.03	Amit et al. (2016)
Leaves	Fiber	15–20	80–85	0	Amit et al. (2016)
Millet husk	Husk	33	27	14	Amit et al. (2016)
Miscanthus	Fiber	40	18	25	Eric et al. (2011)
Napier grass	Fiber	38.75	19.76	26.99	Mohammed et al. (2015)
Newspaper	Fiber	40–55	25–40	18–30	Amit et al. (2016)
Nut shells	Shells	25–30	25–30	30–40	Amit et al. (2016)
Oat straw	Straw	31–37	27–38	16–17	Amit et al. (2016)
Orchard grass (medium maturity)	Fiber	32	40	4.7	Amit et al. (2016)
Pinewood	Wood	39	24	20	Amit et al. (2016)
Poplar wood	Wood	35	17	26	Amit et al. (2016)
Rice husk	Husk	31	24	14	Amit et al. (2016)

(Continued)

TABLE 2.1 (*Continued*)
Composition of LB in Different Residues

Lignocellulosic Feedstock	Source of Residue	Composition (%)			References
		Cellulose	Hemicellulose	Lignin	
Rice straw	Straw	32.1	24	18	Amit et al. (2016)
Rye straw	Straw	33–35	27–30	16–19	Amit et al. (2016)
Sabai grass	Fiber	49.9	23.72	20.88	Amit et al. (2016)
Softwood stem	Stem	45–50	25–35	25–35	Amit et al. (2016)
Sponge gourd fibers	Fibers	63.4	15.3	14.7	Viviane et al. (2017)
Sugarcane bagasse	Bagasse	52	20	24	Radhakumari et al. (2016)
Sugarcane tops	Fibers	39.04	30.53	8.97	Gil-Lopez et al. (2019)
Sugarcane straw	Straw	36	21	16	Saad et al. (2008)
Sunflower husk	Husk	48.4	34.6	17	Perea-Moreno et al. (2018)
Sunhemp residue	Fibers	43.4–48	11.9–13	17.4–18.4	Amit et al. (2016)
Sweet sorghum bagasse	Bagasse	34–45	25–27	18–21	Radhakumari et al. (2016), Khalil et al. (2015)
Switchgrass	Fibers	30–50	10–40	5–20	Shruti & Kalburgi (2016)
Timothy grass	Fibers	34.2	30.1	18.1	Sonil et al. (2014)
Water hyacinth	Whole plants	21	34	7	Deshpande et al. (2008)
Waste papers from chemical pulps	Industrial waste papers	60–70	10–20	5–10	Jahirul et al. (2012)
Wheat bran	Bran	30	50	15	Graminha et al. (2008)
Wheat straw	Straw	30	50	15	Shruti & Kalburgi (2016)

Lignin is a complex amorphous aromatic heteropolymer with a three-dimensional network structure that accounts for 15%–40% of the dry weight and tightly holds cellulose and hemicellulose(Figure 2.1). The main composition is the polymerized phenylpropanoid building units (p-coumaryl, coniferyl, and sinapyl alcohol) linked by ether bonds. The composition of lignin content varies in different plant species as well as within various parts of plants (xylem vessels, tracheary elements, and xylem fibers). Lignin serves as structural support in biomass and is mostly accumulated in secondary cell walls. In addition, it makes the cell wall impermeable and resistant to microbial and oxidative attack. Due to these properties, lignin plays a negative role in biomass conversion (Santos et al. 2018). Lignin contributes strongly to LB recalcitrance, which affects the production of fermentable sugars by limiting the

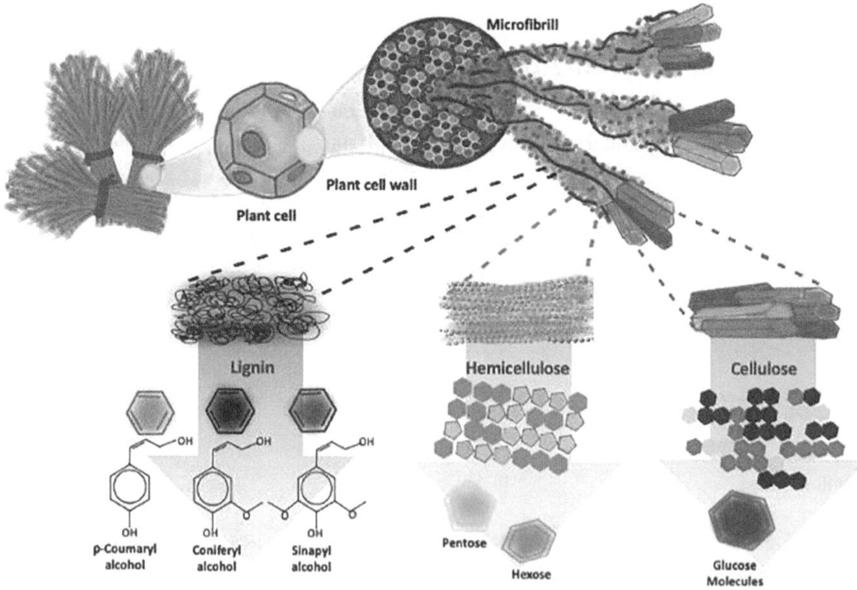

FIGURE 2.1 Structure of LB and its compositions (cellulose, hemicellulose, and lignin).

accessibility of enzymes during hydrolysis and the economic value of plant biomass for biorefinery processes.

LB also contains small amounts of pectin, proteins, extractives, and inorganic compounds. Extractives are those with a low molecular weight and nonstructural components that are soluble in neutral organic solvents or water. It consists of biopolymers such as terpenoids, steroids, resins acids, lipids, waxes, and fats, as well as phenolic constituents in the form of stilbenes, flavonoids, tannins, and lignans. The inorganic matter in LB is regarded as ash content, which consists of major elements (Si, Na, K, Mg, and Ca) and minor elements (Al, Fe, Mn, P, and S) (Nanda et al. 2014, Brandt et al. 2013).

2.3 BIOREFINERY PROCESS

Biorefinery is a system or a group of facilities or processing plants that functions to meet sustainability criteria in terms of environmental, social, and economic aspects. The process consists of upstream, midstream, and downstream steps that produce multiple byproducts or end products that are cost-efficient and marketable. In this scenario, different types of biomass can be used based on availability and productivity (De Jong & Jungmeier 2015, Show & Sriariyanun 2021). To analyze the harmful effect and potential use of the product from the biorefinery process, there is a strategy called LCA. The LCA study of biorefinery systems can evaluate how the product impacts environmental pollution. Based on the study of Dufosse et al. (2017), biorefining of LB from agricultural crops has a low degree of environmental effect, even though there are factors such as agricultural practices, harvesting methods, and product yields that cause adverse effects on environmental pollution. On the contrary, LB provides advantages by lowering the use of nonrenewable energy and CO_2 emissions.

LB is abundantly available in nature and is convertible into value-added chemicals and bioproducts that are renewable, sustainable, eco-friendly, and economically beneficial. The demand for LB has increased over the last few decades because of the elevated price of energy from fossil fuels (Asgher et al. 2013). In addition, accumulations of LB from different sources of residues such as agricultural, forestry, and municipal solid wastes cause hazards to the environment and the loss of valuable materials present in the biomass (Amit, et al. 2016). Usually, in the past, LB was considered as waste that was used for only combustion, which caused the release of fine particles, especially PM10 and PM2.5. Recently, this waste has been considered a raw material to convert into biotechnological products. If this biomass undergoes biotechnological conversion for valorization in different industries, for instance, paper, food, agriculture, biorefinery, animal feeds, enzyme production, and fertilizers, it will help to put forward a step to prevent environmental pollution (global warming, greenhouse effect, etc.) to some extent.

Biorefinery could produce one or a few small scales, however high-esteem, chemical items, and a low-esteem, yet large-scale biofuel, for example, biodiesel or bioethanol. Simultaneously, it can produce power (electricity) and process heat, through combined heat and power (CHP) innovation, for its utilization and possibly enough available to be purchased of power to the neighborhood utility. The high-value products increase revenues, the high-volume fuel helps address energy issues, and the production of electricity assists with bringing down energy costs, reducing GHG release from the traditional power plant and solving the problems of waste management systems. Based on this reality, researchers worldwide are studying and investigating ways to replace energy sources that are potentially sustainable and environmentally friendly. LB is used in lignocellulose biorefineries. To convert LB into sustainable value-added products, it requires multiple biorefinery processing steps. This process includes pretreatment, saccharification, and fermentation to obtain either an intermediate or final bioproduct (low or high volume) that can be used for various purposes or in industries (Xiao et al. 2012, Sriariyanun and Kitsubthawee 2020). The biorefinery process involved in the production of bioethanol is diagrammatically represented in Figure 2.2.

2.3.1 Pretreatment Methods of LB

The initial step of the biorefining process is the pretreatment by different methods that are categorized as physical, chemical, biological, physicochemical, or combined methods. Pretreatment is a necessary step to reduce the size of lignocellulose biomass to handle-sized particles by using physical means, such as cutting or milling. The purposes of pretreatments included the primary fractionation of cellulose, hemicellulose, and lignin; the modification of cellulose to be susceptible form to hydrolysis; and the removal of inhibitory compounds of hydrolysis and fermentation (Cheng et al. 2020, Sriariyanun & Kitsubthawee 2020). Chemical characteristics of LB consist of complexity in the structure, chemical composition, degree of polymerization,

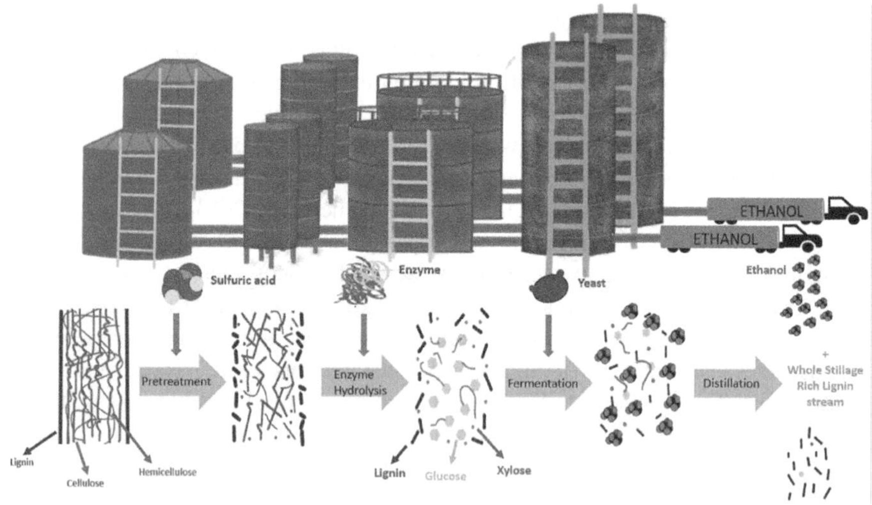

FIGURE 2.2 Diagrammatic representation of steps involved in biorefinery process for the production of bioethanol.

and presence of hydroxyl and acetyl groups. Despite chemical factors, there are physical features that contribute to the recalcitrant nature of LB, such as crystallinity, particle size, accessible surface area (ASA), and volume (pore size) (Aya & Gabriel 2019). Researchers have performed several studies on the crystallinity index of cellulose, which is a macromolecular parameter that acts as a part of crystalline regions in amorphous regions. Studies by Xu et al. (2019), on pretreated corn stover, demonstrated that crystallinity negatively correlates with hydrolysis rate. The significance of the pretreatment process was also shown to be related to the breaking down of crystalline structure and mitigation of inhibitory compounds (Anuj et al. 2018). Particle size is another important factor that enhances the rate of the hydrolysis reaction. Physical methods (grinding, milling, and extrusion) used in the pretreatment process will decrease the particle size for easy breakdown of the complex structure of cellulose in LB (Yu et al. 2019). The significance of particle size on pretreatment efficiency varies depending on plant species. For instance, poplar has less effect on hydrolysis yield when the size is below 400 μm (Chang & Holtzapple 2000), while in wheat straw, the size threshold should be 270 μm (Silva et al. 2012).

According to the study of Liu et al. (2014), the ASA of LB plays a key role in enzymatic saccharification, and it is associated with specific surface area (SSA) and pore volume. These two factors are inversely proportional (low particle size and high pore volume) and have a high impact on ASA. It was demonstrated previously that an increase in ASA enhances hydrolysis efficiency (Torr et al. 2016). As particle size decreases, SSA is increased, leading to a leveling up of the hydrolysis process. Pretreatment enhances the surface area of cellulose so that enzymes can easily act during hydrolysis for better sugar yield (Silvi et al. 2017, Lu et al. 2019). Another significant factor is the pore size or accessible volume of biomass, which is also related to the particle size and shapes of the cellulose fibrils for the hydrolysis process. The cellulose substrate needs a pore size greater than the size of the cellulase enzyme (~5.1 nm) to allow physical and chemical interactions between enzymes and substrates (Herbaut et al. 2018). A similar finding was made to show a solid connection between the size of biomass and the hydrolysis yield of cellulosic substrates (Peciulyte et al. 2015). However, some studies prove there is no such correlation between these factors (Kruyeniski et al. 2019).

These chemical and physical factors of LB influence biomass to become highly unsusceptible and potentially recalcitrant to chemical, microbial, and enzymatic decomposition, which hinders the utilization of LB during valorization to produce bioproducts. Therefore, various pretreatment methods, such as physical, chemical, physicochemical, and biological methods, are required for the degradation of the lignocellulosic complex (Anuj et al. 2018, Sankaran et al. 2020). Further advanced pretreatment methods have been developed to reduce energy and cost. It could be expected that the desired pretreatment method should require low energy, get maximum sugar yield with minimal loss of carbohydrates, be less expensive and be compatible with other downstream processes (Goldy et al. 2018). Each pretreatment has its advantages and disadvantages, as described in Table 2.2.

TABLE 2.2
Types of Pretreatment Methods and Their Pros and Cons

Pretreatments	Pros	Cons	References
	Physical Methods		
Grinding and chipping	Reduction of particle size, breakdown of crystalline structure	High energy usage and machine cost	Siti et al. (2014)
Milling	Reduction of particle size, breakdown of crystalline structure	High energy usage and machine cost	Siti et al. (2014)
	Chemical Methods		
Concentrated acids	Low consumption of acids and temperature, high hydrolysis of cellulose and hemicellulose, change in lignin structure, reuse of chemicals	Hazardous, toxic and corrosive, formation of inhibitors, high equipment cost	Zahid et al. (2014)
Dilute acids	Low consumption of enzymes, high hydrolysis of cellulose	High operation cost, low lignin removal, high solid waste and inhibitors, low sugar yield	Edem & Moses (2013)
Alkali	Removal of hemicellulose and lignin	Not applicable for forest biomass, formation of inhibitors and salts, high recovery process increase cost, long residence time	Goldy et al. (2018)

(Continued)

TABLE 2.2 (*Continued*)
Types of Pretreatment Methods and Their Pros and Cons

Pretreatments	Pros	Cons	References
Organosolv	Degrade LB components, obtain pure form of byproducts	Require high temperature and pressure, costly, cause corrosion and environmental pollution, high energy requirement and recovery process	Javier et al. (2019)
Wet oxidation	Removal of lignin, dissolve hemicellulose and cause dissolution of cellulose	Costly, require high pressure, not applicable for forest residues	Edem & Moses (2013)
Alkaline hydrogen peroxide	Delignification improves enzymatic hydrolysis, releases free radicals, requires low energy, low toxins, recyclable	Decrease the efficiency of sugar yield, high cost of oxidants	Javier et al. (2019)
Sulphite	Removal of lignin, recovery of cellulose and hemicellulose, enhance sugar yield	Sugar breakdown occurs in extreme condition, require large volume of water for washing, high cost of downstream purification chemicals	Edem & Moses (2013)
Glycerol	Remove lignin, high cellulose convertible rates	Low sugar produced when use alone	Edem & Moses (2013)
Inorganic salts	Removes hemicellulose from LB, high sugar recovery	Low sugar recovery, need wastewater treatment	Edem & Moses (2013)
Ionic liquids	Eco-friendly, disrupt lignin and decrystallize cellulose structure, enhance cellulase activity, reusable of chemical	Viscosity of product, high cost	Javier et al. (2019)
Ozonolysis	Lowers lignin content, no inhibitor	High level of Ozone is needed, costly	Goldy et al. (2018)
Pyrolysis	Formation of gaseous and liquid products	High energy and ash content	Goldy et al. (2018)
	Physicochemical Methods		
Extrusion	Particle size reduces, adaptable to many physical parameters, low toxic production	Incomplete breakdown of hemicellulose, low removal of lignin, production of inhibitors	Laura & Vincenza (2016)
Steam explosion	High sugar yield, economically feasible, low toxic, removes hemicellulose, disrupt lignin matrix	Partial degradation of crystalline structure of LB, high inhibitor formation	Sharma et al. (2015)

(*Continued*)

TABLE 2.2 (*Continued*)
Types of Pretreatment Methods and Their Pros and Cons

Pretreatments	Pros	Cons	References
Hydrothermal	Removes hemicellulose efficiently, high sugar recovery, low inhibitors, no addition of catalyst and chemicals	No lignin removal, long residence time	Zahid et al. (2014)
AFEX	Low energy and inhibitor formation, degradation of lignin and hemicellulose	Not suitable for high lignin content biomass, expensive	Zahid et al. (2014)
Supercritical CO_2 explosion	Separates hemicellulose and cellulose, no toxins, high efficiency	No removal of lignin, require high pressure, costly for industrial purpose	Laura & Vincenza (2016)
	Biological Methods		
Microorganisms	Delignification, low energy and reaction parameters, absence of catalyst, chemicals and inhibitors, inexpensive	Low rate of hydrolysis, require large sterile spaces and equipment	Goldy et al. (2018)

2.3.1.1 Physical Methods

Physical method is the initial step needed in the pretreatment process before going to the next step where the bulky size of the biomass is reduced by different mechanical techniques including milling (dry, compression, vibratory ball, and wet milling), chipping, and grinding (Kumari & Singh 2018). One of the merits of physical pretreatment is that it does not produce any inhibitory substances (e.g., furfural and HMF); nevertheless, it requires high energy and has a high operational cost (Veluchamy & Kalamdhad 2017, Dahunsi 2019). These drawbacks can be minimized by combining chemical pretreatment that will reduce energy consumption via a reduction in biomass friction and biomass toughness (Rodiahwati & Sriariyanun 2016). Among physical pretreatment methods, the screw press or extruder is the most selected method due to its high shear, rapid mixing, short residence time, lack of inhibitor formation, and moderate operating conditions. Although a screw press or extruder could alter the structure of biomass to make it more vulnerable to further hydrolysis, the lignin content is not removed. Therefore, a combination of chemical pretreatment such as alkaline addition has been applied to promote the efficiency of the physical pretreatment.

2.3.1.2 Chemical Methods

In general, the chemical method leads to disruption of the chemical structure of LB. Chemical pretreatment primarily employs acids (e.g., HCl, oxalic acid, sulfuric acid, phosphoric acid, acetic acid, citric acid, and tartaric acid), bases (e.g., sodium hydroxide, potassium hydroxide, lime, urea, ammonia, sodium carbonate, calcium hydroxide, and methylamine), and organic solvents. In acid-based pretreatment, there are two main methods that depend on the concentration of acids used. They are high concentration acid and dilute acid treatments. In high concentration acid treatment, a 30%–70% concentration of acid is used and the reaction takes place at a low temperature. As the concentration of acid is high, it causes a corrosive reaction that damages the instrument and leads to an intensive requirement for wastewater treatment. Due to this reason, specialized noncorrosive materials (ceramic or carbon-brick lining) are required, which will increase the cost of production (Sun & Chen 2007). In comparison with high acid concentration, dilute acid treatment is mostly used as it works at low concentrations (0.2%–2.5% v/v) at a high temperature that helps to avoid corrosive effects and toxicity (Badiei et al. 2014). Acid treatments are applied in many agricultural residues, which can effectively solubilize hemicellulose. The main drawback of the procedure is the formation of a secondary product that acts as an inhibitory compound to further steps of the biorefining process, especially hydrolysis and fermentation. Consequently, the operational cost of inhibitor purification is increased (Zahid et al. 2014). Alkaline-based pretreatment involves delignification by increasing the surface area, lowering the degree of polymerization by breaking bonds that link with hemicellulose, and disruption of the crystalline nature of cellulose (Romero-Guiza et al. 2017). This method is mostly used for the LB, such as agricultural residues (wheat and rice straws) and forest residues (corn stover). Alkaline pretreatment with strong operational conditions results in the loss of sugars. Because sodium hydroxide can be used at low concentrations and temperatures, it poses less risk during process operation. However, sodium hydroxide waste causes an impact on environmental pollution. Therefore, researchers came up with alternatives, such as potassium hydroxide and lime treatment (calcium oxide and calcium hydroxide), but these chemicals have a higher cost and lower efficiency (Javier et al. 2019).

Organosolv chemical pretreatment has been developed by using single organic solvent systems (methanol, ethanol, acetone, acetic acid, peracetic acid, etc.) or mixtures of organic solvents utilized along with or without the addition of catalysts (mineral acids and organic acids) (Zhao et al. 2009, Zhang et al. 2016, Amnuaycheewa et al. 2017, Amnuaycheewa et al. 2016). However, the addition of acidic catalyst may cause corrosion problems, and the purification step is required to remove inhibitory substances from biomass, resulting in high utilization of water in the washing process (Edem & Moses 2013). Pretreatment with mixtures of ethanol (solvent) and formic acid (catalyst) was done with the benefit of reducing the process cost and lowering the formation of inhibitory substances to some extent (Javier et al. 2019). Wet oxidation, alkaline hydrogen peroxide, oxidizing agents (ozonolysis), glycerol, sulfite, aqueous N-methylmorpholine-N-oxide (NMMO), and inorganic salts are the other chemical pretreatment methods having a similar mode of action with advantages and limitations (Table 2.2).

An advanced chemical pretreatment that has attracted researchers in the past few years is the use of ionic liquids (ILs), which produce holocellulose from LB matrix. ILs have multiple features, such as being an eco-friendly solvent with a low melting point and low vapor pressure, being nonvolatile and highly thermostable, and having a high efficiency in cellulose and lignin extractions (Gundupalli et al. 2021a). In addition, the high sugar yield obtained from IL pretreatment and the short time consumption of this pretreatment are the characteristics that should draw attention from the research and development sectors. Nevertheless, application of IL pretreatment for industrial use is still not ready due to the high cost of ILs compared to other chemical pretreatments. One of the solutions to reduce IL costs is IL recycling (Akkharasinphonrat et al. 2017). Later, deep eutectic solvents (DESs) have been introduced for pretreatment application because DESs have similar properties with ILs in LB fractionation with reduced cost and toxicity (Xia et al. 2018, Panakkal et al. 2021). DES helps to remove lignin (up to 90%) without affecting cellulose (Liu et al. 2017). Like ILs, DES also has a high viscosity that obstructs its use in pilot-scale industries (Zdanowicz et al. 2018).

2.3.1.3 Physicochemical Methods

The physicochemical pretreatment method overcomes the disadvantages of physical and chemical methods. Even though the operation of physicochemical pretreatment is more complex than other methods, it provides a high energy yield and a high polysaccharide yield in downstream steps. It also has benefits in the mitigation of environmental impacts, a reduction in processing time, and being economically feasible (Theuretzbacher et al. 2015). Extrusion, steam explosion, hydrothermal, ammonia fiber explosion (AFEX), and supercritical CO_2 explosion are the most efficient and eco-friendly physicochemical methods used in the biorefinery of various biomasses (Mupondwa et al. 2017). In the extrusion method, biomass is loaded and passed through an extruder to undergo physical treatment while chemicals, such as NaOH, are added through the barrel (Lamsal et al. 2010). In this scheme, the effects of physical pretreatment and chemical pretreatment could have synergistic effects to improve pretreatment performance. This combined pretreatment leads to changes in the recalcitrant structure of lignocellulose biomass in multiple ways, such as separation of the components, increasing the biomass porosity and surface area to increase enzyme accessibility without any production of inhibitors, so that the enzyme activity increases, which simultaneously increases hydrolysis and product recovery (Yoo et al. 2011).

One of the most widely used pretreatments is the steam explosion, which can be carried out in the presence or absence of a catalyst at a temperature below 260 degrees and a pressure below 50 bar for 30 s to 20 min (Ullah et al. 2018). It was demonstrated that the mechanism of this pretreatment method and the final products obtained were almost similar to extruder pretreatment, where the components of the biomass will be recovered separately. However, the production of acid, thermal energy, and toxic substances affects the midstream and downstream processes (Laura & Vincenza 2016). AFEX is the method where liquid ammonia is loaded under different conditions such as high pressure and low temperature in a short period of time, and it results in a high degree of delignification and increases polysaccharide yield. The important merit of this process is that the ammonia can be retrieved using the ammonia recycle percolation (ARP) process in the aqueous form after pretreatment (Wang et al. 2019). Thus, this helps to decrease the cost and pollution from the process. Supercritical CO_2 (SC-CO_2) explosion is another physicochemical method where supercritical fluids (CO_2) are used; the fluid behaves as a gas or liquid as its temperature and pressure lie above the critical point (Daza et al. 2016). This method functions in the degradation of lignin content in LB and increases the enzyme access to the cellulose substrate. Furthermore, the carbonic acid produced from this process can be again hydrolyzed from biomass to form sugars. The use of SC-CO_2 is increasing at the present for the treatment of LB due to the use of CO_2 adds more advantages to this pretreatment method, such as nontoxic, inexpensive, high sugar yield, reusable, and producing no waste materials. However, for further industrial purposes, it is still relatively expensive and more difficult to scale up (Gu et al. 2013).

2.3.1.4 Biological Methods

When compared to other methods, the biological pretreatment process is considered to be inexpensive, nontoxic, and nonpolluting, requiring less energy input and no additions of reagents or catalysts (Asgher et al. 2012). Wide varieties of microorganisms (fungi and bacteria) are used in this pretreatment due to their abilities to produce enzymes that hydrolyze the structure of LB. Even though this method is inexpensive and does not require complex equipment, it still takes a relatively long duration to complete and consumes more space than chemical and physical pretreatment. It requires monitoring optimum parameters for the growth of microorganisms to achieve a better yield (Liu et al. 2017, Iqbal & Asgher 2013). Some of the microorganisms, such as a white rot fungus called *Phanerochaete chrysosporium*, can degrade lignin to promote dissociation of LB. However, during their growth and metabolism, cellulose and hemicellulose are consumed, and the final yield of the process could be reduced. Alternatively, laccase enzyme produced from bacterial and fungal cultures is applied as a biocatalyst to degrade lignin. However, laccase could be reused by the immobilization method, but the enzyme is sensitive to optimal conditions and relatively costly. Due to these limitations, a biological method is not ready for commercialization. Hence, it is necessary to explore further studies on the optimization of cultural conditions of microbes and reduce operation time to increase the efficiency of pretreatment in order to increase the productivity in large-scale industries.

Because of the current disadvantages of every pretreatment strategy, combinations of pretreatment methods could be advantageous for lowering the cost and increasing the efficiency of pretreatment. For instance, combinations of biological with chemical or physical methods (biological and steam explosion) and a combination of three treatments (mechanical, thermal, and chemical) could promote pretreatment efficiency and reduce cost (Moset et al. 2018). This combination will increase the digestibility of LB when compared with the single process when it takes place alone. There are many factors that affect the selection of appropriate pretreatment methods, including type of feedstock, LB composition, sugar yield, and possible end product. Moreover, it has to consider whether the pros and cons of each method affect the operation cost and productivity. The advanced study of all these factors will help to start up a successive LB biorefinery system.

2.3.2 Hydrolysis

After pretreatment, the next stage of the biorefinery process is hydrolysis, which involves either enzymes or nonenzymes. LB (containing cellulose, hemicellulose, and lignin) is subjected to saccharification for a further breakdown of polysaccharides into monomeric units. These monomeric sugars are taken into the next stage of the biorefinery process, i.e., fermentation, to obtain value-added products. Enzymatic hydrolysis is the most widely used hydrolysis process, which is simple, eco-friendly, high sugar yield, and economically feasible when compared with hydrolysis performed using other agents (e.g., acid hydrolysis) (Anuj et al. 2018). LB contains cellulose, hemicellulose, and lignin, which could be hydrolyzed to fermented sugars and subsequently converted to bioproducts. Modified LB structure by pretreatment allows enzyme accessibility to biomass and enhances the hydrolysis process of that biomass.

Many types of research have been conducted to obtain the optimized conditions for LB hydrolysis by using a live microorganism, enzymes, or chemical/physical reactions. Chandel et al. (2012) found that cellulase enzymes produced from fungus (*Aspegillus* or *Trichoderma species*) are widely used for the LB hydrolysis process due to their characteristics such as low utilization and low production of inhibitors and toxins. Furthermore, it is a noncorrosive process with high sugar output (~90% yield). Cellulases are cocktail enzymes composed of multiple enzymes that break the polymeric chain of cellulose into simple hexose sugars. Some commercial cellulases available in the market, i.e., Cellulast 1.5L, Accellerase 1500, Ctec2, contain hemicellulose, which breaks hemicellulose into hexose and pentose sugars (Gundupalli et al. 2021b). Regardless of significant explorations completed in the past on enzymatic hydrolysis, there are still a few worries, in particular the presence of lignin as a hindrance to fermentation, that need genuine consideration to concoct new, efficient, and cost-effective alternative enzymes for farm-up production in industries (Michailos et al. 2019). According to the study by Gupta et al. (2016), different types of enzymes were tested in hydrolysis and showed that fragmentation of glycosidic bonds in hemicellulose as well as arbitrary cleavage of interior bonds in xylan are caused by endoxylanase, and lignin degradation requires ligninases (Gupta et al. 2016). Masran et al. (2016) demonstrated that the primary downsides for utilizing microorganisms to delignify the LB included long process duration, overconsumption of sugars by the microorganisms, and microorganism tainting issues. More experimental studies are required to improvise and optimize the saccharification process using improved enzymes extracted from microorganisms by using single or enzyme cocktails.

2.3.3 Fermentation Process

The final stage of biorefinery to produce the targeted product after pretreatment and hydrolysis is the fermentation process. In this process, sugars are fermented in the presence of microorganisms, where end bioproducts and intermediate products are obtained. The products from LB include second-generation biofuels and biochemicals. The main disadvantage that occurs during the hydrolysis process is the formation of inhibitors. Usually, saccharification and fermentation are done in two separate steps when the optimal conditions of enzymes used in both processes are different. Such a process is called separated hydrolysis and fermentation (SHF). Later, studies got interested in operating biorefinery process by combining the second and third steps in order to rectify the effect of the inhibitor and also speed up fermentation. Such a complex process is called the simultaneous saccharification and fermentation (SSF) process, where both steps occur in the same fermenter, which depends

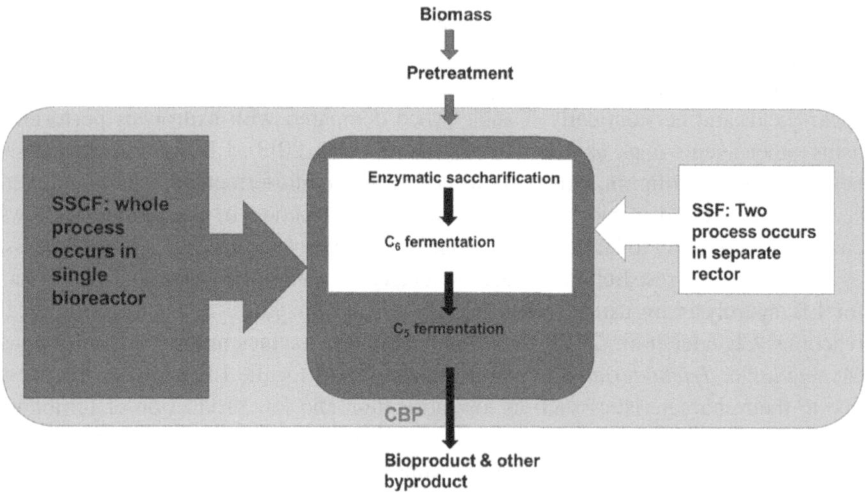

FIGURE 2.3 Schematic representation of CBP in terms of biorefinery from LB.

on working factors and the type of feedstock used (Shadbahr et al. 2017, Singh et al. 2018) (Figure 2.3).

Another strategy used in the consolidated process is presaccharification and simultaneous saccharification and fermentation (PSSF) to halfway defeat the contention in development conditions in SSF (Figure 2.3). Essentially, the ideal conditions for enzymatic saccharification are applied first for a while to improve sugar production, and afterward, the set point is moved to the optimum condition for fermentation with strains (e.g., yeast) added to deliver finished results (e.g., ethanol) (Cavalagilo et al. 2017). It was observed that PSSF has higher productivity when compared to SSF; however, an extra control procedure is required for the optimization of reactor activity. Hydrolysis and fermentation of both cellulose and hemicellulose occur at the same time and in the same reactor with both C_5 and C_6 sugars by using genetically modified microorganisms, and this process is called simultaneous saccharification and co-fermentation (SSCF). This process makes the entire process more reasonable and cost-effective (Zahid et al. 2014, Chen et al. 2018).

As the classical biorefinery process is conducted in multiple steps, the reduction of product yield, as well as other drawbacks related to the formation of inhibitory factors and infeasible costs, could be observed. To rectify these limitations, Khatun et al. (2017) investigated integrating multiple steps of the biorefinery process into a single bioreactor, which they named as consolidated bioprocessing (CBP). This method lowers the complexity of the entire biorefinery process. The microorganisms used for this purpose play an important role, as the selected strains should be capable of completing the entire process (pretreatment, hydrolysis, and fermentation) in a single cell (Kawaguchi et al. 2016). The main characteristics of the microbes for CBP should have the increased rate of saccharification, product tolerance after pretreatment, and suitability for fermentation. The enzymes produced by the CBP microbes should have the capacity to work in both hydrolytic and fermentation processes by either excreting or immobilizing enzymes on the cell surface. This adds high bioconversion activity, sugar release, low contamination, and low inhibitors. Besides, the reutilization of enzymes will decrease the total operation cost and allow the use of whole cell biocatalysts (Liu et al. 2015, Lin & Tao 2017).

Based on the functions of microorganisms used in CBP, two methods are applied: native and recombinant strategies (Jiang et al. 2017). In the native strategy, wild strains (natural type strains) are used that naturally produce various enzymes and utilize different substrates to increase the efficiency of the process and yield bioproducts from LB. For instance, cellulolytic microorganisms, such as fungi or bacteria, were genetically and metabolically modified at the genomic and gene expression levels and improved to scale up the production of cellulases for the valorization of value-added products (Benocci et al. 2019). Similarly, researchers are focusing on consortium (microbial co-cultivation) approaches, which play an advantageous role in enhancing the conversion of substrate and yield of commodity chemicals (Liu et al. 2019). Recombinant strategies focus on the use of recombinant microbial strains that can perform multiple functions at a time, such as hydrolytic strains with fermentation ability or fermentation strains to perform a hydrolytic role. For example, hemicellulose present in LB is composed of xylose with a side chain of acetyl groups, such as uronic acids and arabinose. The degradation of the side chains of hemicellulose required various enzymes based on the types of attached side chains. Mostly used industrial yeast, *Saccharomyces cerevisiae*, neglects to use xylose yet can ferment d-xylulose (an isomer of xylose). By utilizing gene engineering, recombinant *S. cerevisiae* (xylose isomerase bacteria and xylose reductase fungus) converts xylan and improves the fermentation process in bioethanol production (Wertz & Bedue 2013). Hence, while applying microorganisms for biomass conversion, it is necessary to choose a host organism with the ideal qualities, underlining strains that can utilize minimally expensive substrates, provide protection from ecological pressure, and have a high possibility of scaling up the final products (Abrego et al. 2017). CBP, combined with SSF and SSCF, is widely used to scale up the product and reduce the production cost (Goldy et al. 2018). Moreover, studies of genetically engineered microorganisms capable of performing multiple functions in a single reactor will contribute to the advancement of CBP technology in the future by making it economically convenient (Edgar et al. 2020). Still, there are limitations to using CBP for industrial purposes, as when compared with the native strategy method, recombinant technology gives optimum product yield. In that case, the use of genetically manipulated microorganisms is restricted in some countries due to the health and environmental concerns it raises. Furthermore, a long duration of fermentation is required for CBP even if the cost and energy consumption are less (Hasunuma & Kondo 2012).

2.4 VALUE-ADDED PRODUCTS

The final stage after fermentation is the recovery of the biorefinery products through the downstream process. In this case, process handling ought to be compelling, incorporate not many strides quite far to obtain the product without any loss, and be economically suitable. Valdivia et al. (2016) suggested various downstream processing steps, including partitioning of particles, breaking down of cells, extraction, concentration, purification, and drying. Progress in the efficiency of the recovery process also affects the cost, machinery, and waste produced by the system. The availability and utilization of raw materials, enhanced product, and a competitive cost will increase the demand for the product in the global market. In addition, replaced items in the market add to customer demand, for instance, polylactic acid (bioplastic) replacing polyethylene terephthalate (PET; synthetic plastic). The value-added products are grouped based on the composition present in LB (Figure 2.4). Products derived from the fermentation of lignin, cellulose, and hemicellulose, which are the major compositions in LB, are further used for commercial purposes based on the demand in the global market. Biorefinery products include biofuels (ethanol, butanol, methane, and hydrogen), platform chemicals (organic acids), chemical commodities (byproducts), and other building blocks of biomaterials (bioplastics, biopolymers, and biocomposites). According to the global demand prediction for 2020, biofuel production decreases in 2030 due to the development of electric vehicles in the United States and Europe (EEC, 2020). Therefore, the below session discusses some of the value-added bioproducts that have high demand in the market at present and in the future based on current reports.

The terminology of platform chemicals has been invented and applied in parallel to biorefinery due to their potential to be applied in various classes of production lines of finished products in the market. Platform chemicals are considered as small chemical molecules, or building blocks, that function as intermediates or substrates for subsequent chemical reactions or enzymatic reactions to produce more complex chemicals, compounds, or polymers. The global markets for platform chemicals derived from biomass are expected to grow up to $20 billion in 2022. Among those, C_3 platform chemicals are the major players in the market for various industries, as they are obtained from the biodiesel production process in the form of glycerol and its derivatives. Currently, other platform chemicals, such as C_2–C_5 molecules, have received spotlights from various industries, especially in terms of value-added products, such as bioplastics and pharmaceutical compounds. For example, organic acids derived from LB include citric acid, succinic acid, gluconic acid, oxalic acid, lactic acid, malic acid, butyric acid, fumaric acid, levulinic acid, 5-hydroxymethylfurfural (5-HMF), sorbitol, and other solvents. (Figure 2.4). Moreover, xylitol, furfurals, and acetic acid from hemicellulose-derived secondary products and phenol and polymers from lignin are considered as platform chemicals (Figure 2.4). These platform chemicals contain particles with an assortment of functional groups, conveying the capability of getting converted into various value-added bioproducts useful for many industries on the world market (Table 2.3). Some of the important chemicals are discussed below.

2.4.1 Citric Acid

Worldwide, citric acid is the second biggest fermentation product after ethanol, with 1.7 million tons of yearly production. It is the intermediate chemical produced in Kreb's cycle that could be linked to carbohydrate metabolism, synthesis of amino acids, and catabolism of fatty acids in living organisms. It has various applications in biomedicine, cosmetics, nanotechnology, and bioremediation of heavy metals (Amit et al. 2016). Mostly, citric acid is produced through submerged and solid-state

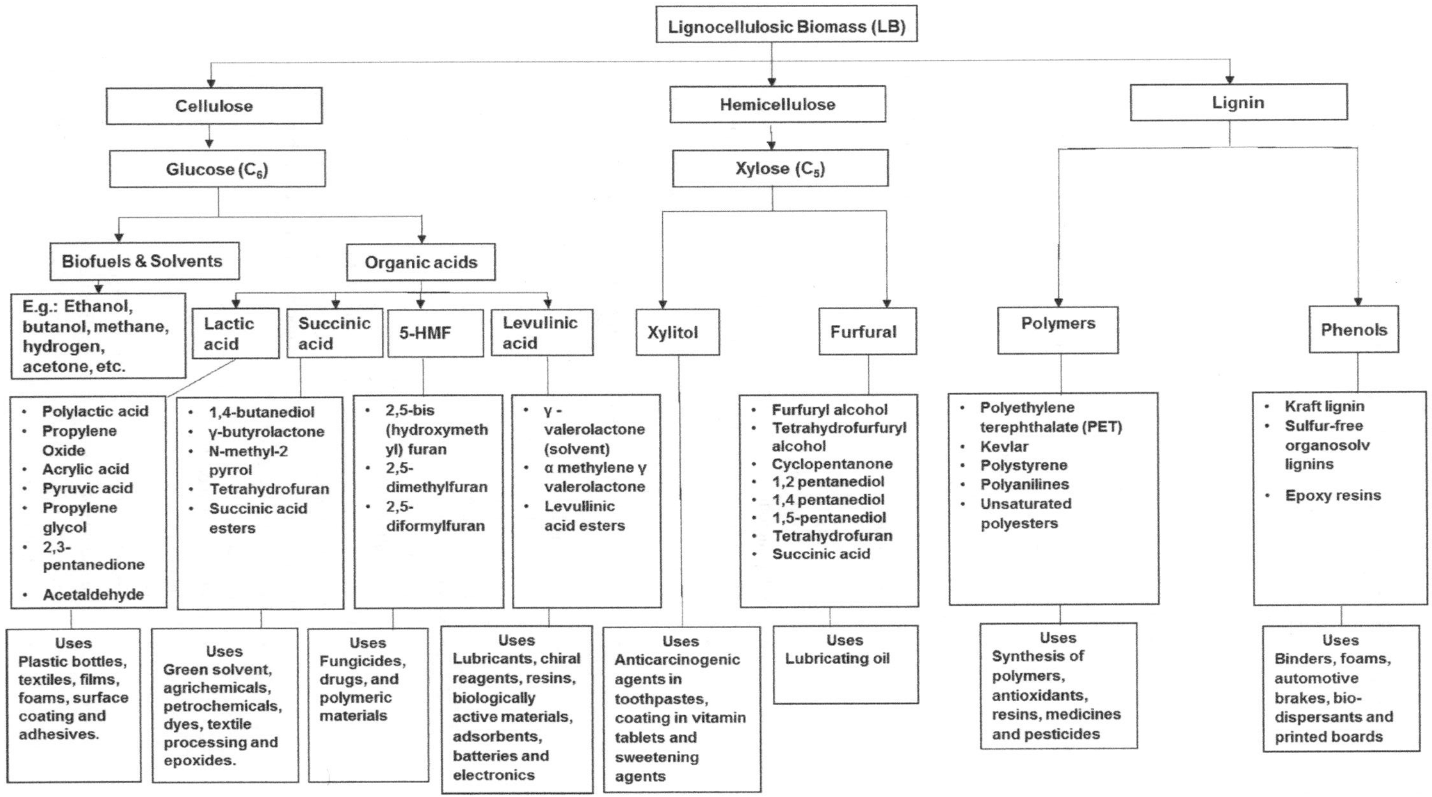

FIGURE 2.4 Value-added products and their possible uses obtained from LB biorefinery platform including cellulose, hemicellulose, and lignin.

TABLE 2.3
Platform Chemicals Obtained from LB and Their Potential in Industries

Platform Chemicals	Derivatives	Uses
Lactic acid	• Polylactic acid • Propylene oxide • Acrylic acid • Pyruvic acid • Propylene glycol • 2,3-Pentanedione • Acetaldehyde	Plastic bottles, textiles, films, foams, surface coating, and adhesives.
Succinic acid	• 1,4-Butanediol • γ-Butyrolactone • N-methyl-2-pyrrolidone • Tetrahydrofuran • Succinic acid esters	Green solvent, application in agrichemicals, petrochemicals, dyes, textile processing, and epoxides.
HMF	• Levulinic acid • 2,5-Bis(hydroxymethyl) furan • 2,5-Dimethylfuran • 2,5-Diformylfuran	Fungicides, drugs, and polymeric materials
Levulinic acid	• γ-Valerolactone • α -Methylene-γ-valerolactone • Levullinic acid esters	Lubricants, chiral reagents, resins, biologically active materials, adsorbents, batteries, and electronics
Xylose	• Xylitol	Anticarcinogenic and sweetening agent
Furfural	• Furfuryl alcohol • Tetrahydrofurfuryl alcohol • Cyclopentanone • 1,2-Pentanediol • 1,4-Pentanediol • 1,5-Pentanediol • Tetrahydrofuran • Succinic acid • Maleic acid • Maleic anhydride	Lubricating oil, gasoline additives, resin, adhesive, flavoring agent
Phenols/polymers	• PET • Kevlar, polystyrene • Polyanilines • Unsaturated polyesters • Kraft lignin • Sulfur-free organosolv lignins • Epoxy resins	Synthesis of polymers, antioxidants, resins, medicines, pesticides, phenol-formaldehyde, wood board and electric board
Bioplastics	• Thermoplastic starch • Cellulose-acetate plastics • Polylactic acid • Polyhydroxy alkanoates	Bundling, dispensable items and clinical applications

fermentation using *Aspergillus niger*. The study by Ping-Ping et al. (2016) used corn stover feedstock as raw material with *A. niger SIIM M288* and obtained a 94.11% citric acid yield without generation of free wastewater throughout the process. This research provided advancements in the scale-up production of citric acid yield from LB. According to the study by Wang et al. (2018), the use of recombinant or mutant strains of *A. niger* improved the titer rate of citric acid production during fermentation. Further studies of genetic engineering and genome analysis are required to study the screening of targeted genes of citric acid metabolisms and develop a new methodology to increase the yield of citric acid. Due to the increase in market demand for citric acid, it is not feasible to produce it from other raw materials (starch and glucose) rather than LB because of the rise in competition for food and feeds. Regardless of whether agricultural residues produce impurities or inhibitors during the pretreatment and fermentation processes, they will acquire significant economic and environmental advantages in the citric acid fermentation industry. Consequently, cost-effective removal of these inhibitors or complex compositions is a need for innovative technology (Wei et al. 2019).

2.4.2 Succinic Acid

Another organic acid that has high market potential in the fields of food, agriculture, and pharmaceuticals is succinic acid, which is produced at a rate of almost 18,000 metric tons per year (Goldy et al. 2018). Currently, Bioamber, Quebec, Canada, is the major producer of succinic acid (Anuj et al. 2018). Cellulose has the potential to be utilized as a substrate for microbial fermentation to produce succinic acid, which undergoes further reaction results to form various value-added bioproducts, such as 1,4-butanediol (polyesters), tetrahydrofuran (polyglycol), and γ-butyrolactone (industrial solvent), through esterification. Fumaric acid (medicinal value) and maleic acid (drug formulation) are the other intermediates produced through the dehydrogenation and cyclization of succinic acid (Kirtika et al. 2019). Jerico et al. (2017) conducted a study that discovered a new approach to producing succinic acid from mildly pretreated LB using the CBP method using lignocellulolytic (*Phanerochaete chrysosporium*) and acidogenic fungal co-cultures (*Aspergillus niger* and *Trichoderma reesei*). This CBP process produced the highest titers of 32.43 g/L after 72 h of batch slurry fermentation and 61.12 g/L after 36 h of the beginning of the slurry fermentation stage. Because of this outcome, this methodology is a promising option in contrast to current bacterial succinic acid creation because of its negligible substrate pretreatment prerequisites, which could reduce production costs. Another study experimented on *Basfa succiniciproducens* BPP7 in SHF (Fed batch) method with *Arundo donax* as raw material, yielding 52% of succinic acid (Donatella et al. 2019).

2.4.3 Lactic Acid

One of the organic acids that is widely produced in the world is lactic acid, with a market size of 472 kilotons per year from the cellulose present in LB. Annually, 40,000 tons of lactic acid are produced all over the world. It is used in food, textile, pharmaceutical, cosmetic, and other chemical industries. The main manufacturers of lactic acid are Chongqing Bofei Biochemical Products & Henan Jindan, HiSun, Wuhan companies in China, as well as Corbion Ltd, USA (Anuj et al. 2018). The demand for lactic acid is expected to reach around 1,960 kilotons by 2025 (Hu et al. 2017). This platform chemical mainly attracted as it can produce polylactic acid, which can replace synthetic plastics (PET). Other derivatives of lactic acid include propylene oxide (foam), acrylic acid (surface coatings and adhesives), pyruvic acid, propylene glycol, 2,3-pentanedione, and acetaldehyde (Kirtika et al. 2019). Lactic acid undergoes many chemical reactions, such as dehydration (acrylic acid and propylene oxide), hydrogenation (propylene glycol), decarboxylation (acetaldehyde), condensation (2,3 pentanedione), esterification (PLA), and oxidation (pyruvic acid) to form its derivatives (Matty 2013).

Microbial fermentation technology is used for large-scale production of lactic acid from LB because of its low cost and nonfood raw material. Agricultural crop residues are utilized for the production of lactic acid through microbial fermentation. Currently, lactic acid is produced in a batch reactor. Recently, Hassan et al. (2020) manufactured lactic acid from bagasse of banana peduncles, sugarcane, or carob by using *Bacillus coagulans* strains for the fermentation process. The pilot scale of 35 L obtained by utilizing carob biomass without yeast extract brought 0.84 g LA/g sugar yield and an efficiency of 2.30 g LA/L/h, which shows an extremely encouraging process for future industrial scale-up of LA.

2.4.4 Hydroxymethylfurfural

One of the significant platform chemicals from cellulose is the 5-HMF in which the valorization results in many derivatives including levulinic acid, 2,5-bis(hydroxymethyl) furan, 2,5-dimethylfuran, and 2,5-diformylfuran. These furan-based chemicals are used for the production of fungicides, medicines, and synthetic polymers (Kirtika et al. 2019). The main manufacturers of 5-HMF are AVA Biochem Ltd, Germany, which produces 0.02 kilotons per year with a theoretical market value of 0.1 million USD per year (Chandel & Silveira 2017). The maximum yield (92%) of 5-HMF is obtained by pretreating cellulose with DES, which is considered to be promising due to this level of conversion yield (Zhao et al. 2014). The crystalline nature of cellulose presented a challenging criterion for the production of 5-HMF. The change of biomass into 5-HMF has been broadly explored by numerous scientists. The viable change of the polysaccharide part of biomass undergoes physical and chemical pretreatment that results in the breakage of the crystalline nature of cellulose. Later, LB is processed through various complex reaction systems containing one or more catalysts (Perez et al. 2019). Hanna et al. (2020) produced 5-HMF (38%) from pectin-free sugar beet pulp. Also, the liquid–liquid extraction method is used to separate the chemical from the reaction mixture. The outcomes are satisfying and reasonable for scaling up production in industries and may turn into motivation to present a harmless production strategy to the ecosystem that is simple, proficient, and utilized to treat waste residues from the sugar refining industry. More refined techniques ought to be created to scale up 5-HMF from substrates, which require effective catalysts for the reactions along with sophisticated reactors to perform the reactions cost-effectively and without harming the environment.

2.4.5 Levulinic Acid

The bioproduct from cellulose, which consists of carboxyl and carbonyl groups, produces different chemical derivatives like γ-valerolactone (solvent), α-methylene γ-valerolactone (acrylic monomer), and levulinic acid esters. These platform chemicals are used for various purposes such as lubricants, resins, chiral reagents, electronics, and so forth. It is produced at a rate of 71.6% globally per year (Goldy et al. 2018). Currently, Segetis Inc., MN, USA, and Zibo Shuangyu Chemicals Ltd, China, are the major producers (Chandel & Silveira 2017). Biorefinery process occurs in two reactors where saccharification and removal of water occur to produce 5-HMF in the first reactor and conversion of levulinic acid occurs in the second reactor, respectively. The yields (60%) were acquired dependent on the monomers present in the substrate. As of now, a few issues such as low yield, high equipment prerequisites, trouble in the downstream process, and pollution caused in the production of levulinic acid from cellulose have still not been rectified. Thus, further studies are needed to find the solution for scaling up the synthesis of levulinic acid, which requires multistep processes and well recovery technology to yield pure chemicals to apply in industries (Kirtika et al. 2019).

2.4.6 Sorbitol

One of the value-added platform chemicals derived from cellulose is used widely in food, cosmetics, medical, and other industrial applications. Production of sorbitol mostly occurs through hydrogenation of glucose in the presence of noble metals or catalysts, such as nickel, platinum, and ruthenium. Sixty percent of the yield is obtained when nickel phosphide (Ni2P) catalyst is utilized for the cellulose conversion (Liang et al. 2012). Another catalyst, the Raney nickel, has a low cost and has been used to increase yield in industries (Aho et al. 2015), but it requires high energy and cause environmental pollution (Gericke et al. 2015). Beatriz et al. (2020) developed an alternative chemical pathway to convert glucose to sorbitol through catalytic transfer hydrogenation using a hydrogen donor, which opens a new method for producing sustainable sorbitol.

2.4.7 Xylitol

Xylitol is produced from LB by the saccharification of hemicellulose to produce xylose. Further breakdown of xylose results in xylitol, which has wide application as a sweetening agent in the food, pharmaceutical, and nutraceutical industries (Amit et al. 2016). It is anticarcinogenic and undergoes insulin-independent metabolism. Xylitol is currently used in oral hygiene products (toothpastes, mouthwashes, or chewing gums) and confectionaries. Xylitol is found in very small amounts in some fruits, vegetables, algae, and mushrooms, but not at a level for commercial extraction (Marie et al. 2018). For the pilot-scale synthesis of xylitol from agricultural feedstock, required acid hydrolysis and chemical reduction of xylose were required. This results in 50%–60% of the final yield; moreover, this process is energy-consuming and causes high costs in the recovery step (Mohamad et al. 2015). Yadav et al. (2012)

first synthesized xylitol by a multistep process, where saccharification and hydrogenation occurred in the presence of a catalyst in different bioreactor. Later, Liu et al. (2016), found the one-pot method using silicone dioxide catalyst along with sulfuric acid. Microbial production of xylitol is a profoundly appealing and elective technique that can produce a low cost, high yield product that is viable because it can be accomplished without other optimum conditions. This includes microbes such as yeasts, fungi, and some bacteria. The main drawback of this method is the formation of inhibitors during acid hydrolysis, which will hinder microbial growth. Removal of toxins requires additional purification steps that will increase the cost of production and also make it difficult to maintain optimum parameters for fermentation. Enzymatic saccharification studies were conducted by Wan and Jahim (2016), but the yield was lower than acid saccharification. Later, Marie et al. (2018) researched the production of xylitol from wheat straw as a raw material using enzymatic hydrolysis. It was hydrolyzed with a xylanase from *Trichoderma viride* for the production of xylose, and a crude XR (crude extracts purified from native or recombinant strains) extract from *Candida guillermondii* was used for the conversion of xylose to xylitol. Thus, it proved that the enzymatic synthesis of xylitol from wheat straw is feasible.

2.4.8 Furfural

Furfural was listed by the US Department of Energy (DOE) as one of the top 12 value-added products obtained from hemicellulose and utilized in the fields of oil refining, plastic, pharmaceutical, and agrochemical industries. China produces around 700,000 tons per year (Mao et al. 2012). Dehydration of xylose produces furfural, the chemical that acts as intermediate for furan, furan alcohols, and tetrahydrofuran. Economically, furfural is created by the acid-catalyzed reaction of C_5 sugars that are first hydrolyzed by sulfuric acid to xylose. Later, it is dehydrogenated, recovered from the fluid stage by steam stripping to avoid additional degradation, and cleansed by double distillation (Mamman et al. 2008). Many studies were conducted to produce furfural through mineral and solid acid catalysts, but their corrosive effects increased the capital cost. Studies of pretreatment using ionic liquid enhanced the chemical yield (Tong et al. 2010), and the development of the one-pot process (Matsagar et al. 2017) reduced the cost of synthesis of furfural to be utilized as lubricating oil in factories. The scale-up synthesis of furfural and its selectivity directly from hemicellulose in LB is challenging. Further studies are required to develop a novel catalyst and solvent system for the extraction of furfural from hemicellulose derivatives effectively and economically.

2.4.9 Acetic Acid

Hydrolysis of the acetyl group of hemicelluloses results in the formation of acetic acid. Acetic acid is mainly used as an application for the production of cellulosic pulp and other household products. Also, it has been used for the fractionation of LB from different feedstocks, such as dhaincha, kash, banana stem, and jute fiber (Amit et al. 2016). The main producers of acetic acid are India (Jubilant Life Science Ltd) and China (Songyuan Ji'an Biochemical). The global market size is approximately

1,357 kilotons annually, with a theoretical market value of 837.3 million USD per year (Chandel & Silveira 2017). A study by Mandana et al. (2016) converted LB sugars to acetic acid by using the strain *Moorella thermoacetica*. After steam pretreatment, the biomass undergoes hydrolysis to obtain water-soluble hydrolysates, which are then used in batch fermentations. Acetic acid yield was obtained 70% from sugarcane straw and 39% from forest residue. For effective bioconversion of lignocellulosic sugars to acetic acid, it is imperative to have a suitable equilibrium of carbohydrates in a hydrolysate. Subsequently, the choice and design of LB and pretreatments are important measures for the industrial application of this process.

2.4.10 Lignin-Based Phenols/Polymers

The polymers derived from lignin consist of PET, Kevlar, polystyrene, polyanilines, and unsaturated polyesters. The Bio-PET can be produced from ethylene glycol and *p*-terephthalic acid. Another value-added product is kraft lignin, sulfur-free organosolv lignins, and epoxy resins, which are used to produce phenol-formaldehyde, wood board, and printed board, respectively. Lignin is a source of phenolic monomers (alkylated phenol and guaiacol) used for the synthesis of polymers, antioxidants, resins, medicines, and pesticides (Jung et al. 2015). Because of the complexity of the structure of lignin, which is hard to separate from LB, the process results in a lower product yield, high treatment parameters, and an advanced downstream process that is expensive. Therefore, the synthesis of these items relies upon the advancement of new lignin valorization mechanisms.

2.4.11 Bioplastics

Every year, new bioplastics are synthesized, with production expected to reach 2.61 metric tons by 2023. Presently, more than half of bioplastics synthesis is occurring in Asia (TMR 2020). Significant instances of biobased and biodegradable plastics are thermoplastic starch, cellulose-acetate plastics, polylactic acid, and polyhydroxy alkanoates. They are essentially utilized for, for example, bundling and dispensing items. They are perceived as naturally well-disposed, and some of them have been utilized for clinical applications because of their lower or zero harmfulness and high biocompatibility (EB 2017). Biobased PE, polytrimethylene terephtalate, polyamides(nylon), polyurethane, and thermosets such as epoxy resins are the other arising bioplastics in the global market (Fabricio et al. 2018).

2.5 CHALLENGES IN THE COMMERCIALIZATION OF LIGNOCELLULOSE BIOREFINERY

For the commercialization of the value-added products on the market, it is necessary to understand the technical and economic issues that arise during the production of biorefinery products from LB. The technology that is used in laboratories may not be suitable for production on a pilot scale, especially in industries. In such a case, technological advancement will be required, and a variety of equipment will be forced to use. This will lead to an increase in the capital costs (inputs), which include energy, equipment, feedstock, chemicals, and manpower. In addition, the steps involved in the biorefinery process (pretreatment, saccharification, fertilization, and recovery processes) to scale up the production should be designed to achieve the targeted products (in terms of yield, efficiency, and cost of the product) (Rahul et al. 2020). Economic analysis depends on the biomass or feedstock utilized and energy yield. In the LB biorefinery, different types of feedstock are used for developing chemical commodities. In both chemical and microbial conversion of biomass, the cost differs based on the treatment exploited to attain maximum yield. Certain countries provide subsidies to the biomass industry that uses it to produce biofuels and bioelectric energy but are not available for green chemicals and commodities. Besides, the ecological effect of the biobased items should be considered, while surveying the future allure of individual items. It is concerned to ensure that all biobased items are more harmless to the ecosystem than their fossil-based partners. Some of the challenges that are related to the commercialization of LB products are briefed below.

2.5.1 Scale-Up Challenges

Large-scale production of bioproducts from the biorefinery process requires minimum infrastructure and manpower, along with regulatory management. According to Sanford et al. (2016), the regulatory management controls both the biorefinery processes and bioproduct recovery. For pilot scale, parameters must be adjusted from laboratory conditions when it comes to an industrial plant. Different biomass processes have different conditions to valorize various products. The availability of biomass in large tons, the huge processing unit required for the plant to scale up the product, and the standardization of protocol are the factors contributing to the prerequisite. Consequently, scale-up challenges include the advancement of precise handling flow sheets, process modeling, measure demonstrating cost affectability investigation, life cycle evaluation, determining and recognizing risk factors (Junqueira et al. 2016). Rahul et al. (2020) and Sanford et al. (2016) proposed some factors that have to be considered when scaling up the bioproducts based on LB. They developed a process design for different types of biomass (nature and composition), expected outcomes, and risk evaluation. Moreover, fundamental information is gathered and compiled by conducting numerous tests at the pilot plant and incorporating whatever number of factors (pretreatment conditions and other factors) as could reasonably be expected to decrease the danger of disappointment at commercial scale production. Thus, obtain a finished product and thoroughly check it before handing it over to customers. Raw materials, water utilization, waste disposal or no waste produced, usage of instruments, and manpower are the other basic elements in starting up the biorefinery.

2.5.2 Technical Challenges

Technical maturity is another challenging factor that influences the effective execution of biorefinery plants internationally. This is due to the failure in parameter execution from the laboratory to the commercial scale. Each biomass processing step in the biorefinery requires technical improvement to ensure the smooth running of the plant. Every processing step is challenging, from biomass collection to the recovery step. The pretreatment process is the initial step where the biomass composition is separated into cellulose, hemicellulose, and lignin. Each biomass has a different composition, as shown in Table 2.1. Therefore, the pretreatment methods will be different for various LB based on the type of feedstock and composition. A decrease in the molecular size is the first physical method utilized in the pretreatment strategy for LB. However, one of the detriments of the physical method is its failure to eliminate the lignin, decrease particle size, and remove other impurities, which is a cost-effective process. Accordingly, diminishing the energy prerequisites and expanding efficiency of mechanical processing of biomass would assist with working on the financial aspects of the entire cycle (Neumann et al. 2016). On account of chemical pretreatment, there are some limitations mentioned in Table 2.2, such as the high cost and corrosive effect of regents; the formation of inhibitory substances reduces the hydrolysate yield. Consequently, accomplishing higher efficiency reagents, non-corrosive reactors, and low formation of toxins by consolidating lower grouping of compound reagents with other pretreatments may help reduce costs (Paudel et al. 2017). In the case of ionic liquid pretreatment, there are more advantages over other solvents, but their purification step again adds cost to the process (Capecchi et al. 2015). Notwithstanding the benefits, biological treatment has a few disadvantages, such as optimum growth conditions, bigger space, longer treatment time, and loss of sugars (Paudel et al. 2017). Still, the formation of inhibitory substances is low when compared with the other two methods.

Pretreated fractions undergo enzymatic saccharification, which is fermented into different sugars. To obtain a high sugar yield, appropriate enzyme cocktails should be added to accelerate the reaction, which is a big concern. CBP (SSF or SSCF) or one-pot processing techniques are promising methods where the entire process can be done in a single reactor to reduce the reactor and enzyme costs. The use of recombinant microorganisms will increase the yield of the product as well (Edger et al. 2020). A few companies such as Genomatica, Global Yeast, Taurus, and Xylome are involved in developing biocatalysts that could be straightforwardly applied in LB biorefineries (Anuj et al. 2018). More advanced research studies are required in the area of recovery and purification of value-added products. After fermentation, the product undergoes various downstream processes and purification steps (distillation, centrifugation, crystallization, concentration, etc.). According to Ragauskas et al. (2014), a byproduct such as lignin is isolated and utilized for energy production. It adds value to LB biorefineries, as various products that can be derived from them have immense applications that require further studies.

2.5.3 Economic Challenges

Investors mainly get attracted to biorefineries because of the biomass availability and cost of production. The main capital cost needed is in the biomass processing, pretreatment, equipment, operations, enzymatic hydrolysis, reactors, reagents, biocatalysts, and purification costs (depending on each product). Combining hydrolysis and fermentation steps can save money to some extent. Lignin can be utilized to produce bioproducts that increase the return on investment and also support a zero-waste policy. Economic research of feedstock is still ongoing. When scaling up the product, the capital cost will also increase (Raud et al. 2019). Karimi and Karimi (2019), who conducted a techno-economic study on biobutanol production (10.48GL), sold out at a price of $1.29/L, where the total cost of production is estimated at around $0.66/L. The yield of a product depends basically on the type of biomass and pretreatment selected. Therefore, the appropriate selection of biomass and pretreatment method will allow reducing the percentage of total processing cost. For instance, biomass, pretreatment, and saccharification account for 65% of the total cost, and the remaining 35% includes fermentation, recovery, and purification processes (Huang et al. 2014).

Every progression that is present in the LB biorefinery process ought to be mechanized and automated to get the desired bioproduct with a high yield. Numerous biorefinery plants have been set up in the last two decades in many countries, but their full operation has not been accomplished so far because of the absence of mechanization and automation of the different steps to processing at pilot scale, from the collection of biomasses to biorefinery processes such as pretreatment, saccharification, later fermentation, and finally downstream processing and product purification. The entire process is systematically regulated and controlled to gain a value-added product. Based on the studies of Haro et al. (2014), automation will increase in the case of producing multiple bioproducts from a single biomass, which includes energy coordination and item broadening that reduces the hurdles of investment.

2.6 CONCLUSION

LB possesses tough and strong physical and chemical structures to protect itself from natural stress. Therefore, the biorefining process to convert lignocellulose to value-added products requires multistep processes, including pretreatment, hydrolysis, fermentation, and separation. This classic theme of the process leads to nonviable and nonfeasible processes for industrial production. The developments in research and technology have been conducted by combining physical, chemical, and biological methods to improve the deconstruction and conversion of lignocellulose to monomeric sugars. The variations of selected microorganisms allow the conversion of LB derivative sugars to various platform chemicals and complex products that are in demand in the global market. The challenges in scale-up, technical limitations, and economic feasibility are still the main targets for improvement to achieve the successful commercialization of these value-added LB derivatives in the global market.

ACKNOWLEDGMENT

The authors would like to thank King Mongkut's University of Technology, North Bangkok (Research University Grant No. KMUTNB-BasicR-64-37) for the financial support of this work.

REFERENCES

Abrego U, Chen Z and Wan C, 2017, Consolidated bioprocessing systems for cellulosic biofuel production. In Yebo Li, Xumeng Ge, eds., *Advances in Bioenergy*. Elsevier, the Netherlands, 2: pp. 143–182. https://www.sciencedirect.com/bookseries/advances-in-bioenergy/vol/2/suppl/C

Aho A, Roggan S, Eranen K, Salmi T and Murzin DY, 2015, Continuous hydrogenation of glucose with ruthenium on carbon nanotube catalysts. *Catalysis Science & Technology*, 5: 953–959.

Akkharasinphonrat R, Douzou T and Sriariyanun. 2017. Development of ionic liquid utilization in biorefinery process of lignocellulosic biomass. *KMUTNB International Journal of Applied Science and Technology*, 10(2): 89–96.

Amit K, Archana G and Dharm D, 2016, Biotechnological transformation of lignocellulosic biomass in to industrial products: An overview. *Advances in Bioscience and Biotechnology*, 7: 149–168. https://doi.org/10.4236/abb.2016.73014.

Amnuaycheewa P, Hengaroonprasan R, Rattanaporn K, Kirdponpattara S, Cheenkachorn K, Sriariyanun M. 2016. Enhancing enzymatic hydrolysis and biogas production from rices-traw by pretreatment with organic acids. *Industrial Crops and Products*, 84: 247–254.

Amnuaycheewa P, Rodiahwati W, Sanvarinda P, Cheenkachorn K, Tawai A, Sriariyanun M. 2017. Effect of organic acid pretreatment on Napier grass (Pennisetum purpureum) straw biomass conversion. *KMUTNB International Journal of Applied Science and Technology*, 10(2): 107–117.

Antonella A, Peter C, Chien-Yuan L, Davinia S and Violeta SN, 2016, Development of lignocellulosic biorefinery technologies: Recent advances and current challenges. *Australian Journal of Chemistry*, 69(11): 1201–1218. https://doi.org/10.1071/CH16022.

Anuj KC, Vijay KG, Akhilesh KS, Felipe AFA and Silvio SDS, 2018, Review: The path forward for lignocellulose biorefineries: Bottlenecks, solutions, and perspective on commercialization. *Bioresource Technology*, 264: 370–381. https://doi.org/10.1016/j.biortech.2018.06.004.

Asgher M, Ahmad Z and Iqbal HMN, 2013, Alkali and enzymatic delignification of sugarcane bagasse to expose cellulose polymers for saccharification and bio-ethanol production. *Industrial Crops and Products*, 44: 488–495.

Asgher M, Iqbal HMN and Asad MJ, 2012, Kinetic characterization of purified laccase produced from *Trametes versicolor* IBL-04 in solid state bio-processing of corncobs. *BioResources*, 7: 1171–1188.

Aya Z and Gabriel P, 2019, Lignocellulosic biomass: Understanding recalcitrance and predicting hydrolysis. *Frontiers in Chemistry*, 7: 874. https://doi.org/10.3389/fchem.2019.00874.

Badiei M, Asim N, Jahim JM and Sopian K, 2014, Comparison of chemical pretreatment methods for cellulosic biomass. *APCBEE Procedia*, 9: 170–174.

BBC Research, 2021, Biorefinery Products: Global Markets. https://www.bccresearch.com/market-research/energy-and-resources/biorefinery-products-markets-report.html.

Beatriz G, Jovita M, Gabriel M, Juan A, Melero and Jose I, 2020, Article: Production of sorbitol via catalytic transfer hydrogenation of glucose. *Applied Sciences*, 10: 1843. https://doi.org/10.3390/app10051843.

Belhadj B, Bederina M, Dheilly RM, Mboumba-Mamboundou LB and Queneudec M, 2020, Evaluation of the thermal performance parameters of an outside wall made from lignocellulosic sand concrete and barley straws in hot and dry climatic zones. *Energy & Buildings*. 225: 110348. https://doi.org/10.1016/j.enbuild.2020.110348

Benocci T, Aguilar-Pontes MV, Kun RS, Lubbers RJM, Lail K, Wang M, Lipzen A, Ng V, Grigoriev IV, Seiboth B, Daly P and de Vries RP, 2019, Deletion of either the regulatory gene ara1 or metabolic gene xki1 in *Trichoderma reesei* leads to increased CAZyme gene expression on crude plant biomass. *Biotechnology for Biofuels*, 12: 81.

Brandt A, Grasvik J, Hallett JP and Welton T, 2013, Deconstruction of lignocellulosic biomass with ionic liquids. *Green Chemistry*, 15(3): 550–583.

Capecchi L, Galbe M, Barbanti L and Wallberg O, 2015, Combined Ethanol and Methane Production Using Steam Pretreated Sugarcane Bagasse. *Industrial Crops and Products*, 74: 255–262.

Cavalagilo G, Gelosia M, Ingles D, Pompili E, D'Antonio S and Cotana F, 2017, Response surface methodology for the optimization of cellulosic ethanol production from Phragmites australis through pre-saccharification and simultaneous saccharification and fermentation. *Industrial Crops and Products*, 83: 431–437.

Chandel AK, Silva SS, Carvalho W and Singh OV, 2012, Sugarcane bagasse and leaves: Foreseeable biomass of biofuel and bio-products. *Journal of Chemical Technology & Biotechnology*, 87: 11–20.

Chandel AK and Silveira MHL, 2017, *Sugarcane Bio-Refinery: Technologies, Commercialization, Policy Issues and Paradigm Shift*. Elsevier Press, Sao Paulo, Brazil.

Chang VS and Holtzapple MT, 2000, Fundamental factors affecting biomass enzymatic reactivity. *Applied Biochemistry and Biotechnology*, 84: 5–37. https://doi.org/10.1007/978-1-4612-1392-5_1.

Chen Y, Wu Y, Zhu B, Zhang G and Wei N, 2018, Co-fermentation of cellobiose and xylose by mixed culture of recombinant Saccharomyces cerevisiae and kinetic modeling. *PLoS One* 13: 199104. https://doi.org/10.1371/journal.pone.0199104.

Cheng YS, Mutrakulcharoen P, Chuetor S, Cheenkachorn K, Tantayotai P, Panakkal EJ and Sriariyanun M, 2020, Recent situation and progress in biorefining process of lignocellulosic biomass: Toward green economy. *Applied Science and Engineering Progress*, 13(4): 299–311.

Cherubini F, 2010, The biorefinery concept: Using biomass instead of oil for producing energy and chemicals. *Energy Conversion and Management*, 51: 1412–1421.

Dahunsi S, 2019, Mechanical pretreatment of lignocelluloses for enhanced biogas production: Methane yield prediction from biomass structural components. *Bioresource Technology*, 280: 18–26.

Danping J, Xueting Z, Xumeng G, Tian Y, Tian Z, Yang Z, Zhiping Z, Chao H, Chaoyang L and Quanguo Z, 2021, Insights into correlation between hydrogen yield improvement and glycerol addition in photo-fermentation of *Arundo donax* L. *Bioresource Technology*, 321: 124467. https://doi.org/10.1016/j.biortech.2020.124467.

Daza SLV, Orrego ACE and Cardona ACA, 2016, Supercritical fluids as a green technology for the pretreatment of lignocellulosic biomass. *Bioresource Technology*, 199: 113–120.

De Jong E and Jungmeier G, 2015, Biorefinery concepts in comparison to petrochemical refineries. *Industrial Biorefineries & White Biotechnology*, 3–33. http://dx.doi.org/10.1016/B978-0-444-63453-5.00001-X.

Deshpande SK, Bhotmange MG, Chakrabarti T and Shastri PN, 2008, Production of cellulase and xylanase by Trichoderma reesei (QM 9414 mutant), *Aspergillus niger* and mixed culture by solid state fermentation (SSF) of water hyacinth (*Eichhornia crassipes*). *Indian Journal of Chemical Technology*, 15(5): 449–456.

Donatella C, Lucio Z, Sergio D, Licia L, Giovanna R, Olimpia P, Vincenza F and Chiara S, 2019, Improved production of succinic acid from *Basfa succiniciproducens* growing on *A. donax* and process evaluation through material flow analysis. *Biotechnology for Biofuels*, 12: 22. https://doi.org/10.1186/s13068-019-1362-6.

Dufosse K, Ben AW and Gabrielle B, 2017, Life-cycle assessment of agricultural feedstock for biorefineries. In *Life-Cycle Assessment of Biorefineries*. Elsevier, pp. 77–96. https://doi.org/10.1016/b978-0-444-63585-3.00003-6, ISBN 9780444635853.

Eastern Economic Corridor (EEC), 2020, Biofuel and Biochemical, Thailand https://www.eeco.or.th/web-upload/fck/editor-pic/files/industry/Biofuel%20and%20Biochemical_2020.pdf.

Edem CB and Moses M, 2013, Review article chemical pretreatment methods for the production of cellulosic ethanol: Technologies and innovations. *International Journal of Chemical Engineering*, 1–21. http://dx.doi.org/10.1155/2013/719607.

Edgar O, Anusuiya S, Rubi C, Raul T and Hector AR, 2020, Review: Consolidated bioprocessing, an innovative strategy towards sustainability for biofuels production from crop residues: An overview. *Agronomy*, 10: 1834. https://doi.org/10.3390/agronomy10111834.

Elieber BB, Danyelle CF, Dayanne DSM, Morsyleide de FR, Joao PSM, Edcleide MA and Renate MRW, 2016, Processing and properties of PCL/cotton linter compounds. *Materials Research*. 1–9. http://dx.doi.org/10.1590/1980-5373-MR-2016-0084.

Eric A, Rebecca A, Matthew M, Adebosola O, Andrew W and Thomas V, 2011, Review growth and agronomy of Miscanthus × giganteus for biomass production. *Biofuels*, 2(2): 167–183.

Erik G, Nadine P and Nina H, 2019, A path transition towards a bioeconomy-the crucial role of sustainability. *Sustainability*, 11(11): 3005. https://doi.org/10.3390/su11113005.

European Bioplastics (EB), 2017, What Are Bioplastics? https://www.european-bioplastics.org/bioplastics/.

Fabricio CP, Carolina BCP and Jonas C, 2018, Prospective biodegradable plastics from biomass conversion processes. *Biofuels - State of Development*, 245–271. http://dx.doi.org/10.5772/intechopen.75111.

FAOSTAT Analytical Brief 23, Gross domestic product and agriculture value added 1970–2019. Global and regional trends. http://www.fao.org/3/cb4651en/cb4651en.pdf

Faruq M, Hamad A and Mohammad J, 2020, *Sustainable Nanocellulose and Nanohydrogels from Natural Sources*, First edition. Elsevier, the Netherlands p. 424.

Fredina D, Rike Y, Yuyun I, Andri H, Yu-I H and Hiroshi U, 2021, Temperature driven structural transition in the nickel-based catalytic graphitization of coconut coir. *Diamond & Related Materials*, 117: 108443. https://doi.org/10.1016/j.diamond.2021.108443.

Gericke D, Ott D, Matveeva VG, Sulman E, Aho A, Murzin DY, Roggan S, Danilova L, Hessel V, Loeb P, et al., 2015, Green catalysis by nanoparticulate catalysts developed for flow processing? Case study of glucose hydrogenation. *RSC Advances*, 5: 15898–15908.

Gil-Lopez DIL, Lois-Correa JA, Sanchez-Pardo ME, Dominguez-Crespo MA, Torres-Huerta AM, Rodriguez-Salazar AE and Orta-Guzman VN, 2019, Production of dietary fibers from sugarcane bagasse and sugarcane tops using microwave-assisted alkaline treatments. *Industrial Crops & Products*, 135: 159–169. https://doi.org/10.1016/j.indcrop.2019.04.042.

Global Bioenergy Statistics, 2020, *World Bioenergy Association*, 1–64. http://www.worldbioenergy.org/uploads/201210%20WBA%20GBS%202020.pdf.

Goldy DB, Ajit KS and Ramkrishna S, 2018, Review: Lignocellulosic biorefinery as a model for sustainable development of biofuels and value added products. *Bioresource Technology*, 247: 1144–1154. https://doi.org/10.1016/j.biortech.2017.09.163.

Graminha EBN, Gonçalves AZL, Pirota RDPB, Balsalobre MAA, Da Silva R and Gomes E, 2008, Review Enzyme production by solid-state fermentation: Application to animal nutrition. *Animal Feed Science and Technology*, 144(1–2): 1–22.

Gu TY, Held MA and Faik A, 2013, Supercritical CO_2 and ionic liquids for the pretreatment of lignocellulosic biomass in bioethanol production. *Environmental Technology*, 34: 1735–1749. https://doi.org/10.1080/09593330.2013.809777.

Gundupalli MP, Rattanaporn K, Chuetor S, Rodiahwati W, Sriariyanun M. 2021a, Biocomposite production from Ionic liquids (IL) assisted processes using biodegradable biomass. In *Toward the Value-Added Biocomposites: Technology, Innovation and Opportunity*. CRC Press. https://doi.org/10.1201/9781003137535-9ss and Biosystems Engineering. https://doi.org/10.1007/s00449-021-02607-6.

Gundupalli MP, Sahithi AST, Cheng Y, Tantayotai P and Sriariyanun M, 2021b, Differential effects of inorganic salts on cellulase kinetics in enzymatic saccharification of cellulose and lignocellulosic biomass. *Bioprocess and Biosystems Engineering*, 44(11): 2331–2344.

Gupta VK, Kubicek CP, Berrin JG, Wilson DW, Couturier M, Berlin A, Filho EXF and Ezeji T, 2016, Fungi enzymes for bio-products from sustainable and waste biomass. *Trends in Biochemical Sciences*, 41: 633–645.

Haiyan C, Yingjuan F, Zhaojiang W and Menghua Q, 2015, Degradation and redeposition of the chemical components of aspen wood during hot water extraction. *Bioresources*, 10(2): 3005–3016.

Hanna P, Małgorzata K, Paweł W, Przemysław S, Agata G, Adrianna Z and Michał P, 2020, Article: Sustainable production of 5-hydroxymethylfurfural from pectin-free sugar beet pulp in a simple aqueous phase system-optimization with doehlert design. *Energies*, 13: 5649. https://doi.org/10.3390/en13215649.

Haro P, Perales LV, Arjona R and Ollero P, 2014, Thermochemical biorefineries with multiproduction using a platform chemical. *Biofuels, Bioproducts and Biorefinery*, 8, 155–170.

Hassan A, Hiba NAT, Roland S, Augchararat K and Joachim V, 2020, Article: Production of lactic acid from carob, banana and sugarcane lignocellulose biomass. *Molecules*, 25: 2956. https://doi.org/10.3390/molecules25132956.

Hasunuma T and Kondo A, 2012, Consolidated bioprocessing and simultaneous saccharification and fermentation of lignocellulose to ethanol with thermotolerant yeast strains. *Process Biochemistry*, 47: 1287–1294.

Herbaut M, Zoghlami A, Habrant A, Falourd X, Foucat L, Chabbert B, et al., 2018, Multimodal analysis of pretreated biomass species highlights generic markers of lignocellulose recalcitrance. *Biotechnology and Biofuels*, 11: 52. https://doi.org/10.1186/s13068-018-1053-8.

Hu Y, Kwan TH, Daoud WA and Lin CSK, 2017, Continuous ultrasonic-mediated solvent extraction of lactic acid from fermentation broths. *Journal of Cleaner Production*, 145: 142–150.

Huang J, Chen D, Wei Y, Wang Q, Li Z, Chen Y and Huang R, 2014, Direct ethanol production from lignocellulosic sugars and sugarcane bagasse by a recombinant *Trichoderma reesei* strain HJ48. *The Scientific World Journal*, 2014: 798683.

Iqbal HMN and Asgher M, 2013, Characterization and decolorization applicability of xerogel matrix immobilized manganese peroxidase produced from Trametes versicolor IBL-04. *Protein & Peptide Letters*, 5: 591–600.

Isikgor FH and Becer CR, 2015, Lignocellulosic biomass: A sustainable platform for the production of bio-based chemicals and polymers. *Polymer Chemistry*, 6: 4497–4559. https://doi.org/10.1039/C5PY00263J.

Jahirul MI, Rasul MG, Chowdhury AA and Ashwath N, 2012, Biofuels Production through Biomass Pyrolysis - A technological review. *Energies*, 5: 4952–5001. https://doi.org/10.3390/en5124952.

Javier S, Maria DC, Nicolas R and Jesus F, 2019, Biomass resources. *The Role of Bioenergy in the Emerging Bioeconomy*: 25–111. https://doi.org/10.1016/B978-0-12-813056-8.00002-9.

Jerico A, Andro M, Logan H and Shaun S, 2017, Article: Direct succinic acid production from minimally pretreated biomass using sequential solid-state and slurry fermentation with mixed fungal cultures. *Fermentation*, 3: 30. https://doi.org/10.3390/fermentation3030030.

Jiang Y, Xin F, Lu J, Dong W, Zhang W, Zhang M, Wu H, Ma J and Jiang M, 2017, State of the art review of biofuels production from lignocellulose by thermophilic bacteria. *Bioresource Technology*. 245: 1498–1506.

Jing Y, Hao X, Jianchun J, Ning Z, Jingcong X, Jian Z, Quan B and Min W, 2020, Itaconic acid production from undetoxified enzymatic hydrolysate of bamboo residues using *Aspergillus terreus*. *Bioresource Technology*, 307: 123208. https://doi.org/10.1016/j.biortech.2020.123208.

Jung KA, Woo SH, Lim SR and Park JM, 2015, Pyrolytic production of phenolic compounds from the lignin residues of bioethanol processes. *Chemical Engineering Journal*, 259: 107–116.

Junqueira TL, Cavalett O and Bonomi A, 2016, The virtual sugarcane biorefinery-A simulation tool to support public policies formulation in bioenergy. *Industrial Biotechnology*, 12(1): 62–67.

Kamm B and Kamm M, 2004, Principles of biorefineries. *Applied Microbiology and Biotechnology*, 64: 137–145.

Karimi AM and Karimi K, 2019, Biobutanol production from corn stover in the US. *Industrial Crops and Products*, 129: 641–653.

Kawaguchi H, Hasunuma T, Ogino C and Kondo A, 2016, Bioprocessing of bio-based chemicals produced from lignocellulosic feedstocks. *Current Opinion in Biotechnology*, 42: 30–39.

Keller V, Lyseng B, English J, Niet T, Palmer-Wilson K, Moazzen I, Bryson R, Peter W and Andrew R, 2018, Coal-to-biomass retrofit in Alberta -value of forest residue bioenergy in the electricity system. *Renewable Energy*, 125: 373–383.

Khalil SRA, Abdelhafez AA and Amer EAM, 2015, Evaluation of bioethanol production from juice and bagasse of some sweet sorghum varieties. *Annals of Agricultural Sciences*, 60(2): 317–324. https://doi.org/10.1016/j.aoas.2015.10.005.

Khatun MM, Yu X, Kondo A, Bai F and Zhao X, 2017, Improved ethanol production at high temperature by consolidated bioprocessing using *Saccharomyces cerevisiae* strain engineered with artificial zinc finger protein. *Bioresource Technology*, 245: 1447–1454.

Kirtika K, Ravindra P and Brajendra KS, 2019, Review bio-based chemicals from renewable biomass for integrated biorefineries. *Energies*, 12: 233. https://doi.org/10.3390/en12020233.

Kondusamy D, Saumya A, Mehak K and Karthik R, 2020, Economics and cost analysis of waste biorefineries. *Refining Biomass Residues for Sustainable Energy and Bioproducts*: 545–565. https://doi.org/10.1016/b978-0-12-818996-2.00025-9.

Kruyeniski J, Ferreira PJ, Carvalho MGVS, Vallejos ME, Felissia FE, Area MC, et al., 2019, Physical and chemical characteristics of pretreated slash pine sawdust influence its enzymatic hydrolysis. *Industrial Crops & Products*, 130: 528–536. https://doi.org/10.1016/j.indcrop.2018.12.075.

Kumar A, Dharm D and Archana G, 2016, Production of crude enzyme from Aspergillus nidulans AKB-25 using black gram residue as the substrate and its industrial applications. *Journal of Genetic Engineering and Biotechnology*, 14: 107–118. http://dx.doi.org/10.1016/j.jgeb.2016.06.004.

Kumari D and Singh R, 2018, Pretreatment of lignocellulosic wastes for biofuel production: A critical review. *Renewable & Sustainable Energy Reviews*, 90: 877–891.

Lamsal B, Yoo J, Brijwani K and Alavi S, 2010, Extrusion as a thermomechanical pre-treatment for lignocellulosic ethanol. *Biomass Bioenergy,* 34: 1703–1710. http://dx.doi.org/10.1016/j.biombioe.2010.06.009.

Laura C and Vincenza F, 2016, Green methods of lignocellulose pretreatment for biorefinery Development. *Applied Microbiology and Biotechnology*, 100: 9451–9467. http://dx.doi.org/10.1007/s00253-016-7884-y.

Lenka B, Maros S, Alica B and Maros S, 2017, Review utilization of waste from coffee production. *Faculty of Materials Science and Technology in Trnava Slovak University of Technology in Bratislava*, 25(40): 91–101.

Leu SY and Zhu J, 2013, Substrate-related factors affecting enzymatic saccharification of lignocelluloses: Our recent understanding. *Bioenergy Research*, 6: 405–415. http://dx.doi.org/10.1007/s12155-012-9276-1.

Liang G, Cheng H, Li W, He L, Yu Y and Zhao F, 2012, Selective conversion of microcrystalline cellulose into hexitols on nickel particles encapsulated within ZSM-5 zeolite. *Green Chemistry*, 14: 2146–2149.

Lin B and Tao Y, 2017, Whole-cell biocatalysts by design. *Microbial Cell Factories*, 16: 106.

Liu J, Shi P, Ahmad S, Yin C, Liu X, Liu Y, Zhang H, Xu Q, Yan H and Li QX, 2019, Co-culture of *Bacillus coagulans* and *Candida utilis* efficiently treats Lactobacillus fermentation wastewater. *AMB Express*, 9: 15.

Liu S, Okuyama Y, Tamura M, Nakagawa Y, Imai A and Tomishige K, 2016, Selective transformation of hemicellulose (xylan) into n-pentane, pentanols or xylitol over a rhenium-modified iridium catalyst combined with acids. *Green Chemistry*, 18: 165–175.

Liu Y, Chen W, Xia Q, Guo B, Wang Q, Liu S, et al., 2017, Efficient cleavage of lignin-carbohydrate complexes and ultrafast extraction of lignin oligomers from wood biomass by microwave-assisted treatment with deep eutectic solvent. *ChemSusChem*, 10: 1692–1700.

Liu Z, Inokuma K, Ho S-H, Haan RD, Hasunuma T, van Zyl WH, et al., 2015, Combined cell-surface display- and secretion-based strategies for production of cellulosic ethanol with *Saccharomyces cerevisiae*. *Biotechnology for Biofuels*, 8: 162.

Liu ZH, Qin L, Li BZ and Yuan YJ, 2014, Physical and chemical characterizations of corn stover from leading pretreatment methods and effects on enzymatic hydrolysis. *ACS Sustainable Chemistry & Engineering*, 3: 140–146. http://dx.doi.org/10.1021/sc500637c.

Lu M, Li J, Han L and Xiao W, 2019, An aggregated understanding of cellulase adsorption and hydrolysis for ball-milled cellulose. *Bioresource Technology*, 273: 1–7. http://dx.doi.org/10.1016/j.biortech.2018.10.037.

Lv S, Yu Q, Zhuang X, Yuan Z, Wang W, Wang Q, et al., 2013, The influence of hemicellulose and lignin removal on the enzymatic digestibility from sugarcane bagasse. *Bioenergy Research*, 6: 1128–1134. http://dx.doi.org/10.1007/s12155-013-9297-4.

Madhumitha J, 2021, U.S. Biomass Energy – Statistics and facts. *Energy and Environment*. https://www.statista.com/topics/1000/biomass-energy/#dossierSummary.

Mamman S, Lee JM, Kim YC, Hwang IT, Park NJ, Hwang YK, Chang JS and Hwang JS, 2008, Furfural: Hemicellulose/xylose derived biochemical. *Biofuels, Bioproducts and Biorefining*, 2: 438–454.

Mandana E, Azra VS and Renata B, 2016, Fermentation of lignocellulosic sugars to acetic acid by *Moorella thermoacetica*. *Journal of Industrial Microbiology and Biotechnology*, 43(6): 807–816. http://dx.doi.org/10.1007/s10295-016-1756-4.

Mao LY, Zhang L, Gao NB and Li AM, 2012, $FeCl_3$ and acetic acid co-catalyzed hydrolysis of corncob for improving furfural production and lignin removal from residue. *Bioresource Technology*, 123: 324–331.

Maria H and Gerfried J, 2019, Biorefineries. *The Role of Bioenergy in the Bioeconomy*: 179–222. https://doi.org/10.1016/B978-0-12-813056-8.00005-4.

Marie KW, Hussein FK and Khalida AS, 2018, Production of Xylitol from Agricultural Waste by Enzymatic Methods. *American Journal of Agricultural and Biological Sciences*, 13(1): 1–8. https://doi.org/10.3844/ajabssp.2018.1-8.

Matsagar BM, Hossain SA, Islam T, Alamri HR, Alothman ZA, Yamauchi Y, Dhepe PL and Wu KCW, 2017, Direct production of furfural in one-pot fashion from raw biomass using Bronsted acidic ionic liquids. *Scientific Reports*, 7: 13508.

Masran R, Zanirun Z, Bahrin EK, Ibrahim MF, Yee PL and Abd-Aziz S, 2016, Harnessing the potential of ligninolytic enzymes for lignocellulosic biomass pretreatment. *Applied Microbiology and Biotechnology*, 100: 5231–5246.

Matty J, 2013, Market potential of biorefinery products. *Systems Perspectives on Biorefineries*, 30–41. https://doi.org/10.13140/2.1.1157.6007.

Merklein K, Fong SS and Deng Y, 2016, Biomass utilization. *Biotechnology for Biofuel Production and Optimization*, 291–324. https://doi.org/10.1016/B978-0-444-63475-7.00011-X.

Michailos S, Parker D and Webb C, 2019, Design, sustainability analysis and multiobjective optimisation of ethanol production via syngas fermentation. *Waste Biomass Valorization*, 10: 865–876. https://doi.org/10.1007/s12649-017-0151-3.

Mohamad NL, Kamal SMM and Mokhtar MN, 2015, Xylitol bioproduction: A review of recent studies. *Food Reviews International*, 31: 74–89. https://doi.org/10.1080/87559129/2014/961077.

Mohammad WA, Jameel SAA, Sawsan I, Ghadeer Q, Marc M and Omar SA, 2021, Potential use of corn leaf waste for biofuel production in Jordan (physio-chemical study). *Energy*, 214: 118863.

Mohammed IY, Yousif AA, Feroz KK, Suzana Y, Ibraheem A and Soh AC, 2015, Comprehensive characterization of napier grass as a feedstock for thermochemical conversion. *Energies*, 8: 3403–3417.

Moset V, Xavier CDAN, Feng L, Wahid R and Moller HB, 2018, Combined low thermal alkali addition and mechanical pre-treatment to improve biogas yield from wheat straw. *Journal of Cleaner Production*, 172: 1391–1398.

Mupondwa E, Li X, Tabil L, Sokhansanj S and Adapa P, 2017, Status of Canada's lignocellulosic ethanol: Part I: Pretreatment technologies. *Renewable Sustainable Energy Review*, 72: 178–190.

Nanda S, Mohammad J, Reddy S, Kozinski J and Dalai A, 2014, Pathways of lignocellulosic biomass conversion to renewable fuels. *Biomass Conversion and Biorefinery*, 4: 157–191. http://dx.doi.org/10.1007/s13399-013-0097-z.

Neumann P, Pesante S, Venegas M and Vidal G, 2016, Developments in pre-treatment methods to improve anaerobic digestion of sewage sludge. *Reviews in Environmental Science and Bio/Technology*, 15: 173–211.

Nyika J, Adediran AA, Olayanju A, Adesina OS, Edoziuno FO, 2020, The potential of Biomass in Africa and the debate on its carbon neutrality. In *Biotechnological Applications of Biomass*. InTech Open. http://dx.doi.org/10.5772/intechopen.93615.

Pan X, Gilkes N and Saddler JN, 2006, Effect of acetyl groups on enzymatic hydrolysis of cellulosic substrates. *Holzforschung*, 60: 398–401. http://dx.doi.org/10.1515/HF.2006.062.

Panakkal EJ, Cheng YS, Phusantisampan T, Sriariyanun M. 2021. Deep eutectic solvent mediated process for productions of sustainable polymeric biomaterials. In *Toward the Value-Added Biocomposites: Technology, Innovation and Opportunity*. CRC Press. http://dx.doi.org/10.1201/9781003137535-10.

Patthra P, Chakrit T, Pornpan P, Prathana K, Rattiya W, Akihigo K and Khanok R, 2020, One - step biohydrogen production from cassava pulp using novel enrichment of anaerobic thermophilic bacteria community. *Biocatalysis and Agricultural Biotechnology*, 27: 101658. https://doi.org/10.1016/j.bcab.2020.101658.

Paudel SR, Banjara SP, Choi OK, Park KY, Kim YM and Lee JW, 2017, Pretreatment of agricultural biomass for anaerobic digestion: Current state and challenges. *Bioresource Technology*, 245: 1194–1205.

Peciulyte A, Karlstrom K, Larsson PT and Olsson L, 2015, Impact of the supramolecular structure of cellulose on the efficiency of enzymatic hydrolysis. *Biotechnology and Biofuels*, 8: 56. http://dx.doi.org/10.1186/s13068-015-0236-9.

Perea-Moreno M, Francisco M and Alberto-Jesus P, 2018, Sustainable energy based on sunflower seed husk boiler for residential buildings. *Sustainability*, 10: 3407. http://dx.doi.org/10.3390/su10103407.

Perez GP, Mukherjee A and Dumont MJ, 2019, Insights into HMF catalysis. *Journal of Industrial and Engineering Chemistry*, 70: 1–34.

Ping-Ping Z, Jiao M and Jie B, 2016, Fermentative production of high titer citric acid from corn stover feedstock after dry dilute acid pretreatment and biodetoxification. *Bioresource Technology*, 224: 563–572. http://dx.doi.org/10.1016/j.biortech.2016.11.046.

Radhakumari M, Suresh KB, Satyavathi B and Andrew SB, 2016, A review on 1st and 2nd generation bioethanol production-recent progress. *Journal of Sustainable Bioenergy Systems*, 6(3): 72–92. http://dx.doi.org/10.4236/jsbs.2016.63008.

Ragauskas AJ, Beckham GT, Biddy MJ, Chandra R, Chen F, Davis MF, et al., 2014, Lignin valorization: Improving lignin processing in the biorefinery. *Science*, 344: 1246843. http://dx.doi.org/10.1126/science.1246843.

Rahul S, Carlos SO, Krishnamoorthy H, Satinder KB, Sara M, Pierre V and Antonio A, 2020, Lignocellulosic biomass-based biorefinery: An insight into commercialization and economic standout. *Current Sustainable/Renewable Energy Reports*, 7: 122–136. http://dx.doi.org/10.1007/s40518-020-00157-1.

Raud M, Kikas T, Sippula O and Shurpali NJ, 2019, Potentials and challenges in lignocellulosic biofuel production technology. *Renewable & Sustainable Energy Reviews*, 111: 44–56.

Robak K and Balcerek M, 2018, Review of second generation bioethanol production from residual biomass. *Food Technology Biotechnology*, 56: 174–187. http://dx.doi.org/10.17113/ftb.56.02.18.5428.

Rodiahwati W and Sriariyanun M, 2016, Lignocellulosic biomass to biofuel production: Integration of chemical and extrusion (screw press) pretreatment. *KMUTNB International Journal of Applied Science & Technology*, 9(4): 289–298.

Romero-Guiza MS, Wahid R, Hernandez V, Moller H and Fernandez B, 2017, Improvement of wheat straw anaerobic digestion through alkali pre-treatment: Carbohydrates bioavailability evaluation and economic feasibility. *Science of the Total Environment*, 595: 651–659.

Saad MBW, Oliveira LRM, Candido RG, Quintana G, Rocha GJM and Gonçalves AR, 2008, Preliminary studies on fungal treatment of sugarcane straw for organosolv pulping. *Enzyme and Microbial Technology*, 43(2): 220–225.

Samaneh K, Paridah MT, Ali K, Alain D and Ali A, 2013, Kenaf bast cellulosic fibers hierarchy: A comprehensive approach from micro to nano. *Carbohydrate Polymers*, 101: 878–885. http://dx.doi.org/10.1016/j.carbpol.2013.09.106.

Sanford K, Chotani G, Danielson N and Zahn JA, 2016, Scaling up of renewable chemicals. *Current Opinion in Biotechnology*, 38: 112–122.

Sankaran R, Parra CRA, Pakalapati H, Show PL, Ling TC, Chen WH and Tao Y, 2020, Recent advances in the pretreatment of microalgal and lignocellulosic biomass: A comprehensive review. *Bioresource Technology*, 298: 122476–122489.

Santos VTO, Siqueira G, Milagres AMF and Ferraz A, 2018, Role of hemicellulose removal during dilute acid pretreatment on the cellulose accessibility and enzymatic hydrolysis of compositionally diverse sugarcane hybrids. *Industrial Crops & Products*, 111: 722–730. http://dx.doi.org/10.1016/j.indcrop.2017.11.053.

Saowalak A, Awanwee P, Donato G, Vito D, Wonnop V and Giuseppina A, 2020, The effectiveness of durian peel as a multi-mycotoxin adsorbent. *Toxins*, 12: 108.

Shadbahr J, Khan F and Zhang Y, 2017, Kinetic modeling and dynamic analysis of simultaneous saccharification and fermentation of cellulose to bioethanol. *Energy Conversion and Management*, 141: 236–243. http://dx.doi.org/10.1016/j.enconman.2016.08.025.

Sharma S, Kumar R, Gaur R, Agrawal R, Gupta RP, Tuli DK and Das B, 2015, Pilot scale study on steam explosion and mass balance for higher sugar recovery from rice straw. *Bioresource Technology*, 175: 350–357. http://dx.doi.org/10.1016/j.biortech.2014.10.112.

Show PL, Sriariyanun M. 2021. Prospect of liquid biphasic system in microalgae research. *Applied Science and Engineering Progress*, 14(3): 295–296. http://dx.doi.org/10.14416/j.asep.2020.12.001.

Shruti AB and Kalburgi PB, 2016, Production of bioethanol from waste newspaper. *Procedia Environmental Sciences*, 35: 555–562.

Silva GG, Couturier M, Berrin JG, Buleon A and Rouau X, 2012, Effects of grinding processes on enzymatic degradation of wheat straw. *Bioresource Technology*, 103: 192–200. http://dx.doi.org/10.1016/j.biortech.2011.09.073.

Silveira MLL, Furlan SA and Ninow JL, 2008, Development of an alternative technology for the oyster mushroom production using liquid inoculum. *Food Science and Technology (Campinas)*, 28(4): 858–862.

Silvi ORP, Arsa PIDG and Tatang HS, 2017, Determining the enzyme accessibility of ammonia pretreated lignocellulosic substrates by Simon's Stain method. *Journal of Engineering and Applied Sciences*, 12(18): 5307–5312.

Singh S, Chakravarty I, Pandey KD and Kundu S, 2018, Development of a process model for simultaneous saccharification and fermentation (SSF) of algal starch to third-generation bioethanol. *Biofuels*, 1–9. http://dx.doi.org/10.1080/17597269.2018.1426162.

Siti NCK, Muhammad SJ, Amizon A, Nor SMS and Ahmad RMD, 2014, Mechanical pretreatment of lignocellulosic biomass for biofuel production. *Applied Mechanics and Materials*, 625: 838–841.

Somerville C, Youngs H, Taylor V, Davis SC and Long SP, 2010, Feedstocks for Lignocellulosic Biofuels. *Science*, 329: 790–792.

Sonil N, Ajay KD and Janusz AK, 2014, Butanol and ethanol production from lignocellulosic feedstock: Biomass pretreatment and bioconversion. *Energy Science and Engineering*, 2(3): 138–148. http://dx.doi.org/10.1002/ese3.41.

Sriariyanun M and Kitsubthawee K, 2020, Trends in lignocellulosic biorefinery for production of value-added biochemicals. *Applied Science and Engineering Progress*, 13(4): 283–284.

Sun FB and Chen HZ, 2007, Evaluation of enzymatic hydrolysis of wheat straw pretreated by atmospheric glycerol autocatalysis. *Journal of Chemical Technology and Biotechnology*, 82(11): 1039–1044.

Theuretzbacher F, Lizasoain J, Lefever C, Saylor MK, Enguidanos R, Weran N, Gronauer A and Bauer A, 2015, Steam explosion pretreatment of wheat straw to improve methane yields: Investigation of the degradation kinetics of structural compounds during anaerobic digestion. *Bioresource Technology*, 179: 299–305.

Thorsell S, Epplin FM, Huhnke RL and Taliaferro CK, 2004, Economics of a coordinated biorefinery feedstock harvest system: Lignocellulosic biomass harvest cost. *Biomass Bioenergy*, 27: 327–37.

Tong X, Ma Y and Li Y, 2010, Biomass into chemicals: Conversion of sugars to furan derivatives by catalytic processes. *Applied Catalysis A: General*, 385: 1–13.

Torr KM, Love KT, Simmons BA and Hill SJ, 2016, Structural features affecting the enzymatic digestibility of pine wood pretreated with ionic liquids. *Biotechnology and Bioengineering*, 113: 540–549. http://dx.doi.org/10.1002/bit.25831.

Transparency Market Research (TMR), 2020, Bio-based Platform Chemicals Market - Global Industry Analysis, Size, Share, Growth, Trends and Forecast 2018–2026. https://www.transparencymarketresearch.com/bio-based-platform-chemicals-market.html.

Ullah K, Sharma VK, Ahmad M, Lv P, Krahl J and Wang Z, 2018, The insight views of advanced technologies and its application in bio-origin fuel synthesis from lignocellulose biomasses waste, a review. *Renewable Sustainable Energy Review*, 82: 3992–4008.

Valdivia M, Galan JL, Laffarga J and Ramos JL, 2016, Biofuels 2020: Biorefineries based on lignocellulosic materials. *Microbial Biotechnology*, 9: 585–594.

Veluchamy C and Kalamdhad AS, 2017, Influence of pretreatment techniques on anaerobic digestion of pulp and paper mill sludge: A review. *Bioresource Technology*, 245: 1206–1219.

Viviane AE, Elen BAVP, Ana MFS, Monica ACSB, Antonio GS and Leila LYV, 2017, Study of natural fibers from waste from sponge gourd, peach palm tree and papaya pseudo stem. *International Journal of Environmental & Agriculture Research*, 3(2): 11–24.

Wan ANI and Jahim JM, 2016, Enzymatic hydrolysis of pretreated Kenaf using a recombinant xylanase: Effects of reaction conditions for optimum hemicellulose hydrolysis. *American Journal of Agricultural and Biological Sciences*, 11: 54–66. http://dx.doi.org/10.3844/ajabssp.2016.54.66.

Wang D, Xin Y, Shi H, Ai P, Yu L, Li X and Chen S, 2019, Closing ammonia loop in efficient biogas production: Recycling ammonia pretreatment of wheat straw. *Biosystems Engineering*, 180: 182–190.

Wang SZ, Sun XX and Yuan QP, 2018, Strategies for enhancing microbial tolerance to inhibitors for biofuel production. *Bioresource Technology*, 258: 302–309.

Wei H., Wen-jian L, Hai-quan Y and Ji-hong C, 2019, Current strategies and future prospects for enhancing microbial production of citric acid. *Applied Microbiology and Biotechnology*, 103: 201–209. https://doi.org/10.1007/s00253-018-9491-6.

Wertz JL and Bedue O, 2013, *Lignocellulosic Biorefineries*. CRC Press, Spain, pp. 366–375.

Xia Q, Liu Y, Meng J, Cheng W, Chen W, Liu S, Liu Y, Li J and Yu H, 2018, Multiple hydrogen bond coordination in three-constituent deep eutectic solvents enhances lignin fractionation from biomass. *Green Chemistry*, 20(12): 2711–2721.

Xiao W, Wang Y, Xia S and Ma P, 2012, The study of factors affecting the enzymatic hydrolysis of cellulose after ionic liquid pretreatment. *Carbohydrate Polymers*, 87: 2019–2023.

Xu H, Che X, Ding Y, Kong Y, Li B and Tian, et al., 2019, Effect of crystallinity on pretreatment and enzymatic hydrolysis of lignocellulosic biomass based on multivariate analysis. *Bioresource Technology*, 279: 271–280. https://doi.org/10.1016/j.biortech.2018.12.096.

Yadav M, Mishra DK and Hwang JS, 2012, Catalytic hydrogenation of xylose to xylitol using ruthenium catalyston NiO modified TiO_2 support. *Applied Catalysis A: General*, 425: 110–116.

Yoo J, Alavi S, Vadlani P and Amanor-Boadu V, 2011, Thermo-mechanical extrusion pretreatment for conversion of soybean hulls to fermentable sugars. *Bioresource Technology*, 102: 7583–7590. https://doi.org/10.1016/j.biortech.2011.04.092.

Yu H, Xiao W, Han L and Huang G, 2019, Characterization of mechanical pulverization/phosphoric acid pretreatment of corn stover for enzymatic hydrolysis. *Bioresource Technology*, 282: 69–74. https://doi.org/10.1016/j.biortech.2019.02.104.

Zahid A, Muhammad G and Muhammad I, 2014, Agro-industrial lignocellulosic biomass a key to unlock the future bio-energy: A brief review. *Journal of Radiation Research and Applied Sciences*, 7: 163–173.

Zdanowicz M, Wilpiszewska K and Spychaj T, 2018, Deep eutectic solvents for polysaccharides processing. A review. *Carbohydrate Polymer*, 200: 361–380.

Zhang K, Pei Z and Wang D, 2016, Organic solvent pretreatment of lignocellulosic biomass for biofuels and biochemicals: A review. *Bioresource Technology*, 199: 21–33.

Zhao Q, Sun Z, Wang S, Huang G, Wang X and Jiang Z, 2014, Conversion of highly concentrated fructose into 5-hydroxymethylfurfural by acid-base bifunctional HPA nanocatalysts induced by choline chloride. *RSC Advances*, 4: 63055–63061.

Zhao X, Cheng K and Liu D, 2009, Organosolv pretreatment of lignocellulosic biomass for enzymatic hydrolysis. *Applied Microbiology and Biotechnology*, 82: 815–827.

Zhao X, Zhang L and Liu D, 2012, Biomass recalcitrance Part I: The chemical compositions and physical structures affecting the enzymatic hydrolysis of lignocellulose. *Biofuels, Bioproducts and Biorefining*, 6: 465–482. http://dx.doi.org/10.1002/bbb.1331.

3 Inhibitors and Microbial Tolerance during Fermentation of Biofuel Production

Muneera Lateef
Nigde Omer Halisdemir University

Ziaul Hasan and Asimul Islam
Jamia Millia Islamia
Shared first authorship

CONTENTS

DOI: 10.1201/9781003203452-3

3.1 INTRODUCTION

Currently, concerns related to climate change and global warming have increased. As expected to the increasing prices and enough exploitation of fossil fuels, there is a legislative restriction on the use of fossil fuel energy sources that leads to options and more discoveries for biofuel production. According to statistics, the global consumption of petroleum was 86 million barrels per day in 2008, and it is expected to be 98 million barrels per day in 2020 and 112 million barrels per day in 2035 (USEIA, 2011). The main idea behind this whole campaign is to stop using fossil fuels excessively and lean towards renewable sources such as lignocellulose biomass, which is capable of producing biofuel (Sarangi and Nanda 2018). Lignocellulosic biomasses are composed of residual agricultural waste such as wood, feedstocks (which consist of agricultural and forestry remnants) (Nanda et al., 2014a), municipal trash, and glop material, and have the extraordinary ability to improve biofuel manufacturing, thereby gaining energy security and decreasing greenhouse gases (Nanda et al., 2014b).

Because of the lower price and sufficient access, lignocellulosic biomass can be an alternative to biofuel. It is reported that about 40 million litres of lignocellulose biomass are generated worldwide (Sanderson 2011). Hemicellulose, cellulose, and lignin are part of the complicated lignocellulose. Glucose polymer is an integral part of cellulose that gives strength to a plant. On the other hand, lignin ensures the strength of the whole structure of the plant (Limayem and Ricke 2012). This kind of feedstock needs a lot of physical and labour-intensive work, which is expensive, and before being converted into biofuel, it needs to go through several steps such as pretreatment for hydrolysis to take place, followed by fermentation and distillation.

Pretreatment is one of the most significant and crucial steps due to its expensive and sensitive nature for any kind of inhibition. Pretreatment involves many steps, including (i) making it amorphous to facilitate hydrolysis, (ii) increasing porosity through chemical and enzymatic hydrolysis and (iii) converting hemicellulose and lignin to cellulose (Tran et al., 2019). This hydrolysis process has many types, among which chemical hydrolysis can lead to hazardous environmental impacts. Pretreatment has many types, and physical pretreatment is one of them. The main idea is to provide available surface space for lignocellulose so that hydrolytic enzymes can work by shrinking or deforming the structure. This kind of work can be done under sheer stress. Traditionally, physical treatment includes processes such as chipping, grinding, and milling (Taherzadeh and Karimi 2008).

Chemical pretreatment comprises the usage of different chemicals to shear the lignocellulose structure, such as acids, alkalis (ILs), oxidants, and organic solvents. One of the more favourable methods for industrial purposes is dilute acid technology (Carvalheiro et al., 2008). Oxygen, ozone, hydrogen peroxide, and chlorine dioxide are also used for lignin removal and oxidative fermentation (Sun et al., 2016). The next step after the pre-treatment process of lignocellulose biomass is fermentation. That carbohydrate component, cellulose, and hemicellulose are now ready to become fermentable (Karimi, 2015). Yeast (*Saccharomyces cerevisiae*) is chosen for fermentation due to its high production of ethanol and high tolerance towards inhibitors. There are other microbes, such as bacteria, that are also used for fermentation for industrial purposes (Adegboye et al., 2021).

Although lignocellulose is one of the favourite materials for conversion into biofuel, due to its tough texture, pre-treatment is rigorous and expensive, which could also increase the price of biofuel (Shafiei et al., 2011; Shafiei et al., 2013; Shafiei et al., 2014). Pre-treatment of feedstock also produces inhibitors in hydrolysates such as hydroxymethyl-furfural and lignin derivatives (Kudahettige-Nilsson et al., 2015). And these inhibitors affect microbes' growth, like *Clostridium*, and that leads to a diminishing yield of product (Cai et al., 2013).

Butanol toxicity is also one of the issues in fermentation because *Clostridium spp.* barely tolerates 2% of butanol in acetone, butanol, and ethanol (ABE) fermentation. So, the bottleneck in the fermentation process is inhibition by some by-products that are produced during the pre-treatment process and sometimes a lower tolerance of microbes. In this chapter, the fermentation process, including microbes and all those inhibitory compounds produced during the process, is explained. To overcome the issue of inhibitors, microbial tolerance has been explained, and how we can exploit new biotechnology techniques and generate some breakthroughs for the manufacturing of biofuel has been explained. In the direction of the manufacturing of biofuel and microbial tolerance against inhibitors, the future outlook is also mentioned (Kabir et al., 2015; Wang et al., 2018).

3.2 BREAKDOWN OF CELLULOSE AND PRODUCTION OF BIOFUEL

Generally, fermentation is defined as the breakdown of chemical substances like glucose in the absence of air by bacteria, fungi, or another microorganism while making wine and beer for ages. The glycolysis process in the cytosol leads to the conversion of hexose (fructose and glucose) into pyruvate (Weusthuis et al., 1994) and the formation of ATP and NADH (Zamora, 2009). Sugar molecules move through facilitated diffusion inside the cell. Gal2, Hxt1, Hxt2, Hxt3, Hxt4, Hxt6, and Hxt7 are the glucose transporters. First, fructose 1,6-biphosphate is formed from glucose (Maier et al., 2002; Ratledge, 1991). To incorporate three steps, two molecules of ATP are required (Bellou et al., 2014).

In the second stage, phosphate is formed by the reaction of glyceraldehyde-3-phosphate and dihydroxyacetone (Aggelis, 2007). Transfer of 1,3-biphosphoglycerate from glyceraldehyde-3-phosphate takes place. Glyceraldehyde-3-phosphate dehydrogenase catalyses the production of one mole of NADH while the reaction takes place. At 3-phosphoglycerate, formed by the conversion of 1,3-biphosphoglycerate (Aggelis, 2007), phosphoglycerate kinase catalyses the reaction that results in the simultaneous release of one mole of ATP (Festel, 2008; Arshad et al.,

2014). Lastly, pyruvate is formed by 3-phosphoglycerate, the last glycolysis product, occurring immediately with the production of another mole of ATP. In this process, a single mole of glucose is produced via glycolysis to produce NADH with four moles of ATPAs and two moles of pyruvic acid. To activate one mole of hexose, two moles of ATP are required. Stable energy is obtained for the cell during glycolysis, with only two ATP consumed per hexose. Yeasts utilise pyruvate, produced by glycolysis, in a variety of metabolic pathways. NAD+ from NADH reproduces NAD+ with the help of microbes to bring back the potential oxidation–reduction potential of the cell, which can be accomplished by respiration or fermentation (Zamora, 2009). Glycolysis products and processing come to an end here. Furthermore, the ability to progress by fermentation of alcohol, respiration, and glycerol-pyruvic fermentation is dependent on a variety of factors (Ribéreau-Gayon et al., 2006). The NADH-reducing ability produced by glycolysis must be 20 M in anaerobic conditions and sent to an electron acceptor to revive the NAD+ utilised by glycolysis. The operation is known as alcohol fermentation, and it happens in the cytoplasm, where electrons are accepted by acetaldehyde (Ratledge, 2004). Two other enzyme reactions are involved in alcohol fermentation. With the help of magnesium and the cofactor, thiamine pyrophosphate, the conversion of pyruvate into acetaldehyde is facilitated. Acetaldehyde gets reduced into ethanol by alcohol dehydrogenase, which utilises zinc as a cofactor (Salwan and Sharma, 2018), recycling NADH to NAD+. Carbon dioxide and ethanol, the end products of alcoholic fermentation, are normally thrown out of the cell. (Arshad et al., 2011). In biofuel production, the same fermentation is exploited to convert lignocellulose biomass into ethanol. The second step is cellulose degradation, which leaves lignin solid. As described above, hexoses and pentoses break down into monomer sugar molecules, but pentose breakdown is more difficult than hexose. Pentose is primarily composed of xylose, and there has been more than 15 years of research into xylose-fermenting microorganisms (Moreno et al., 2017).

Three types of fermentation procedures exist: the first is separate hydrolysis and fermentation (SHF) (Karimi, 2015), the second is consolidated bioprocessing (CBP), the third is simultaneous saccharification, and the last is fermentation (SSF). The involvement of the actinobacteria group has importance because it produces compounds such as xylanases, chitinases, cellulases, proteases, and laccases that help in the further degradation of biomass (Salwan and Sharma, 2018). Overall, fermentation is such a huge and lengthy process that, after pretreatment, includes the method of fermentation that depends on which one has much higher production efficiency and then the big role of microorganism selection, which has several criteria (Lewin et al., 2016; Adegboye et al., 2018). In fermentation, *S. cerevisiae* and its different strains are used. These microorganisms are chosen according to their tolerance and efficiency in producing ethanol. *S. cerevisiae* is one of the microorganisms that is preferred for fermentation because of its inhibition against products that are produced during the formation and its tolerance to them. For a better yield of bioethanol, now there is also a genetically modified microorganism present. It is reported that *S. cerevisiae* and *E. coli* are engineered strains that do not convert other sugars in the presence of glucose (Romaní et al., 2015; Jessop-Fabre et al., 2016; Milne et al., 2016; Pasotti et al., 2017).

TABLE 3.1
Microbes Involved in the Production of Biofuel

S. No.	Microbes	Substrate	Fermentation Product	Inhibited by	References
1.	*Clostridium spp.*	Wood pulp	Butanol	High concentration of butanol	Li et al. (2014)
2.	*Streptomyces spp.*	Terpene biosynthetic pathway	Bisabolene (advanced biofuel precursor)	Vanillin	Bhatia et al. (2019)
3.	*Escherichia coli*	Sugars	Ethanol	Furfural	Wang et al. (2012)
4.	*Saccharomyces cerevisiae*	Sugars	Ethanol	High concentration of ethanol, HMF, furfural, vanillin	Carmona-Gutierrez et al. (2012); Yang et al. (2013)
5.	*Zymomonasmobilis*	Sugars	Ethanol	High concentration of ethanol	Carmona-Gutierrez et al. (2012); Yang et al. (2013)

These genetic changes take place in transcriptional regulators and transporters. Here are some bacteria and fungi listed below (Table 3.1) that take part in ethanol bioconversion. Lastly, during the fermentation process (Mendez-Perez et al., 2017), distillation takes place, in which bioethanol goes under purification to separate the pure bioethanol from the fermented broth. Several sugars rise after the pretreatment process, determining the amount of bioethanol.

3.3 INHIBITION AND ITS ROLE IN THE FERMENTATION OF BIOFUEL

The act of stopping or slowing down the process and desired product formation due to some undesirable compounds or an undesired concentration of by-products is known as inhibition. Because bacterial species do not readily consume lignin–cellulose–hemicellulose biomass complexes, pretreatment is required for their degradation. Some of the hemicellulose and cellulose are fermented into fermented sugar during pretreatment. These are then transformed into desired compounds (biofuel) (Abubaker et al., 2012). During the conversion of hemicellulose into sugars, some unwanted sugars are produced that inhibit the fermentation process and further lead to a lower yield of biofuels such as bioethanol (e.g., ABE). Streamlining pretreatment process parameters, like using inhibitor-tolerant strains with high-yielding performance and continuous recovery of fermented products throughout microbial fermentation for biofuel production, can decrease inhibition. (Koppolu and Vasigala, 2016; Cheah et al., 2020).

3.4 TYPE OF INHIBITORS

Compounds such as weak acids, derivatives, and phenolic compounds hinder microbial strains (Oliet et al., 2002), and even higher concentrations of products in the fermentation broth are toxic to microbial species (Li et al., 2013; Lyu et al., 2017). As shown in Figure 3.1, inhibitors are classified into two types: process inhibitors and inherent inhibitors (López et al., 2004). The formation of various sugars and lignin degradation products is primarily determined by the pretreatment method and process parameters (such as temperature and time). Process inhibitors are compounds that form because of the pretreatment and neutralisation processes (Trinh et al., 2014; Xie et al., 2015, Shibuya et al., 2017). Inherent inhibitors are produced during the fermentation process, such as butanol, as a by-product that produces toxicity.

3.4.1 Process Inhibitors (Derived from Pretreatment)

3.4.1.1 Short-Chain Aliphatic Acids

Lignocellulose hydrolysate, acetic levulinic acid, and formic acid are examples of short-chain aliphatic acids (Zhang et al., 2016). While the acetyl group of hemicellulose is being hydrolysed, acetic acid is produced. Formic acid and levulinic acid, on the other hand, are produced during the 5-hydroxymethyl-furfural (HMF) reaction (Ulbricht et al., 1984). It has an impact on microbe growth rate and biomass yield, as well as the growth lag phase (Pampulha and Loureiro-Dias, 2000). It affects microbe growth rate and biomass yield, as well as the growth lag phase (Pampulha and Loureiro-Dias, 2000). A variety of factors influence acid toxicity for microbial inhibition, including exposed concentration and chain length, with longer chain acids becoming more hydrophobic (Wilbanks and Trinh, 2017).

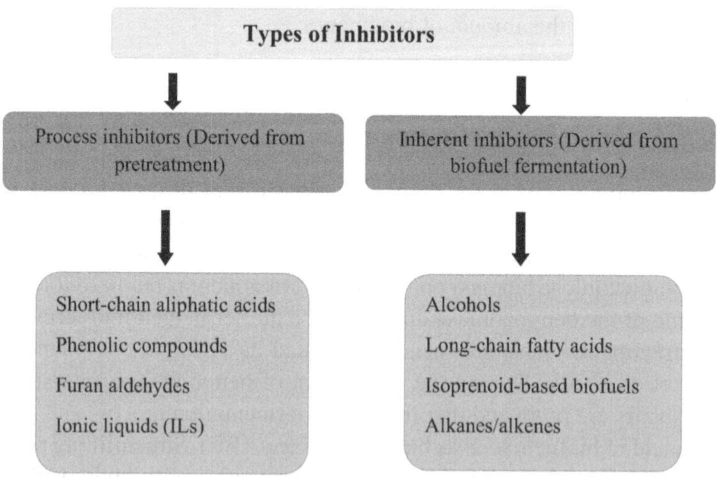

FIGURE 3.1 Lignocellulose degradation and generation of inhibitors.

3.4.1.2 Phenolic Compounds

Some phenolics, such as vanillic acid, ferulic acid, syringic acid, and 4-hydroxy benzoic acid, are formed in one of the pretreatment methods of acid-base, hydrothermal, alkaline, and oxidative degradation of lignin, as well as some alcohols, such as guaiacol, catechol, and vanillyl alcohol, and some aldehydes, such as vanillin and syringic aldehyde. During the fermentation, if these kinds of compounds are produced, the lag phase becomes prolonged and affects the production of ethanol, lactic acid, and xylitol (Jönsson and Martín, 2016; Lin et al., 2015; Wang et al., 2017; Zhang et al., 2016). Certain effects can be seen when phenolic compounds are produced. They interfere with membrane fluidity by entering it and distorting the hydrophobicity of the membrane. Production of toxic compounds certainly changes the ratio of proteins to lipids and interferes with the function of the selective barrier (Keweloh et al., 1990). Furthermore, the presence of these phenolics may increase reactive oxygen species, which can lead to cytoskeleton damage, programmed cell death, and DNA mutagenesis (Ibraheem and Ndimba, 2013).

3.4.1.3 Furan Aldehydes

In lignocellulose hydrolysate, some furan aldehydes are found, which are HMF and furfural (Park et al., 2011). These are derivatives of hexose and pentose in the process of fermentation (Chheda et al., 2007). The formation of furan depends on how severe the pretreatment is and what kind of biomass species are present (Arshad et al., 2018). The toxicity of furan aldehyde impacts glycolytic and fermentative enzymes such as acetaldehyde, pyruvate, alcohol dehydrogenase, and the breakage of a single strand of double-stranded DNA (Hadi and Rehman, 1989; Wang et al., 2017). The membrane is distorted by furan aldehyde, which also distorts, causing an outflow of intracellular magnesium and affecting the membrane's hydrophobicity. Furfural also contributes to the increased pernicious effects of acetic acid and phenol in lignocellulose hydrolysates (Zaldivar et al., 1999).

3.4.1.4 Ionic Liquids

Currently, a new method based on ionic liquids (ILs) has gained a lot of attention due to its potential for many types of lignocellulosic biomass and its ability to decrystallise cellulose (Konda et al., 2014). In an earlier study, numerous washes were needed to remove the IL, even if ethanol was used as a solvent (Li et al., 2013). After pretreatment, it was discovered that some amounts of toxicity to microbes remained in the biomass slurry. A major problem with so many inhibitors in IL-based pretreatment is IL toxicity. In so many of the recently investigated ILs, choline-based and imidazolium-based ILs are the highly impactful and extra-consumed reagents for the pretreatment of a wide range of sources of biomass (Shafiei et al., 2011). The most extensively researched choline based ILs for pretreatment are cholinium amino acid, chloride, and choline acetate. Cholinium acetate and choline chloride were found to be less toxic to bacteria than cholinium amino acid. Gram-negative bacteria are said to be more vulnerable than gram-positive bacteria (Hou et al., 2013).

3.4.2 Inherent Inhibitors (Derived from Biofuel Fermentation)

3.4.2.1 Alcohols

The number of alcohols produced by engineered microorganisms includes ethanol and several others, such as propanol, butanol, and isopentenyl. The production of these alcohols is contributed by the glycolytic pathway, CoA-dependent oxidation, and keto acid. The toxicity of alcohol leads to a diminished yield (Woodruff et al., 2013). If a longer carbon chain of alcohol breaks in the membrane and breaks the bond between the lipid tails, which is a hydrogen bond, the membrane can be distorted (Ly and Longo, 2004). Alcohols are toxic to organic acids if the carbon number is greater than 4. The only distinction between their hindrances is the steric effect and polarity (Wilbanks Cong and, 2017). An increase in n-butanol concentration causes inner membrane intrusion via peptidoglycan (Fletcher et al., 2016).

3.4.2.2 Long-Chain Fatty Acids

Some functions such as cell division and phosphate uptake in cells may be inhibited by low concentrations of fatty acids, and fatty acids that contain high concentrations may diminish oxidation of the substrate (Hundová and Fencl, 1977). An earlier study has shown that the diminishing abilities of fatty acids grow with the length of the chain and level of unsaturation (Kelsey et al., 2006). Additionally, *E. coli* treated with short-chain fatty acids had a relatively low hydrophobicity of the membrane 12, a higher saturated/unsaturated lipid ratio, and a relatively long lipid length (Royce et al., 2013). Reduced fatty acid concentrations can slow growth, cell division, and phosphate absorption, whereas higher fatty acid concentrations can hinder substrate oxidation (Wilbanks and Cong, 2017). According to past research, the inhibition properties of fatty acids increase with the length of the chain and the degree of unsaturation (Kelsey et al., 2006). The toxic effects of fatty acids as a product will not be the only thing that determines the cellular physiology of engineered strains, as inherent genetic changes and heterologous pathways can also modify the structure of lipids and middle metabolites (Ruffing and Jones, 2012).

3.4.2.3 Alkanes/Alkenes

The main determinants of natural gas and petroleum are alkenes and alkanes, which are also generated by microbial species via the following routes. Aldehyde decarboxylase transformed fatty acyl–acyl transport protein into fatty aldehyde, which was then transformed by acyl ACP reductase into alkanes and alkenes (Ruffing and Jones, 2012). These alkanes and alkenes may be impeded in some sense throughout fermentation, which could give rise to toxicity.

3.5 MECHANISM OF INHIBITION

The presence of numerous inhibitors shows a synergistic inhibitory effect. Sugar degradation compounds formed during the pre-treatment of lignocellulosic biomass include weak acids and furan derivatives. The lower concentration of these compounds did not inhibit the growth of microbes, but as their concentration increases,

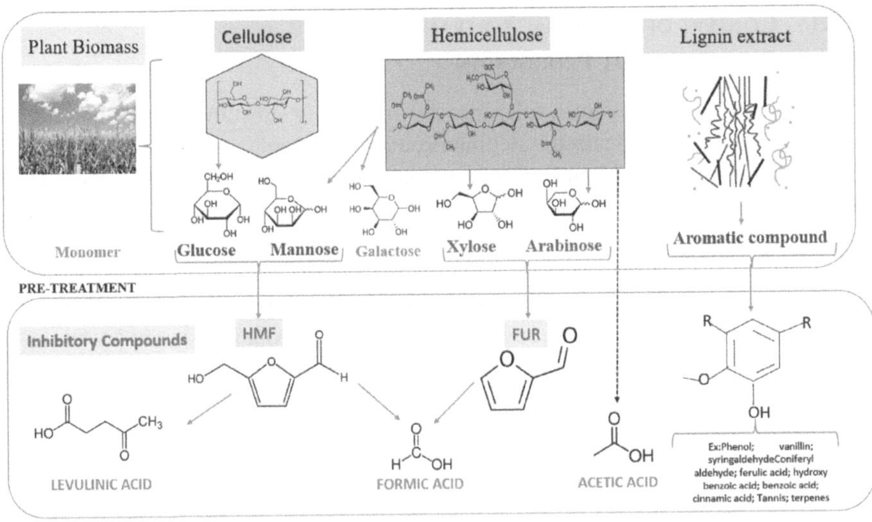

FIGURE 3.2 Type of Inhibitors.

a rapid decline in pH takes place and results in growth inhibition. During the degradation of lignin, numerous compounds such as aromatic, polyaromatic, phenolic, and aldehydic compounds are created. Recent research on inhibitory substances has focused on phenolics. Just like previously stated, the most toxic chemical for *Clostridium spp.* is ferulic acid, followed by coumaric acid (Kang and Nielsen, 2017). Following phenolics, furan derivatives and weak acids were reported to be toxic to cell growth, 2,3-butanediol generation, and enzyme activity (Wang et al., 2018). Figure 3.2 illustrates the diminishment of cellulose, hemicellulose, and lignin separation. Additionally, inhibitory substances such as levulinic acid, hydroxymethyl-furfural, furan, acetic acid, and formic acid were generated. These substances are critical in inhibiting biofuel yield by deteriorating the cell membrane. The existence of these substances also has an effect on the processes of cellular enzymes (Tramontina et al., 2020).

3.6 MICROBIAL TOLERANCE

Previously, several kinds of microbial strain inhibitors were involved in the process of pretreatment and fermentation of biofuel. These inhibitors put an end to microbial growth, reducing the yield of biofuel and elevating the cost of fermentation. The production of biofuel from sustainable sources such as lignocellulose biomass highly depends on strong and tolerant microbial strains. The latest molecular advancement has given ground-breaking techniques by which desirable tolerance strains for further biofuel production pathways can be obtained (Trastoy et al., 2018). So, in this part, we will go through the primary understanding of microbial tolerance and different ways of enhancing microbial tolerance.

3.6.1 Concept of Microbial Tolerance and Its Role in Biofuel Production

It is critical to increase strain tolerance for inhibiting substances, metabolic reaction products, and the desired outcome products. Multiple ways have been applied to overcome these problems to achieve an optimal yield of biofuel production (Rühl et al., 2009). It is already explained in the above paragraph that these inhibiting compounds are produced during the pretreatment or fermentation process. One of them is excessive butanol and ethanol toxicity. To overcome these issues, scientists went through several ways, which will be described in further paragraphs. Microbial tolerance consists of making microbes resistant to stress or toxic compounds. Metabolic engineering is applied to check pathways for the production of biofuel. All these techniques and methods are used to check the tolerance and increase the yield of biofuel. It is a powerful platform to provide tolerant strains for biofuel production, but it also costs a lot of money (Kang and Nielsen, 2017).

3.6.2 Mechanism and Strategies for Enhancing Microbial Tolerance

Numerous response mechanisms in bacteria are powered up and governed in harsh circumstances by proteins whose expression is controlled by regulator genes. It is plausible for these procedures to engage, enabling a rapid and organised response to a variety of stimulations (Sridhar et al., 2002). Additionally, tolerance patterns differ among both species and strains. Whereas most bacteria are hindered by 1%–2% (v/v) butanol, some evolved *Pseudomonas* strains can endure up to 6% (v/v) butanol (Yu et al., 2017). Even so, what is effective for biofuel may not be effective for the other, and the current strain may only withstand a subcategory of engineering strategies. The findings imply that tolerance mechanisms could be converted into a more efficient production strain. An appropriate server is needed for an organism to be thoroughly studied using effective genetic tools capable of being altered for both biofuel production and tolerance. Extensive attempts have been made to recognise strains with enhanced tolerance and biofuel methods of production, which are quoted in this chapter.

3.6.2.1 Random Mutagenesis

To obtain the right phenotype, random mutagenesis is among the classic approaches for recognising microorganisms with opioid tolerance and is appropriate for biofuels. Nowadays physical mutagen (e.g., X-rays, UV radiation, particle radiation) and chemical mutagens (e.g., ethyl methane sulfonate, N-methyl-N′-nitro-N-nitrosoguanidine) are used often. Nowadays, it is being used to enhance cellulosic ethanol fermentation output and boost microbes' tolerance to lignocellulosic inhibition (Suzuki et al., 2015; Zhang et al., 2014). Because of the numerous plasma effects of heat, neutral reactive species, charged particles, free radicals, high electric fields, and UV radiation, atmospheric and room temperature plasma has a potent implementation in microbial breeding as a rapidly evolving mutational strategy (Qin et al., 2016).

3.6.2.2 Adaptive Laboratory Evolution

By collecting random genetic changes under positive selection, adaptive laboratory evolution is a valuable tool for choosing microbes with the preferred trait (Mohamed et al., 2017). The batch cultivation procedure was being used to perform adaptive laboratory evolution by increasing the strain, which further resulted in microbial tolerance testing. Batch culture is well known for requiring a significant amount of time and energy because it is performed in tube culture, flask culture, and plate culture. Presently, well-plate culture by plate reader has piqued the interest of researchers because it has the potential for high throughput and online monitoring (Matsusako et al., 2017).

In the biofuel brewing process, adaptive laboratory evolution has been used to enhance microbial tolerance to ILs, lignocellulosic hydrolysate, ethanol, n-butanol, furfural ethanol, long-chain fatty acids, acetic acid, and phenol (Wang et al., 2018). The tolerance gene can be tested by comparing parent and transformed strains using a whole-genome resequencing process (Wright et al., 2011; Park et al., 2011). To keep improving microbial tolerance and sensitivity to an inhibitor generated throughout pretreatment, adaptive laboratory evolution was used. Despite being adjusted to grow in an acetic acid medium without acetic acid stress, the strain destroyed its own capacity to grow in high concentrations of acetic acid (Dong et al., 2013). Although strains are genetically stable, phenotypic tolerance will only increase the involvement of prolonged stress.

3.6.2.3 In Situ Detoxification

Upregulation of the NADH-dependent oxidoreductase FucO in *E. coli* may assist in furfural reduction. As per scientific data, heterologous expression of ADH1 from *C. tropicalis* enhanced furfural tolerance and deterioration in *E. coli* (Wang et al., 2018). The expression of aldehyde dehydrogenase 6 enhanced cell formation and ethanol generation in *Saccharomyces cerevisiae* by lowering the inhibitory activity of furan aldehydes (Brynildsen and Liao 2009). ALD6, or aldehyde dehydrogenase 6, is involved in both the direct oxidation of furan aldehydes and the supply of nicotinamide adenine dinucleotide phosphate (NADPH) to degradation processes. There is a gene that is encoded by fdh and is involved in the production of formate dehydrogenase in *Saccharomyces cerevisiae*. It was expressed heterologously and enhanced formic acid meltdown (Rutherford et al., 2010).

3.6.2.4 Heat Shock Proteins

Heat shock protein refers to molecules that aid in protein synthesis and their subsequent actions such as transport, folding, and destruction. Under stress, they aid in protein refolding and prevent protein aggregation; under normal conditions, they act as housekeepers. Heat shock proteins have repeatedly been found to be one of those that are increased through responding to solvent pressure in genetic analysis, and there is evidence that they may be implementing new targets for enhancing biofuel tolerance and generation. In several recent studies, heat shock proteins are continuously upregulated in actions involving solvent stress (Arshad et al., 2011). It exposed *E. coli* to isobutanol stress and analysed changes in the transcriptional response network, discovering that RpoH, a heat shock sigma element, was powerfully stimulated (Tomas et al., 2004). In a study done, changes in transcription in *E. coli* were evaluated under n-butanol stress, and innumerable genes related to heat shock protein and protein misfolding were found, including ibpAB, dnaJ, htpG, and rpoH (Dong et al., 2013). They discovered that when *C. acetobutylicum* was exposed to external butanol stress, several of the known heat shock proteins (hsp18, hsp90, dnaKJ, and groESL) were overexpressed (Minty et al., 2011). Grol, the chaperone, was also altered in isobutanol-tolerant *E. coli* strains (Alsaker et al., 2010). A comprehensive study of butanol, butyrate, and acetate stress in *C. acetobutylicum* produced a similar outcome (Tomas et al., 2003). Several other genes, including dnaK, groES, groEL, hsp90, and hasp18, have increased expression as a result of all three metabolites. Heat shock protein expression patterns change in response to solvent stress. Changing the pattern of heat shock protein expression can also aid in the overexpression of biofuel tolerance. Overexpression of GroESL is said to boost tolerance in *C. acetobutylicum* (Mukhopadhyay, 2015). Under stress, they aid in protein refolding and inhibit protein aggregation; under ordinary operation, they act as housekeepers. Heat shock proteins have repeatedly been found to be one of those that are enhanced in responding to solvent stress in genetic analysis, and there is evidence that they may be effective engineering targets for improving biofuel tolerance and generation. According to many recent studies, heat shock proteins are continuously highly expressed in response to solvent stress. Brynildsen and Liao (2009) exposed *E. coli* to isobutanol stress and examined transcriptional reaction connectivity adjustments, discovering that RpoH, a heat shock sigma factor, was strongly activated (Zhang et al., 2014). Rutherford et al. (2010) measured transcriptional changes in *E. coli* under n-butanol stress and discovered many genes associated with heat shock and protein misfolding, such as rpoH, dnaJ, htpG, and ibpAB (Tomas et al., 2004). Most of those recognised as heat shock proteins (hsp90, hsp18, dnaKJ, and groESL) were found to be overexpressed in *C. acetobutylicum* when exposed to external butanol stress (Minty et al., 2011). In evolving isobutanol-tolerant *E. coli* strains, the chaperone GroL was also modified (Alsaker et al., 2010). A major analysis of butyrate, acetate, and butanol stress in *C. acetobutylicum* yielded similar results (Tomas et al., 2003). Several other genes, including groES, groEL, hsp90, dnaK, and hasp18, have increased expression as a result of all three metabolites. Heat shock protein expression profiles respond differently to solvent stress. Updating the trend of heat shock protein expression can also aid in the upregulation of biofuel tolerance. GroESL upregulation increases tolerance in *C. acetobutylicum* (Dunlop, 2011).

3.6.2.5 Efflux Pumps

Efflux pumps are designed to keep dangerous substances out of the cell by using proton motive force. They contribute to cell sustenance by transferring a wide range of substrates, such as antimicrobial medicines, solvents, and bile salts (Mukhopadhyay, 2015). The five efflux pump family members are the small multidrug resistance, resistance nodulation division efflux pumps, multidrug and toxic compound extrusion, main facilitator superfamily, and ATP-binding cassette (ABC transporter) (Turner and Dunlop, 2015). In recent years, a large number of efflux pumps have been proclaimed to be efficient in the export of organic solvents such as ILs, α-pinene, geranyl acetate, geraniol, alkanes, isopentenol, and biodiesel (Wang et al., 2018). On the other hand, overexpression of efflux pumps can be detrimental to cell development due to membrane insertion machinery overload and membrane component modification (Ruegg et al., 2014). As a result, the expression efflux pump must be calibrated (Kiran et al., 2004). A natural auto-inducible efflux system has been developed to reduce efflux pump inhibition and improve IL tolerance. The concentration of ILs regulated efflux pump expression, making regulation more efficient, reducing negative effects, and eliminating the need for costly inducing molecules.

3.6.2.6 Membrane Modifications

Solvent invasion in solvent-tolerant bacteria can be prevented by changing the composition of the membrane. In nontolerant strains, solvents change the cellular membrane configuration, which has a major effect on biological functions and ultimately causes cell death. To mitigate this adverse effect, tolerant strains may alter the fatty acid composition of their membrane in order to avoid solvent access. Another good example of this method is the conversion of cis to trans unsaturated fatty acids, which can be facilitated by the cis–trans isomerase (Mukhopadhyay, 2015). A rise in the proportion of trans to cis fatty acids is linked to a reduction in membrane fluidity and a corresponding increase in solvent tolerance (Isken and de Bont, 1998). To maintain membrane stability over time, cells can change the ratio of saturated to unsaturated fatty acids. Some alteration in head groups or length of phospholipid has been shown to have some solvent tolerance (Baer et al., 1987; Pinkart and White, 1997). Membrane modifications that reduce permeability may not be useful as stand-alone solutions for increasing biofuel manufacturing speed. Because fuel molecules may become trapped within the cell, they may be beneficial when combined with other mechanisms. Individual lipid bilayer modifications are helpful to enhance resistance to external solvents but not production (Majidian et al., 2018). Natural examples from solvent-tolerant bacteria suggest that a combined technique involving membrane changes and other approaches could be a useful engineering approach. Membrane alterations, for example, could be beneficial when used along with export pumps. On the other hand, fatty acid biosynthesis timing and rate might be amended in response to production levels. Increased tolerance will result from the ability to improve membrane composition. However, manufacturing and tolerance must be attentively evaluated in this method (Ramos et al., 2002).

3.6.2.6.1 Future Directions

There is a crisis of sustainable energy sources facing human beings that is threatening. Fossil fuel sources are on the edge now, so biofuels emerged a few years ago, and nowadays, different types of biofuels are on the market with lower emissions of greenhouse gases. Among biofuels, several types exist, such as biodiesel, biogas, acetone, butanol, and ethanol (ABE). Earlier, feedstocks such as corn and soybeans were the main sources for biofuel production, which caused food insecurity. In terms of sustainable production of biofuel, lignocellulosic mass has come, and it doesn't need any more feedstock (Sarangi and Nanda, 2018). After gradually replacing this feedstock, waste lignocellulose biomass is now being used.

Lignocellulosic biomasses include all the agricultural waste materials that are used for biofuel production, especially ABE production. Pretreatment is considered a crucial and expensive process. When it comes to the presence of harmful substances, pretreatment is the most crucial step that causes the production of multiple toxic compounds, such as acetic acids, aldehydes, and phenolics. For better production of biofuel in the future, this pretreatment process should be less expensive and controlled to lessen the production of toxic by-products (Sarangi and Nanda, 2018). Another major issue is the yield of ethanol and butanol, which is hindered by multiple by-products, and the cost recovering of biofuel is way more expensive. Therefore, more technological advancements are required for high and low-cost recovery of bioethanol. Most functional pretreatment methods require reducing expensive enzyme loading for delignification. Some compounds, such as ethyl acetates, CaO, KOH, $CaCO_3$, NaOH, and $Ca(OH)_2$, can be used to neutralise inhibitors (Ramos et al., 2002). Research can be directed toward the production of dihydrogen, which is currently gaining popularity due to photo fermentation. An inhibitory product like furfural or polyphenol-like compounds can inhibit the production of dihydrogen. Identification of functional strains of microbes is very crucial, but it is also in demand in the present scenario. Metabolic engineering and pathways related to it must be understood by a synthetic biologist. Different molecular techniques are evolving to improve and enhance the productivity of microbes. Clustered, regularly interspaced, short palindromic repeats (CRISPR/Cas9) are one of the advanced techniques being used to step up the productivity of biofuel by editing the genomes of microbes (Wang et al., 2017). This gene-editing method, CRISPR-Cas9 technique, is used in prokaryotes for defence against viruses or any other alien particle. Downregulation or upregulation of genes is done by this technique, which makes it easy to understand the metabolic pathways by using a single-guided RNA for the target sequence that has 20 nucleotides to complementarily allow the editing in the genome. The Cas9 protein contains endonuclease activity and cuts the target DNA sequence. This editing in the microbe's genome enhances the work of useful genes and increases tolerance against inhibitors. According to the literature, this has been well explained in *Streptococcus pyogenes*. The insertion of novel genes and gene knockout methods can be adopted for the benefit of biofuel producing genes (Nanda et al., 2014a). The CRISPR-interference technique is used to knock out unwanted genes, and in this technique, the endonuclease is deactivated, which combines with sgRNA to inhibit transcription of RNA polymerase. For example, zinc finger nucleases and transcription activator-like effector nucleases (TALEN) were earlier used for gene editing (TALENs) (Wang et al., 2016). Several gene-editing techniques exist, including zinc finger nuclease and transcription activator-like effector nucleases. CRISPR is a new gene-editing technique that is widely used these days. *Corynebacterium*, *Clostridium*, *Lactobacillus*, *Mycobacterium*, *Escherichia coli*, *Pseudomonas*, *Staphylococcus*, *Streptomyces*, and *Bacillus* have all been modified with CRISPR to allow for efficient biofuel production (Cho et al., 2018).

A dual operon-based synthetic method was also used for the *E. coli* strain where the ethanol-producing gene has been replaced by a new strain that produces 19 g/L. It is reported that this efficiency was enhanced by using PJ23119. Apart from the above techniques, high-efficiency techniques are used such as proteomics, transcriptomics, and metabolomics, in which they elaborate on the expression and different pathways of genes in high-yielding strains of microbes. Computational and mathematical tools and nanoparticles could be used for mechanistic understanding (Limayem and Ricke, 2012; Zhang et al., 2020).

End-product tolerance has now been proven in enough studies to improve production and strain robustness. Pathway optimisation research, on the other hand, has followed a pattern like that of tolerance improvement (Alonso-Gutierrez et al., 2015). The most required measure will be to blend sufferance traits into formulation strains as new highly productive strains become accessible. Furthermore, most tolerance engineering efforts have focused on improving a host microbe's resident gene role and potential. As more heterologous genes are introduced into dummy hosts such as *E. coli*, the number of dummy varieties accessible for genetic manipulation will increase (Abdelaal et al., 2019).

3.7 CONCLUSION

In this chapter, we have summarised the biomass pretreatment and biofuel fermentation processes in detail. During pretreatment processes, the production of inhibitory compounds takes place, which limits the yield of biofuel. To overcome the inhibitor-sponsored limitations, strategies for improving microbial tolerance were suggested, which included host strain modification, genetic engineering, random mutagenesis, and so on. Tolerance technology and evolutionary engineering are very useful platforms for changing host strains to have desired traits such as tolerance. Nonetheless, the strength of microorganism strains is based on a multicomplex tolerance system rather than one or two tolerance genes. Tolerance modification should not add undue stress or expense, such as the metabolic burden of protein production, efflux pump toxicity, or the cost of an inducer for expression. Currently, research has been focused more on increasing tolerance in microbes against inhibitors as well as enhancing the yield of biofuel production.

REFERENCES

Abdelaal, A.S., Jawed, K. and Yazdani, S.S., 2019. CRISPR/Cas9-mediated engineering of Escherichia coli for n-butanol production from xylose in defined medium. *Journal of Industrial Microbiology and Biotechnology*, *46*(7), pp. 965–975.

Abubaker, H.O., Sulieman, A.M.E. and Elamin, H.B., 2012. Utilization of Schizosaccharomyces pombe for production of ethanol from cane molasses. *Journal of Microbiology Research. (Rosemead Calif.)*, *2*, pp. 36–40.

Adegboye, M.F., Lobb, B., Babalola, O.O., Doxey, A.C. and Ma, K., 2018. Draft genome sequences of two novel cellulolytic streptomyces strains isolated from South African rhizosphere soil. *Genome Announcements*, *6*(26), pp. e00632–18.

Adegboye, M.F., Ojuederie, O.B., Talia, P.M. and Babalola, O.O., 2021. Bioprospecting of microbial strains for biofuel production: metabolic engineering, applications, and challenges. *Biotechnology for Biofuels*, *14*(1), pp. 1–21.

Aggelis, G., 2007. *Microbiology and Microbial Technology*. Stamoulis Publishers, Athens, Greece.

Alonso-Gutierrez, J., Kim, E.M., Batth, T.S., Cho, N., Hu, Q., Chan, L.J.G., Petzold, C.J., Hillson, N.J., Adams, P.D., Keasling, J.D. and Martin, H.G., 2015. Principal component analysis of proteomics (PCAP) as a tool to direct metabolic engineering. *Metabolic Engineering*, *28*, pp. 123–133.

Alsaker, K.V., Paredes, C. and Papoutsakis, E.T., 2010. Metabolite stress and tolerance in the production of biofuels and chemicals: gene-expression-based systems analysis of butanol, butyrate, and acetate stresses in the anaerobe Clostridium acetobutylicum. *Biotechnology and Bioengineering*, *105*(6), pp. 1131–1147.

Arshad, M., Adil, M., Sikandar, A. and Hussain, T., 2014. Exploitation of meat industry by-products for biodiesel production: Pakistan's perspective. *Pakistan Journal of Life and Social Sciences*, *12*, pp. 120–125.

Arshad, M., Zia, M.A., Asghar, M. and Bhatti, H., 2011. Improving bio-ethanol yield: using virginiamycin and sodium flouride at a Pakistani distillery. *African Journal of Biotechnology*, *10*(53), pp. 11071–11074.

Arshad, M., Zia, M.A., Shah, F.A. and Ahmad, M., 2018. An overview of biofuel. *Perspectives on Water Usage for Biofuels Production*, pp. 1–37.

Baer, S.H., Blaschek, H.P. and Smith, T.L., 1987. Effect of butanol challenge and temperature on lipid composition and membrane fluidity of butanol-tolerant Clostridium acetobutylicum. *Applied and Environmental Microbiology*, *53*(12), pp. 2854–2861.

Bellou, S., Baeshen, M.N., Elazzazy, A.M., Aggeli, D., Sayegh, F. and Aggelis, G., 2014. Microalgal lipids biochemistry and biotechnological perspectives. *Biotechnology Advances*, *32*(8), pp. 1476–1493.

Bhatia, S.K., Gurav, R., Choi, T.R., Han, Y.H., Park, Y.L., Park, J.Y., Jung, H.R., Yang, S.Y., Song, H.S., Kim, S.H. and Choi, K.Y., 2019. Bioconversion of barley straw lignin into biodiesel using Rhodococcus sp. YHY01. *Bioresource Technology*, *289*, p. 121704.

Brynildsen, M.P. and Liao, J.C., 2009. An integrated network approach identifies the isobutanol response network of Escherichia coli. *Molecular Systems Biology*, *5*(1), p. 277.

Cai, D., Zhang, T., Zheng, J., Chang, Z., Wang, Z., Qin, P.Y. and Tan, T.W., 2013. Biobutanol from sweet sorghum bagasse hydrolysate by a hybrid pervaporation process. *Bioresource Technology*, *145*, pp. 97–102.

Carmona-Gutierrez, D., Sommer, C., Andryushkova, A., Kroemer, G. and Madeo, F., 2012. A higher spirit: avoiding yeast suicide during alcoholic fermentation. *Cell Death & Differentiation*, *19*(6), pp. 913–914.

Carvalheiro, F., Duarte, L.C. and Gírio, F., 2008. Hemicellulose biorefineries: a review on biomass pre-treatments. *Journal of Scientific & Industrial Research*, *67*(11), pp. 849–864.

Cheah, W.Y., Sankaran, R., Show, P.L., Ibrahim, T.N.B.T., Chew, K.W., Culaba, A. and Jo-Shu, C., 2020. Pretreatment methods for lignocellulosic biofuels production: current advances, challenges and future prospects. *Biofuel Research Journal*, *7*(1), p. 1115.

Chheda, J.N., Román-Leshkov, Y. and Dumesic, J.A., 2007. Production of 5-hydroxymethylfurfural and furfural by dehydration of biomass-derived mono-and poly-saccharides. *Green Chemistry*, *9*(4), pp. 342–350.

Cho, S., Shin, J. and Cho, B.K., 2018. Applications of CRISPR/Cas system to bacterial metabolic engineering. *International Journal of Molecular Sciences*, *19*(4), p. 1089.

Dong, H.W., Fan, L.Q., Luo, Z., Zhong, J.J., Ryu, D.D. and Bao, J., 2013. Improvement of ethanol productivity and energy efficiency by degradation of inhibitors using recombinant Zymomonasmobilis (pHW20a-fdh). *Biotechnology and Bioengineering*, *110*(9), pp. 2395–2404.

Dunlop, M.J., 2011. Engineering microbes for tolerance to next-generation biofuels. *Biotechnology for Biofuels*, *4*(1), pp. 1–9.

Festel, G.W., 2008. Biofuels–economic aspects. *Chemical Engineering & Technology: Industrial Chemistry-Plant Equipment-Process Engineering-Biotechnology*, *31*(5), pp. 715–720.

Fletcher, E., Pilizota, T., Davies, P.R., McVey, A. and French, C.E., 2016. Characterization of the effects of n-butanol on the cell envelope of *E. coli*. *Applied Microbiology and Biotechnology*, *100*(22), pp. 9653–9659.

Hadi, S.M. and Rehman, A., 1989. Specificity of the interaction of furfural with DNA. *Mutation Research Letters*, *225*(3), pp. 101–106.

Hou, X.D., Liu, Q.P., Smith, T.J., Li, N. and Zong, M.H., 2013. Evaluation of toxicity and biodegradability of cholinium amino acids ionic liquids. *PloS One*, *8*(3), p. e59145.

Hundová, Z. and Fencl, Z., 1977. Toxic effects of fatty acids on yeast cells: dependence of inhibitory effects on fatty acid concentration. *Biotechnology and Bioengineering*, *19*(11), pp. 1623–1641.

Ibraheem, O. and Ndimba, B.K., 2013. Molecular adaptation mechanisms employed by ethanologenic bacteria in response to lignocellulose-derived inhibitory compounds. *International Journal of Biological Sciences*, *9*(6), p. 598.

Isken, S., & de Bont, J. A. 1998. Bacteria tolerant to organic solvents. *Extremophiles*, 2(3), 229–238.

Jessop-Fabre, M.M., Jakočiūnas, T., Stovicek, V., Dai, Z., Jensen, M.K., Keasling, J.D. and Borodina, I., 2016. EasyClone-MarkerFree: a vector toolkit for marker-less integration of genes into Saccharomyces cerevisiae via CRISPR-Cas9. *Biotechnology Journal*, *11*(8), pp. 1110–1117.

Jönsson, L.J. and Martín, C., 2016. Pretreatment of lignocellulose: formation of inhibitory by-products and strategies for minimizing their effects. *Bioresource Technology*, *199*, pp. 103–112.

Kabir, M.M., Forgács, G. and Horváth, I.S., 2015. Biogas from lignocellulosic materials. In Keikhosro Karimi (ed) *Lignocellulose-Based Bioproducts* (pp. 207–251). Springer, Cham.

Karimi, K., Tabatabaei, M., Sárvári Horváth, I. and Kumar, R., 2015. Recent trends in acetone, butanol, and ethanol (ABE) production. *Biofuel Research Journal*, *2*(4), pp. 301–308.

Kang, M.K. and Nielsen, J., 2017. Biobased production of alkanes and alkenes through metabolic engineering of microorganisms. *Journal of Industrial Microbiology and Biotechnology*, *44*(4–5), pp. 613–622.

Kelsey, J.A., Bayles, K.W., Shafii, B. and McGuire, M.A., 2006. Fatty acids and monoacylglycerols inhibit growth of Staphylococcus aureus. *Lipids*, *41*(10), pp. 951–961.

Keweloh, H., Weyrauch, G. and Rehm, H.J., 1990. Phenol-induced membrane changes in free and immobilized Escherichia coli. *Applied Microbiology and Biotechnology*, *33*(1), 66–71.

Kiran, M.D., Prakash, J.S.S., Annapoorni, S., Dube, S., Kusano, T., Okuyama, H., Murata, N. and Shivaji, S., 2004. Psychrophilic Pseudomonas syringae requires trans-monounsaturated fatty acid for growth at higher temperature. *Extremophiles*, *8*(5), pp. 401–410.

Koppolu, V. and Vasigala, V.K., 2016. Role of Escherichia coli in biofuel production. *Microbiology insights*, *9*, pp. MBI-S10878.

Konda, N.M., Shi, J., Singh, S., Blanch, H.W., Simmons, B.A. and Klein-Marcuschamer, D., 2014. Understanding cost drivers and economic potential of two variants of ionic liquid pretreatment for cellulosic biofuel production. *Biotechnology for Biofuels*, *7*, pp. 1–11.

Kudahettige-Nilsson, R.L., Helmerius, J., Nilsson, R.T., Sjöblom, M., Hodge, D.B. and Rova, U., 2015. Biobutanol production by Clostridium acetobutylicum using xylose recovered from birch Kraft black liquor. *Bioresource Technology*, *176*, pp. 71–79.

Lewin, G.R., Carlos, C., Chevrette, M.G., Horn, H.A., McDonald, B.R., Stankey, R.J., Fox, B.G. and Currie, C.R., 2016. Evolution and ecology of Actinobacteria and their bioenergy applications. *Annual Review of Microbiology*, *70*, pp. 235–254.

Li, C., Tanjore, D., He, W., Wong, J., Gardner, J.L., Sale, K.L., Simmons, B.A. and Singh, S., 2013. Scale-up and evaluation of high solid ionic liquid pretreatment and enzymatic hydrolysis of switchgrass. *Biotechnology for Biofuels*, *6*(1), pp. 1–13.

Li, J., Baral, N.R. and Jha, A.K., 2014. Acetone–butanol–ethanol fermentation of corn stover by Clostridium species: present status and future perspectives. *World Journal of Microbiology and Biotechnology*, *30*(4), pp. 1145–1157.

Limayem, A. and Ricke, S.C., 2012. Lignocellulosic biomass for bioethanol production: current perspectives, potential issues and future prospects. *Progress in Energy and Combustion Science*, *38*(4), pp. 449–467.

Lin, R., Cheng, J., Ding, L., Song, W., Zhou, J. and Cen, K., 2015. Inhibitory effects of furan derivatives and phenolic compounds on dark hydrogen fermentation. *Bioresource Technology*, *196*, pp. 250–255.

López, M.J., Nichols, N.N., Dien, B.S., Moreno, J. and Bothast, R.J., 2004. Isolation of microorganisms for biological detoxification of lignocellulosic hydrolysates. *Applied Microbiology and Biotechnology*, *64*(1), pp. 125–131.

Ly, H.V. and Longo, M.L., 2004. The influence of short-chain alcohols on interfacial tension, mechanical properties, area/molecule, and permeability of fluid lipid bilayers. *Biophysical Journal*, *87*(2), pp. 1013–1033.

Lyu, H., Lv, C., Zhang, M., Liu, J., Meng, F. and Geng, Z.F., 2017. Kinetic studies of the strengthening effect on liquid hot water pretreatments by organic acids. *Bioresource Technology*, *235*, pp. 193–201.

Maier, A., Völker, B., Boles, E. and Fuhrmann, G.F., 2002. Characterisation of glucose transport in Saccharomyces cerevisiae with plasma membrane vesicles (countertransport) and intact cells (initial uptake) with single Hxt1, Hxt2, Hxt3, Hxt4, Hxt6, Hxt7 or Gal2 transporters. *FEMS Yeast Research*, *2*(4), pp. 539–550.

Majidian, P., Tabatabaei, M., Zeinolabedini, M., Naghshbandi, M.P. and Chisti, Y., 2018. Metabolic engineering of microorganisms for biofuel production. *Renewable and Sustainable Energy Reviews*, *82*, pp. 3863–3885.

Matsusako, T., Toya, Y., Yoshikawa, K. and Shimizu, H., 2017. Identification of alcohol stress tolerance genes of Synechocystis sp. PCC 6803 using adaptive laboratory evolution. *Biotechnology for Biofuels*, *10*(1), pp. 1–9.

Mendez-Perez, D., Alonso-Gutierrez, J., Hu, Q., Molinas, M., Baidoo, E.E., Wang, G., Chan, L.J., Adams, P.D., Petzold, C.J., Keasling, J.D. and Lee, T.S., 2017. Production of jet fuel precursor monoterpenoids from engineered Escherichia coli. *Biotechnology and Bioengineering*, *114*(8), pp. 1703–1712.

Milne, N., Wahl, S.A., Van Maris, A.J.A., Pronk, J.T. and Daran, J.M., 2016. Excessive by-product formation: a key contributor to low isobutanol yields of engineered Saccharomyces cerevisiae strains. *Metabolic Engineering Communications*, *3*, pp. 39–51.

Minty, J.J., Lesnefsky, A.A., Lin, F., Chen, Y., Zaroff, T.A., Veloso, A.B., Xie, B., McConnell, C.A., Ward, R.J., Schwartz, D.R. and Rouillard, J.M., 2011. Evolution combined with genomic study elucidates genetic bases of isobutanol tolerance in Escherichia coli. *Microbial Cell Factories*, *10*(1), pp. 1–38.

Mohamed, E.T., Wang, S., Lennen, R.M., Herrgård, M.J., Simmons, B.A., Singer, S.W. and Feist, A.M., 2017. Generation of a platform strain for ionic liquid tolerance using adaptive laboratory evolution. *Microbial Cell Factories*, *16*(1), pp. 1–15.

Moreno, A.D., Alvira, P., Ibarra, D. and Tomás-Pejó, E., 2017. Production of ethanol from lignocellulosic biomass. In Zhen Fang, Richard L. Smith, Jr., Xinhua Qi (Eds.) *Production of Platform Chemicals from Sustainable Resources* (pp. 375–410). Springer, Singapore.

Mukhopadhyay, A., 2015. Tolerance engineering in bacteria for the production of advanced biofuels and chemicals. *Trends in Microbiology*, *23*(8), pp. 498–508.

Nanda, S., Dalai, A.K. and Kozinski, J.A., 2014a. Butanol and ethanol production from lignocellulosic feedstock: biomass pretreatment and bioconversion. *Energy Science & Engineering*, *2*(3), pp. 138–148.

Nanda, S., Mohammad, J., Reddy, S.N., Kozinski, J.A. and Dalai, A.K., 2014b. Pathways of lignocellulosic biomass conversion to renewable fuels. *Biomass Conversion and Biorefinery*, *4*(2), pp. 157–191.

Oliet, M., Garcıa, J., Rodrıguez, F. and Gilarranz, M.A., 2002. Solvent effects in autocatalyzed alcohol–water pulping: comparative study between ethanol and methanol as delignifying agents. *Chemical Engineering Journal*, *87*(2), pp. 157–162.

Pampulha, M.E. and Loureiro-Dias, M.C., 2000. Energetics of the effect of acetic acid on growth of Saccharomyces cerevisiae. *FEMS Microbiology Letters*, *184*(1), pp. 69–72.

Park, S.E., Koo, H.M., Park, Y.K., Park, S.M., Park, J.C., Lee, O.K., Park, Y.C. and Seo, J.H., 2011. Expression of aldehyde dehydrogenase 6 reduces inhibitory effect of furan derivatives on cell growth and ethanol production in Saccharomyces cerevisiae. *Bioresource Technology*, *102*(10), pp. 6033–6038.

Pasotti, L., Zucca, S., Casanova, M., Micoli, G., De Angelis, M.G.C. and Magni, P., 2017. Fermentation of lactose to ethanol in cheese whey permeate and concentrated permeate by engineered Escherichia coli. *BMC Biotechnology*, *17*(1), pp. 1–12.

Pinkart, H.C. and White, D.C., 1997. Phospholipid biosynthesis and solvent tolerance in Pseudomonas putida strains. *Journal of Bacteriology*, *179*(13), pp. 4219–4226.

Qin, D., Hu, Y., Cheng, J., Wang, N., Li, S. and Wang, D., 2016. An auto-inducible Escherichia coli strain obtained by adaptive laboratory evolution for fatty acid synthesis from ionic liquid-treated bamboo hydrolysate. *Bioresource Technology*, *221*, pp. 375–384.

Ramos, J.L., Duque, E., Gallegos, M.T., Godoy, P., Ramos-Gonzalez, M.I., Rojas, A., Terán, W. and Segura, A., 2002. Mechanisms of solvent tolerance in gram-negative bacteria. *Annual Reviews in Microbiology*, *56*(1), pp. 743–768.

Ratledge, C., 1991. Yeast physiology—a micro-synopsis. *Bioprocess Engineering*, *6*(5), pp. 195–203.

Ratledge, C., 2004. Fatty acid biosynthesis in microorganisms being used for single cell oil production. *Biochimie*, *86*(11), pp. 807–815.

Ribéreau-Gayon, P., Dubourdieu, D., Donèche, B. and Lonvaud, A. eds., 2006. *Handbook of Enology*, Volume 1: The Microbiology of Wine and Vinifications (Vol. 1). John Wiley & Sons, England.

Romaní, A., Pereira, F., Johansson, B. and Domingues, L., 2015. Metabolic engineering of Saccharomyces cerevisiae ethanol strains PE-2 and CAT-1 for efficient lignocellulosic fermentation. *Bioresource Technology*, *179*, pp. 150–158.

Royce, L.A., Liu, P., Stebbins, M.J., Hanson, B.C. and Jarboe, L.R., 2013. The damaging effects of short chain fatty acids on Escherichia coli membranes. *Applied Microbiology and Biotechnology*, *97*(18), pp. 8317–8327.

Ruegg, T.L., Kim, E.M., Simmons, B.A., Keasling, J.D., Singer, S.W., Lee, T.S. and Thelen, M.P., 2014. An auto-inducible mechanism for ionic liquid resistance in microbial biofuel production. *Nature Communications*, *5*(1), pp. 1–7.

Ruffing, A.M. and Jones, H.D., 2012. Physiological effects of free fatty acid production in genetically engineered Synechococcus elongatus PCC 7942. *Biotechnology and Bioengineering*, *109*(9), pp. 2190–2199.

Rühl, J., Schmid, A. and Blank, L.M., 2009. Selected Pseudomonas putida strains able to grow in the presence of high butanol concentrations. *Applied and Environmental Microbiology*, *75*(13), pp. 4653–4656.

Rutherford, B.J., Dahl, R.H., Price, R.E., Szmidt, H.L., Benke, P.I., Mukhopadhyay, A. and Keasling, J.D., 2010. Functional genomic study of exogenous n-butanol stress in Escherichia coli. *Applied and Environmental Microbiology*, *76*(6), pp. 1935–1945.

Salwan, R. and Sharma, V., 2018. The role of actinobacteria in the production of industrial enzymes. In *New and Future Developments in Microbial Biotechnology and Bioengineering* (pp. 165–177). Elsevier, Himachal Pradesh, India.

Sanderson, K., 2011. Lignocellulose: a chewy problem. *Nature*, 474, pp. S12–S14.

Sarangi, P.K. and Nanda, S., 2018. Recent developments and challenges of acetone-butanol-ethanol fermentation. In Prakash Kumar, Sarangi, Sonil Nanda, Pravakar Mohanty (Eds.) *Recent Advancements in Biofuels and Bioenergy Utilization* (pp. 111–123). Springer, Singapore.

Shafiei, M., Kabir, M.M., Zilouei, H., Horváth, I.S. and Karimi, K., 2013. Techno-economical study of biogas production improved by steam explosion pretreatment. *Bioresource Technology*, *148*, pp. 53–60.

Shafiei, M., Karimi, K. and Taherzadeh, M.J., 2011. Techno-economical study of ethanol and biogas from spruce wood by NMMO-pretreatment and rapid fermentation and digestion. *Bioresource Technology*, *102*(17), pp. 7879–7886.

Shafiei, M., Karimi, K., Zilouei, H. and Taherzadeh, M.J., 2014. Economic impact of NMMO pretreatment on ethanol and biogas production from pinewood. *BioMed Research International*, *2014*, pp. 2–14.

Shibuya, M., Sasaki, K., Tanaka, Y., Yasukawa, M., Takahashi, T., Kondo, A. and Matsuyama, H., 2017. Development of combined nanofiltration and forward osmosis process for production of ethanol from pretreated rice straw. *Bioresource Technology*, *235*, pp. 405–410.

Sridhar, M., Sree, N.K. and Rao, L.V., 2002. Effect of UV radiation on thermotolerance, ethanol tolerance and osmotolerance of Saccharomyces cerevisiae VS1 and VS3 strains. *Bioresource Technology*, *83*(3), pp. 199–202.

Sun, S., Sun, S., Cao, X. and Sun, R., 2016. The role of pretreatment in improving the enzymatic hydrolysis of lignocellulosic materials. *Bioresource Technology*, *199*, pp. 49–58.

Suzuki, T., Seta, K., Nishikawa, C., Hara, E., Shigeno, T. and Nakajima-Kambe, T., 2015. Improved ethanol tolerance and ethanol production from glycerol in a streptomycin-resistant Klebsiella variicola mutant obtained by ribosome engineering. *Bioresource Technology*, *176*, pp. 156–162.

Taherzadeh, M.J. and Karimi, K., 2008. Pretreatment of lignocellulosic wastes to improve ethanol and biogas production: a review. *International Journal of Molecular Sciences*, *9*(9), pp. 1621–1651.

Tomas, C.A., Beamish, J. and Papoutsakis, E.T., 2004. Transcriptional analysis of butanol stress and tolerance in Clostridium acetobutylicum. *Journal of Bacteriology*, *186*(7), pp. 2006–2018.

Tomas, C.A., Welker, N.E. and Papoutsakis, E.T., 2003. Overexpression of groESL in Clostridium acetobutylicum results in increased solvent production and tolerance, prolonged metabolism, and changes in the cell's transcriptional program. *Applied and Environmental Microbiology*, *69*(8), pp. 4951–4965.

Tramontina, R., Brenelli, L.B., Sodré, V., Cairo, J.P.F., Travália, B.M., Egawa, V.Y., Goldbeck, R. and Squina, F.M., 2020. Enzymatic removal of inhibitory compounds from lignocellulosic hydrolysates for biomass to bioproducts applications. *World Journal of Microbiology and Biotechnology*, *36*(11), pp. 1–11.

Tran, T.T.A., Le, T.K.P., Mai, T.P. and Nguyen, D.Q., 2019. Bioethanol production from lignocellulosic biomass. *Alcohol Fuels-Current Technologies and Future Prospect*, pp. 1–14.

Trastoy, R., Manso, T., Fernández-García, L., Blasco, L., Ambroa, A., Perez Del Molino, M.L., Bou, G., García-Contreras, R., Wood, T.K. and Tomás, M., 2018. Mechanisms of bacterial tolerance and persistence in the gastrointestinal and respiratory environments. *Clinical Microbiology Reviews*, *31*(4), pp. e00023–18.

Trinh, L.T.P., Kundu, C., Lee, J.W. and Lee, H.J., 2014. An integrated detoxification process with electrodialysis and adsorption from the hemicellulose hydrolysates of yellow poplars. *Bioresource Technology*, *161*, pp. 280–287.

Turner, W.J. and Dunlop, M.J., 2015. Trade-offs in improving biofuel tolerance using combinations of efflux pumps. *ACS Synthetic Biology*, *4*(10), pp. 1056–1063.

Ulbricht, R.J., Northup, S.J. and Thomas, J.A., 1984. A review of 5-hydroxymethylfurfural (HMF) in parenteral solutions. *Toxicological Sciences*, *4*(5), pp. 843–853.

United States Energy Information Administration (USEIA) 2011. International energy outlook 2011. http://www.eia.gov/forecasts/ieo/pdf/0484(2011).pdf.

Wang, J., Zhang, Y., Chen, Y., Lin, M. and Lin, Z., 2012. Global regulator engineering significantly improved Escherichia coli tolerances toward inhibitors of lignocellulosic hydrolysates. *Biotechnology and Bioengineering*, *109*(12), pp. 3133–3142.

Wang, S., Dong, S., Wang, P., Tao, Y. and Wang, Y., 2017. Genome editing in Clostridium saccharoperbutylacetonicum N1–4 with the CRISPR-Cas9 system. *Applied and Environmental Microbiology*, *83*(10), pp. e00233–17.

Wang, S., He, Z. and Yuan, Q., 2017. Xylose enhances furfural tolerance in Candida tropicalis by improving NADH recycle. *Chemical Engineering Science*, *158*, pp. 37–40.

Wang, S., Sun, X. and Yuan, Q., 2018. Strategies for enhancing microbial tolerance to inhibitors for biofuel production: a review. *Bioresource Technology*, *258*, pp. 302–309.

Wang, Y., Chen, L. and Zhang, W., 2016. Proteomic and metabolomic analyses reveal metabolic responses to 3-hydroxypropionic acid synthesized internally in cyanobacterium Synechocystis sp. PCC 6803. *Biotechnology for Biofuels*, *9*(1), pp. 1–15.

Weusthuis, R.A., Pronk, J.T., Van Den Broek, P.J. and Van Dijken, J.P., 1994. Chemostat cultivation as a tool for studies on sugar transport in yeasts. *Microbiological Reviews*, *58*(4), pp. 616–630.

Wilbanks, B. and Trinh, C.T., 2017. Comprehensive characterization of toxicity of fermentative metabolites on microbial growth. *Biotechnology for Biofuels*, *10*(1), pp. 1–11.

Woodruff, L., Pandhal, J., Ow, S.Y., Karimpour-Fard, A., Weiss, S.J., Wright, P.C., Gill, R.T. 2013. Genome-scale identification and characterization of ethanol tolerance genes in Escherichia coli. *Metabolic Engineering*, *15*(1), 124–133.

Wright, J., Bellissimi, E., de Hulster, E., Wagner, A., Pronk, J.T. and van Maris, A.J., 2011. Batch and continuous culture-based selection strategies for acetic acid tolerance in xylose-fermenting Saccharomyces cerevisiae. *FEMS Yeast Research*, *11*(3), pp. 299–306.

Xie, R., Tu, M., Carvin, J. and Wu, Y., 2015. Detoxification of biomass hydrolysates with nucleophilic amino acids enhances alcoholic fermentation. *Bioresource Technology*, *186*, pp. 106–113.

Yang, S., Pan, C., Tschaplinski, T.J., Hurst, G.B., Engle, N.L., Zhou, W., Dam, P., Xu, Y., Rodriguez Jr, M., Dice, L. and Johnson, C.M., 2013. Systems biology analysis of Zymomonas mobilis ZM4 ethanol stress responses. *PLoS One*, *8*(7), p. e68886.

Yu, H., Guo, J., Chen, Y., Fu, G., Li, B., Guo, X. and Xiao, D., 2017. Efficient utilization of hemicellulose and cellulose in alkali liquor-pretreated corncob for bioethanol production at high solid loading by Spathasporapassalidarum U1–58. *Bioresource Technology*, *232*, pp. 168–175.

Zaldivar, J., Martinez, A. and Ingram, L.O., 1999. Effect of selected aldehydes on the growth and fermentation of ethanologenic Escherichia coli. *Biotechnology and Bioengineering*, *65*(1), pp. 24–33.

Zamora, F., 2009. Biochemistry of alcoholic fermentation. In M. Victoria Moreno-Arribas, M. Carmen Polo (Eds.) *Wine Chemistry and Biochemistry* (pp. 3–26). Springer, New York.

Zhang, L., Li, X., Yong, Q., Yang, S.T., Ouyang, J. and Yu, S., 2016. Impacts of lignocellulose-derived inhibitors on l-lactic acid fermentation by Rhizopus oryzae. *Bioresource Technology*, *203*, pp. 173–180.

Zhang, S., Guo, F., Yan, W., Dai, Z., Dong, W., Zhou, J., Zhang, W., Xin, F. and Jiang, M., 2020. Recent advances of CRISPR/Cas9-based genetic engineering and transcriptional regulation in industrial biology. *Frontiers in Bioengineering and Biotechnology*, *7*, p. 459.

Zhang, X., Zhang, X.F., Li, H.P., Wang, L.Y., Zhang, C., Xing, X.H. and Bao, C.Y., 2014. Atmospheric and room temperature plasma (ARTP) as a new powerful mutagenesis tool. *Applied Microbiology and Biotechnology*, *98*(12), pp. 5387–5396.

4 The Role of Metabolic Engineering in the Development of 2G Biofuels (Both in Conversion and Fermentation)

Vinod Kumar Nathan
SASTRA Deemed University

Kalirajan Arunachalam
Mulungushi University

Aparna Ganapathy Vilasam Sreekala
SASTRA Deemed University

CONTENTS

DOI: 10.1201/9781003203452-4

4.1 INTRODUCTION

Second-generation (2G) biofuels or advanced biofuels are formulated by mixing with petroleum-based fuels or can be directly used in modified vehicles, demand the use of various technologies for the extraction of energy from different feedstocks such as lignocellulosic biomass or woody crops, agricultural residues, industrial wastes such as wood chips, pulp or skins after fruit pressing, animal wastes, other crops that cannot be used for food purposes (non-food crops), cereals that bear little grain and so forth (IEA, 2010); whereas, the first-generation category of biofuels was produced from food biomasses (Inderwildi and King, 2009). The 2G biofuels such as cellulosic ethanol and Fischer-Tropsch fuels have a much larger array of feedstock options, a greater energy output yield than fossil fuels (Carriquiry et al., 2011), and significantly reduce the competition on land with carbon-rich feedstock for first-generation biofuel production, thus having less environmental impacts.

Research recently has focused on lignocellulosic biomass feedstocks for their biological degradation to synthesize biofuels such as biodiesel, bioethanol and biohydrogen, and has sparked interest in the hunt for fuels that are more sustainable. Efficient substrate utilization must be achieved by the model strains in order to obtain better fermentation products by adopting better routes for sugar transportation and to withstand the inhibition caused by chemicals as well as fermentation end products, thereby boosting the metabolic fluxes. The engineering of microbes opens a new possibility for making the biofuel production process more economical by utilizing lignocellulosic biomass. Metabolic engineering results in more efficient microbial cell factories, which is a critical component of the next-generation bio-economy. It has been utilized in a variety of microorganisms, including natural and modified ones, to manipulate the biosynthetic pathway to manufacture targeted products. Engineering of metabolic pathways as well as optimizing endogenous metabolism by metabolic engineers has resulted in the development of a wide spectrum of novel chemicals (Adegboye et al., 2021).

Within the fuel sector, biofuel generation from renewable and sustainable resources is becoming increasingly essential. Bioethanol has been the most extensively utilized biofuel as a gasoline additive. Bioethanol may be made from cellulosic biomass, which is a hopeful natural resource. Cellodextrins, xylose, glucose, cellobiose and arabinose, among other sugars found in plant biomass hydrolysates, are poorly fermented by bacteria. Although these modified strains that use xylose have been designed, *Saccharomyces cerevisiae*, the most widely used ethanol producer, can normally use glucose only as a carbon source (Li et al., 2020). Higher alcohols can be mixed at a larger volume than ethanol, resulting in a diminishing of greenhouse gas (GHG) emissions without the requirement of adjusting the existing fuel infrastructure (Liang et al., 2020). Furthermore, research has shown that integrated processes such as concurrent fermentation and saccharification, as well as integrated bioprocessing, increase the yield of bioethanol using cereal crops and residues. Moreover, genome editing and metabolic engineering approaches could help improve the biorefinery theme, resulting in low-cost technology, as well as increasing the bioethanol content (Kumar et al., 2020). Several yeasts are used to produce bioethanol from monomeric sugar. However, yeast strains have significant drawbacks, including limited ethanol tolerance, toxic inhibitors and high sugar concentrations. The aforesaid restrictions can be solved through the engineering of suitable yeast strains at both genetic as well as metabolic levels (Adebami et al., 2021).

Controlled and data-driven metabolic engineering has been allowed by systems-level understanding on several scales – from the genome, transcriptome, proteome and metabolome to the (sub)cellular and bioreactor scales – as well as the development of advanced synthetic biological tools. Meyer (2021) describes the techniques and innovative concepts for bioengineers to understand the versatility of filamentous fungi as cell factories, which will provide them with the knowledge and tools they need to rationally rewire fungal metabolic and regulatory networks for new, smarter and better products that will help to secure a sustainable bioeconomy. In synthetic biology, manufacturing several types of biofuels is possible by engineering the metabolic pathways of microbial hosts, including yeast and heterotrophic as well as photoautotrophic bacteria (Singhal et al., 2020). Metabolic engineering of the system provides for the rapid generation of high-performing bacterial strains for long-term chemical and material synthesis. Machine-learning techniques have been actively enforced across different stages in the metabolic engineering of systems, including metabolic pathway reconstruction, choosing of host strains, optimizing of metabolic flux, as well as fermentation, as the availability of biological big data, such as omics data, has increased in recent years (Kim et al., 2020). Even though the adoption of machine learning in response to the growing amount of bio-big data is already actively employed in most of the steps in systems metabolic engineering, some future prospects for successful machine-learning applications include further upgradation by automating processes using robots, generation of high-quality datasets, availability of experimental datasets in a machine-learning format and so forth.

Recent advances in omics technology, computer modelling, and simulation, systems and synthetic biology are assisting in the effective manipulation and re-designing of microbes at the cellular level via metabolic engineering; for example, how it helps to build and modify pathways for the production of ethanol in *Zymomonas mobilis* (Zhang et al., 2019). As a result, sensible engineering procedures are required, particularly when targeting particular genes and pathways. Metabolic engineering techniques are increasingly being employed to boost the generation of bioethanol and other types of biofuels, especially those of microbial origin. For example, various metabolic engineering procedures have been applied to widen Z. *mobilis*'s substrate range, eliminates competing pathways and enhance the level of tolerance to ethanol and lignocellulosic hydrolysate inhibitors. Using genetic and metabolic engineering techniques, enzymes and microbes for biomass metabolizing as well as bioethanol synthesis have been developed and deployed as reactors suitable for consolidated bioprocessing. In consolidated bioprocessing, hydrolysis and fermentation take place in the same reaction vessel (Banerjee et al., 2019). As a result of these developments in metabolic engineering along with synthetic biology, biotechnologists will have new tools to create a cost-effective biocatalyst with desirable phenotypes (Bilal et al., 2018).

4.2 METABOLIC ENGINEERING FOR BIOFUEL PROCESSES

4.2.1 Bacterial Metabolic Engineering

Lignocellulosic biomass, which contains cellulose, hemicelluloses and lignin, is one of the most plentiful organic renewable resources in the world. These renewable energy resources are being researched for the enhanced generation of ethanol fuel (Nielsen and Keasling, 2011). The main purposes of metabolic engineering are to increase productivity, gain higher yield and widen the substrate and product spectra for biofuel production in an eco-friendly way that leads to a reduction of fossil fuel dependence (Banerjee et al., 2010).

4.2.2 Molecular Biology in Bacterial Metabolic Engineering

Current molecular biology techniques can efficiently boost the flux towards biofuel synthesis by altering enzyme levels. Molecular biology techniques are employed in a variety of ways. Some common strategies employed for accomplishing this include promoter engineering, plasmid selection, copy numbers, synthetic scaffolds, optimization of codons, directed evolution of important enzymes, ribosome binding site enhancement and suppression of competitive pathways (Dueber et al., 2009; Carneiro et al., 2013; Nowroozi et al., 2014). RNA interference, CRISPRs and TALENs are new technologies that help to alter genes involved in microbial metabolic pathways (Pratt and MacRae, 2009; Sun and Zhao, 2013).

4.2.3 Importance and Significance of Bacterial Metabolic Engineering in Biomass Conversion

Biomass conversion by microorganisms is a crucial process in second-generation biofuel production, in which microbes facilitate the breakdown of carbohydrate polymers present in biomass to be converted into simple molecules. Fermentation technology using yeast cells makes it easier to produce biofuels (Ruffing and Chen, 2005). Microorganisms, especially bacterial species, are extensively employed in biomass conversion, which impacts the efficient conversion rate of biomass into biofuels. Therefore, it's imperative to improve the metabolic pathways of microbes or introduce exogenous pathways derived from any organism using the above-mentioned molecular biology techniques. The development of biocatalysts for the hemicellulose conversion found in biomass into biofuels is the focus of metabolic strategies (Aristidou and Penttila, 2000).

Metabolic engineering enhanced product diversity, yield, and ethanol concentration in order to enhance fuel production and simplify downstream recovery of biofuel (De Bhowmick et al., 2015), and engineering microorganisms to utilize low-cost substrates can lower production costs (Gustavsson and Lee, 2016). This metabolic engineering approach in biofuel production indirectly benefits the recycling of agricultural based wastes into biofuels at a higher rate than traditional microbial conversion of biomass, which utilized ordinary microbes in the conversion of biomass to biofuel production. Remarkable impact and employment of metabolic engineering are not only increasing the rate of production of biofuels through the conversion of biomass by microbes but also making it more cost-effective as it facilitates the microbes to use cheap substrates.

4.2.4 Bioethanol Production Using Bioengineered Bacterial Strains

Zymomonas mobilis, an anaerobic bacterium, is the producer of bioethanol, which converts glucose into bioethanol (Ajit et al., 2017), but this bacterium is incapable of converting pentose sugars present in lignocellulosic feedstocks to ethanol. The metabolic engineering in *Z. mobilis* for pentose utilization ability has solved this difficulty. Without utilizing vitamins or amino acids, an engineered *Z. mobilis* strain, TMY-HFPX, with better properties was processed to reach a concentration of 136 g/L ethanol in a solution with glucose estimated at about 295 g/L. This recombinant strain sheltered multiple gene components *xylA/xylB/tktA/talB* operon for the utilization of xylose (Wang et al., 2016).

Through metabolic engineering, *Escherichia coli* was the pioneering microorganism to be genetically modified to make ethanol. The PET operon, which encodes pyruvate decarboxylase as well as alcohol dehydrogenase from *Z. mobilis*, was introduced into an *E. coli* wild strain to produce a recombinant *E. coli* K011 that produces ethanol at a theoretical 95% yield in a complex medium while also showing superior ethanol tolerance to the original strain (Ajit et al., 2017). Complex dietary supplements were required for this recombinant *E. coli* K011, which escalates the ethanol production cost. By deleting the lactate dehydrogenase gene and replacing it with the pyruvate formate lyase along with the PET operon from *Z. mobilis*, recombinant *E. coli* was transformed into SZ110, which could produce ethanol in a less nutritionally demanding environment (Ajit et al., 2017). Under the influence of the promoter, which is strongly light-driven psbAII, the genes for alcohol dehydrogenase II (adh) and pyruvate decarboxylase (pdc) from *Z. mobilis* were blended into the chromosome of the cyanobacterium *Synechocystis* sp. PCC 6803, resulting in a 5.2 mmol per OD730 unit/L/d increase in ethanol productivity (Dexter and Fu, 2009). When Lönn et al. (2003) overexpressed the multicopy xylA gene from Thermus thermophiles, they discovered xylitol synthesis via an endogenous aldose reductase (GRE3), which inhibited the pentose pathway and xylulokinase (XK). The recombinant strain's ethanol productivity was increased by removing gre3 and overexpressing an additional copy of XK. In another study, Lo et al. (2017) altered electron metabolism in *Clostridium thermocellum*, resulting in improved bioethanol production.

4.2.5 Butanol Production Using Bioengineered Bacterial Strains

Another commonly used biofuel is butanol, which is superior to ethanol owing to its lower vapour pressure, high energy density, and being less hygroscopic. Its similarity and compatibility with gasoline have the potential to substitute gasoline, which has been used to fuel automobile engines for a long time (Xu et al., 2017). *Clostridium* strains and bioengineered *E. coli* strains have been used as natural producers of butanol (Huang et al., 2010). Harris et al. (2000) metabolically altered *Clostridium acetobutylicum* PJC4BK to increase butanol concentration by deleting butyrate kinase, which is encoded by the gene *buk* and is participated in the butyrate synthesis pathway. Compared to *Clostridium acetobutylicum* wild type, which produced butanol of about 11.7 g/L, this novel strain generated up to 16.7 g/L of butanol (Wilson et al., 2016).

Abdelaal et al. (2019) reported the development of a dual-operon-based approach in the genome of *E. coli* strain MG1655 that produced 5.4 g/L of n-butanol in a glucose medium and was subsequently replicated in *E. coli* strain SSK42 to yield butanol from xylose via a redox-balanced pathway. Hanai et al. (2007) overexpressed the *thl* and *adc* genes from *C. acetobutylicum* ATCC 824 and the endogenous *atoDA* genes producing CoA transferase in a bioengineered *E. coli* strain to synthesize isopropanol (approximately 5 g/L). The Ehrlich, or 2-keto acid pathway, is an alternate method for the generation of butanol-like alcohols. This route converts keto acids, which are the immediate amino-acid precursors, to aldehydes, which are then reduced to alcohols. Increased iso-butanol production may be due to overexpression of 2-ketoisovalerate biosynthetic genes, ablation of multiple pyruvate-consumption pathways and replacement of the endogenous acetolactate synthase with one that has stronger pyruvate selectivity (Roy et al., 2020).

4.3 METABOLIC ENGINEERING OF SOME COMMON MODEL ORGANISMS

4.3.1 *Clostridium cellulolyticum*

Guedon et al. (2002) reported the case of *C. cellulolyticum* ATCC 35319, a mesophilic bacterium with cellulase activity, which has superior control over carbohydrate metabolism on mineral salt media compared to the media that have a complex composition due to their unfavorability caused by high-concentration substrates. As a result, toxic metabolic products collect inside the bacteria's cells (NADH and pyruvate), and *C. cellulolyticum* is unable to consume carbon sources in excess as a result of this accumulation, necessitating genetic manipulation for carbon source breakdown. The accumulated pyruvate was removed by a mode of metabolic engineering that resulted in a substantial increase in the cellulose fermentation rate of *C. cellulolyticum*. Biotechnological applications such as enhanced biofuel generation are all possible with these transformed organisms. The cellulose degradation route driven by cellulolytic enzymes generated by a recombinant strain of *C. cellulolyticum* is depicted in Figure 4.1.

Similar to the above-mentioned aspect, since *Clostridium cellulolyticum* ATCC 35319 cannot withstand high concentrations of pure cellulosic substrates, nutrients begin to accumulate in dangerous quantities. The rate of lactate dehydrogenase and pyruvate ferredoxin oxidoreductase (PFO) processing is faster than the rate of carbon flux through glycolysis (LDH). As a result, the accumulation of these inhibitory

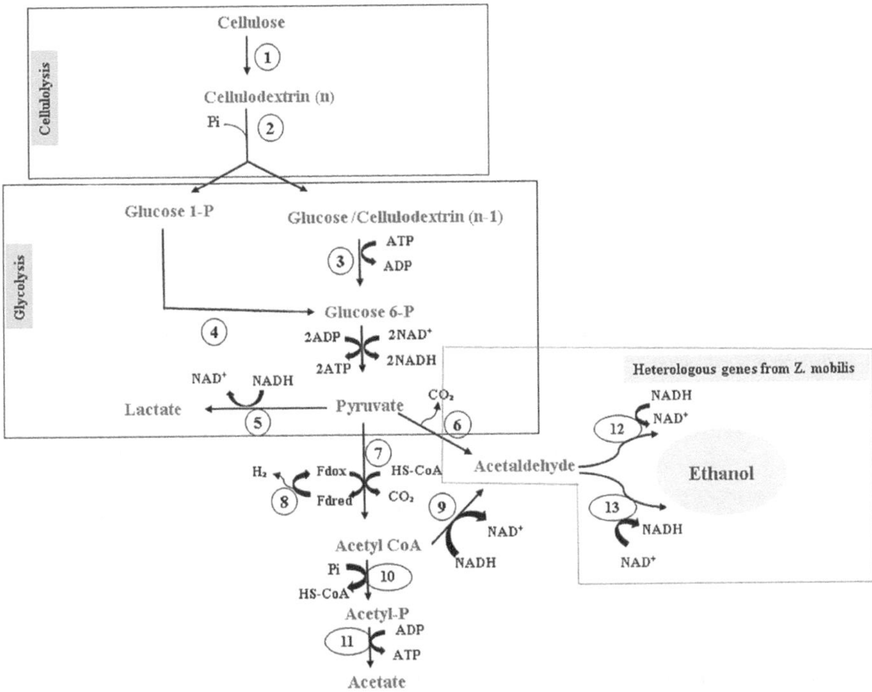

FIGURE 4.1 Execution of cellulose conversion into fermentation products by an engineered *C. cellulolyticum* (strain CC-pMG8). (1) Cellulosome hydrolysis; (2) cellodextrin phosphorylase (EC 2.4.1.49); (3) glucokinase; (4) phosphoglucomutase; (5) L-LDH; (6) PDC; (7) PFO; (8) hydrogenase; (9) acetaldehyde dehydrogenase; (10) phosphotransferase; (11) acetate kinase; (12) ADHII; (13) ADH; (P) phosphate; (Fd) ferredoxin; (ox) oxidation; (red) reduction; (HS-CoA) coenzyme A. (Adapted from Guedon et al., 2002; Senthilkumar and Gunasekaran, 2005.)

chemicals stops the cells from growing. As a result, *C. cellulolyticum* is intolerant to excess carbon sources, making it unsuitable for the production of fuel ethanol from cellulosic biomass. Senthilkumar and Gunasekaran (2005) devised an expression system (pMG8) that consists of the *pdc* and *adh* genes from *Z. mobilis* under the control of a powerful ferredoxin gene promoter of *Cellulolyticum pasteurianum* to reduce pyruvate synthesis. On cellulose medium, the novel recombinant strain CC-pMG8 functioned well, with a specific growth rate (0.049 g/l/h) that was greater than the parent strain's (0.044 g/l/h) and produced two times more ethanol than the parent strain (20 mM). This is also parallelly depicted in Figure 2.1.

4.3.2 *Klebsiella pneumoniae*

Chen et al. (2015) transformed *K. pneumoniae* HR526 into a superior 2,3-butanediol-yielding strain for 2-butanol production by extending the bacterium's 2,3-butanediol synthesis pathway and introducing alcohol dehydrogenases and diol dehydratases. The pathway was optimized, and the diol dehydratase was engineered using protein

TABLE 4.1
Metabolically Modified Bacterial Strains for Biofuel Production

Organisms	Products	Substrate	Reference
Clostridium autoethanogenum	Ethanol	Synthetic medium	Lewin et al. (2016)
E. coli	Fatty alcohol	Synthetic medium	Liu et al. (2014)
Klebsiella pneumoniae	2-Butanol	Glucose	Chen et al. (2015)
C. cellulolyticum	Ethanol	Microcrystalline cellulose	Li et al. (2012)
C. cellulovorans	Ethanol	Cellulose	Yang et al. (2015)
Clostridium thermocellum	Ethanol	Cellulose	Argyros et al. (2011)
Thermoanaerobacterium saccharolyticum	Ethanol	Cellulose	Argyros et al. (2011)
K. oxytoca	Ethanol	Xyl 100	Ohta et al. (1991)
Z. mobilis	Ethanol	Xyl 160	Mohagheghi et al. (2004)
C. cellulomonas	Ethanol	Cellulose 18	Guedon et al. (2002)
C. japonicas MSB28	Ethanol	Microcrystalline cellulose	Gardner and Keating (2010)

engineering, resulting in a higher 2-butanol yield (1,030 mg/L). The capacity of the optimum strain to utilize lignocellulosic biomass feedstock with pentose and hexose sugars for biofuel generation, as well as its survivability in inhibitory chemicals formed during the pretreatment process, is highlighted by Kim et al. (2010). As a result, developing a modified strain using a metabolic engineering approach that can outperform native bacterial strains in terms of pentose and hexose utilization from lignocellulosic and other waste biomasses, as well as improved survival against inhibitory compounds produced during the pretreatment process, is critical. Hence, when designing strains for biofuel production, it is critical to improve inhibitor tolerance (Ling et al., 2014). Table 4.1 shows some metabolically modified bacterial strains that have been successfully employed to produce biofuel.

4.3.3 *Lactobacillus casei*

For ethanol production, Ohta et al. (1991) used *Klebsiella oxytoca* as well as *Erwina chrysanthemi*. Despite the fact that the strains produced less ethanol than *E. coli*, significant progress was made in developing superior strains of *K. oxytoca* that can convert cellulose to ethanol. The pet operon-containing metabolic engineering bacteria *K. oxytoca* M5A1 produced ethanol (>90%) of the fermentation products.

4.3.4 Actinobacteria

Actinobacteria members are involved in the manufacture of a variety of important enzymes, including xylanases (Salwan and Sharma, 2018) and cellulases (Techapun et al., 2003), which aid in the decomposition of cellulose, lignocelluloses, lignin and plant wastes. Because these microbes are a potential source of essential enzymes involved in lignocellulose biomass conversion, the use of metabolic engineering will improve the capacity of bacteria belonging to the actinobacteria group for enhancing the biomass conversion rate of lignocellulosic and waste biomass derived from agriculture crop waste. In recent years, several approaches such as site-directed mutagenesis, random mutagenesis, heterologous protein expression, clustered regularly interspaced short palindromic repeats (CRISPR-Cas) system, and genome and metabolic engineering have been used to improve the expression of enzymes by microbial strains in order to make the impression of biorefineries feasible and economically viable (Fujii et al., 2018; Misra et al., 2019).

Recently, gene-editing technologies such as CRISPRa and CRISPRi have been employed to up-regulate or down-regulate metabolic pathways essential for biofuel production. CRISPR–Cas9 technology helps prokaryotic microbes as a self-protective mechanism against any foreign nucleic acids (Mobolaji Felicia Adegboye et al., 2021). Advanced metabolic engineering at the gene level opens up a lot of possibilities for strain enhancement in order to scale up biofuel production using fermentation technology.

4.4 FUNGAL METABOLIC ENGINEERING

Similar to bacterial metabolic engineering, fungal cells are also engineered for either hypersecretion of fungal enzymes for lignocellulosic biomass pretreatment or ethanol production. In recent years, many filamentous fungi have been subjected to metabolic engineering to intensify enzyme production (Zhao et al., 2018). Over the decades, there has been a thorough understanding of the regulatory pathway for cellulase and other similar lignocellulosic enzymes' expression. This has enabled the engineering of *Trichoderma reesei*, a promising fungus for cellulase production at the metabolic pathway level (Kubicek et al., 2009). Apart from gene-level modification, promoter engineering is also carried out for the enhancement of enzyme expressions. In *T. reesei*, scientists targeted the *cbh1* and *cbh2*, two strongly inducible promoters, for the overexpression of β-glucosidase and other cellulase components (Ma et al., 2011; Li et al., 2017).

Cellulase synthesis is known to be regulated at the transcriptional level, and in filamentous fungi, such regulators are abundantly found (Benocci et al., 2017). Under inducing conditions, overexpression of *xyr1* increased glycoside hydrolase synthesis, whereas deletion of gene *cre1* reduced CCR, but catabolic de-repression was insufficient to expand the production of glycoside hydrolase, implying that hyperproduction is largely dependent on inducers (Wang et al., 2013; Nakari-Setala et al., 2009). Artificial transcription factors were used apart from natural transcription factors to increase cellulase synthesis in *T. reesei*. Trvib-1 gene overexpression improved cellulase synthesis in *T. reesei* Rut-C30 (Zhang et al., 2018a), and the significance of this gene in cellulase regulation has also been documented (Ivanova et al., 2017). As a result, analysing the mutant containing the artificial transcription factor can uncover novel regulatory pathways. Many artificial transcription factors will need to be employed to influence cellulase synthesis in filamentous fungi in the future.

TABLE 4.2
Metabolically Modified Fungal Strains for Biofuel Production

Organisms	Genome editing method	Reference
T. reesei	Artificial transcription factors synthesis for increased cellulase production	Zhang et al. (2018)
Saccharomyces cerevisiae	Gene knockout approach for enhanced bioethanol production.	Divate et al. (2017)
Myceliophthora sp.	CRISPR-Cas9 system for transcription factor alterations.	Liu et al. (2017)
A. niger	sgRNA synthesis and transformation into *A. niger* strain carrying the Cas9 plasmid	Novy et al. (2016)
P. chrysogenum	Transforming the CRISPR-Cas9 ribonucleoproteins (RNPs)	Pohl et al., (2016)
T. reesei	In vitro transcription of gRNA, and transformation into *T. reesei* overexpressing codon-optimized Cas9	Liu et al. (2015)

The type II CRISPR-Cas9 technology was recently designed to enable quick genome editing in numerous organisms, including filamentous fungal species (Doudna and Charpentier, 2014; Fuller et al., 2015; Liu et al., 2015; Nødvig et al., 2015; Matsu-Ura et al., 2015; Pohl et al., 2016; Liu et al., 2017). In addition to Cas9, the *Cpf1* effector protein created was used for editing the genome (Zetsche et al., 2015). Although the use of *Cpf1* has not been used in filamentous fungi, it is certain that this will enhance the efficiency of engineering the genome in most fungi. As a result, CRISPR systems can be employed for both gene insertion and deletion (Pohl et al., 2016; Liu et al., 2017), leading to the disruption of marker-free genes through insertion or deletion (Doudna and Charpentier, 2014). Liu et al. (2015) constructed CRISPR-Cas9 in the filamentous fungus *T. reesei*, which was the first example that was subjected to genome editing. The CRISPR-Cas9 technique offers considerable advantages over traditional genetic editing technology used in filamentous fungi. Firstly, it is simple, and second, it is efficient. Cas9 may be used to target many genes, with the exception of sgRNA. The CRISPR-Cas9 method may be employed to edit several locations at once as well as manipulate a single gene quickly. The third is its flexibility. This offers the possibility of changing several genes without the use of any markers. Fourth, it is not toxic. When Cas9 was expressed in cells, there was no difference in cell proliferation or sporulation (Liu et al., 2017). The off-target effect of CRISPR-Cas9 is one of its drawbacks. As a result, high-fidelity CRISPR-Cas9 nucleases (Kleinstiver et al., 2016) and optimization of designing sgRNA will lead to more precise genome editing (Doench et al., 2016). In filamentous fungi, the possibility of editing is restricted to deletion or replacement. This approach will also be used to either activate certain genes or repress their expression, as well as epigenome editing used in mammalian systems, which was quite successful (Hilton et al., 2015; Konermann et al., 2015).

Saccharomyces cerevisiae, the most extensively employed starter, is exposed to a variety of environmental challenges during bioethanol production, including aldehydes, ethanol, glucose, high temperatures, acid, alkaline and osmotic pressure. A genetically modified *S. cerevisiae* starter that was resistant to a variety of environmental stressors was developed. In *S. cerevisiae*, when the *nth1* gene (neutral trehalase gene) coding for an enzyme that degrades trehalose was deleted, the aldehyde reductase gene (*ari1*) and the gene trehalose-6-phosphate synthase (*tps1*) were co-overexpressed. The modified strain could tolerate about 14% ethanol, whereas the wild strain's development was slowed by 6% ethanol (Divate et al., 2017).

Bioethanol production using yeast results in glycerol production, a by-product that helps to maintain cellular redox balance and osmoregulation in *S. cerevisiae*. It has been proven that restoring the glycerol pathway with an engineered mechanism for NAD+-dependent acetate reduction improves ethanol yields and aids in the detoxification of media containing acetate. However, glycerol non-producing strains' osmo-sensitivity limits their use in high-osmolarity commercial applications. A study looked into engineering options for lowering glycerol production while maintaining osmotolerance in acetate-reducing strains. In low-osmolarity cultures, replacing both native G3PDHs with an archaeal NADP+-preferring enzyme and deleting ALD6 resulted in an acetate-reducing strain with a phenotype that mirrored that of a glycerol-negative gpd1 gpd2 strain. In high-osmolarity cultures, it produced 13% more ethanol than an acetate-reducing strain with the native glycerol pathway.

Although direct cellobiose fermentation could eliminate glucose suppression in biomass fermentation, employing engineered *S. cerevisiae* for cellobiose is currently restricted due to its protracted lag phase. For metabolic engineering, a filamentous cellulolytic fungus called *Myceliophthora thermophila*, which is thermophilic, produces ethanol from glucose and cellobiose. After introducing ScADH1 into the wild-type strain, ethanol production rose by 57% from glucose but not from cellobiose. Overexpression of GLT-1 (glucose transporter) or the cellodextrin transport system (CDT-1/CDT-2) in *N. crassa* boosted ethanol production by 131% and 200%, respectively, from glucose and cellobiose. Decreasing the expression of the *pyc* gene resulted in a final modified strain with a 23% increase in ethanol production, reaching 11.3 g/L on cellobiose (Li et al., 2020).

Furthermore, the engineering of global transcription machinery can be employed to enhance tolerance (Atsumi et al., 2008). To enhance the tolerance of *S. cerevisiae* to glucose/ethanol, Alper et al. (2006) designed gTME, a global transcription machinery engineering system that involves reprogramming the gene transcription in order to achieve cellular phenotypes that were required for the technological processes. A factor that affects transcription, Spt15p, was mutated, resulting in an increase in tolerance and better efficiency in the conversion of glucose to bioethanol. The phenotype was caused by a fusion of three mutations in the *Spt15* gene, which included Tyr195His, Phe177Ser and Lys218Arg (Alper et al., 2006). As a result, gTME can facilitate access to complicated phenotypic features that are difficult to access using traditional methods. El-Rotail et al. (2017), who used the gTME technique to build an SPT15 mutagenesis library in *S. cerevisiae*, recently validated this.

Saccharomyces cerevisiae is known as an appropriate host for intensified bioprocessing since it is a conventional ethanol-producing microbe. However, heterologous cellulase expression increases yeast metabolic burden, resulting in low breakdown efficiency of cellulose, which might be due to low cellulase activity. In this study, yeast strains with cellulase expression that are capable of degrading a variety of cellulosic substrates were produced by adjusting cellulase ratios using a POT1-mediated integration method. This system also featured metabolic engineering methodologies such as codon use optimization, promoter and signal peptide optimization (Song et al., 2018).

The unfolded protein response was validated by heterologous cellulase production in cellulosic yeast. The endoplasmic reticulum chaperone protein BiP and the disulfide isomerase Pdi1p were overexpressed to improve protein folding capacity, whereas the Golgi membrane protein Ca^{2+}/Mn^{2+} ATPase Pmr1p was disrupted to reduce cellulase glycosylation. With carboxymethyl cellulose, the resulting strain, SK18-3, was able to synthesize 5.4 g/L ethanol. With phosphoric acid swollen cellulose hydrolysis, strain SK12-50 produced 4.7 g/L ethanol. When Avicel was utilized as the substrate, SK13-34 produced 3.8 g/L ethanol (75% of the theoretical maximum yield) (Song et al., 2018).

4.5 CHALLENGES IN SCALE-UP FERMENTATION

In scale-up fermentation, there are some challenges that affect bioethanol production. Temperature, pH, salt concentration, aeration rate, ethanol concentration and carbohydrate content all affected the fermentation process for bioethanol production, according to Arora et al. (2017) and Selim et al. (2018). As a result, some unique features of selected microbial strains should have a wide range of substrate utilization, the ability to tolerate high concentrations of sugar, ethanol and by-products produced following the pretreatment step, and minimal by-product generation to achieve maximum yield of bioethanol (Lugani and Sooch, 2018). However, most native strains of microbes used for alcoholic fermentation can only ferment hexose sugars, which results in very low ethanol yields, and wild-type pentose sugar fermenting microbial strains such as *P. tannophilus*, *P. stipitis* and *C. shehatae* are sensitive to high ethanol concentrations, low pH and inhibitors (Hahn-Hägerdal et al., 2007). Another important factor to consider when scaling up fermentation is the microbial strain's tolerance to inhibiting chemicals and metabolic intermediates. This must be overcome in order to obtain the maximum possible yield from the fermentation process, which involves the conversion of lignocellulosic biomass and waste biomass from the agricultural sector. Some of the hazardous substances found in lignocellulosic biomass include furans, weak organic acids and phenols. Because microbial cell development throughout the fermentation process affects the amount of biofuel produced, it is critical to develop strains that are resistant to inhibitors. Furthermore, during microbial fermentation, end products and by-products accumulate, inhibiting microbe development and hence lowering production levels for biofuel production. As a result, when developing strains for biofuel production, it is critical to advance inhibitor tolerance (Mukhopadhyay, 2015).

The ability of microorganisms in the hydrolysis of lignin and cellulose is low compared to physicochemical and enzymatic processes, and controlling by-product formation, which affects the fermentation process while using lignocellulosic biomass, is another challenge in scaling up fermentation towards the production of biofuel. The co-existence of lignocellulosic enzyme producers and ethanol producers is dependent on the process's growth rate and operating conditions, as well as the growth rate of other microbes. Lignocellulosic biomass hydrolysis by microbes is quite low, and reactor configuration and process parameters such as hydraulic retention and solid retention time affect reactant formation. It is critical to have knowledge of metabolic engineering of microbial hosts to help with strain improvement for successful production of second-generation biofuels using lignocellulosic biomasses and biowaste-based biomasses in order to solve the challenges encountered in the fermentation process that slow down the rate of biofuel production. According to Park et al. (2008), genomics and system biology have supplied knowledge for microbial host metabolic engineering. Nonetheless, Chubukov et al. (2013) indicated that it is not appropriate to link "omics" to actual enzyme functioning because they are governed by complicated post-transcriptional regulatory mechanisms, particularly allosteric modulation. Increases in mRNA levels, for example, may not result in increased protein levels, yet enzyme levels can remain stable despite dynamic carbon flow changes (Gygi et al., 1999). 13C-MFA (metabolic flux analysis) is utilized for the direct assessment of enzyme reaction rates during large-scale fermentations to overcome difficulties encountered during fermentation.

Other common occurrences throughout the fermentation process include genetic mutations and metabolic shifts, which result in subpopulations in the cultures and a decrease in ethanol output. According to Rühl et al. (2011), the Sauer group presented a 13C-MFA based on "reporter proteins" to investigate the metabolism of subpopulations. This specialized protein, which was generated by a specific subgroup, stored the subgroup's 13C-labelling information, allowing 13C-MFA to distinguish diverse metabolic flux phenotypes. This concept was evolved utilizing genetically modified *E. coli* (Shaikh et al., 2008; Rühl et al., 2011). Based on the labelled signals from amino acids (hydrolysates) of the green fluorescent protein, 13C-MFA examined the subpopulation metabolism (GFP producers) during fermentation using *E. coli*.

In addition, 13C-MFA can be combined with microbial kinetics to show how cell metabolism varies during the cultivation phase. The Sauer group used a dynamic 13C-MFA method in a fed-batch fermentation to screen the dynamic changes in intracellular fluxes during different growth stages, which revealed that in a riboflavin-producing bacterium's physiology, there was a remarkable shift from an overflow metabolism to an exclusively maintenance metabolism at the late fermentation stage (Rühl et al., 2010). Due to mass transfer restrictions, it is critical to determine responses of microorganisms to industrial bioreactor shapes and culture heterogeneity. Cell physiology of bacteria employed in industrial fermentation can be studied using a metabolic flux analysis and hydrodynamics technique at various sites inside a large bioreactor. However, such metabolic flow analysis tools have not yet reached their full potential.

Fermentation of pentose sugars, which make up a large portion of biomass, is another significant issue in the manufacture of cellulosic bioethanol. The ability of

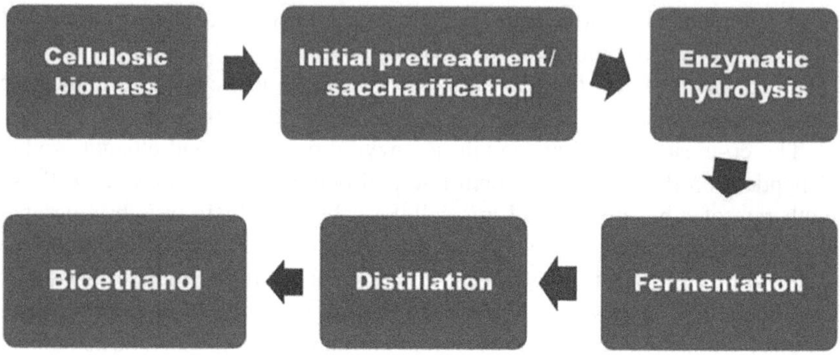

FIGURE 4.2 Utilization of cellulosic biomass for bioethanol production by downstream processing.

many bacteria to use xylose is limited. These issues can be addressed by employing metabolic engineering to build superior strains that can produce considerable amounts of bioethanol from the fermentation of pentose sugars while also lowering costs. The production of bioethanol is schematically described in Figure 4.2.

4.6 PRESENT STATUS AND FUTURE PROSPECTS OF BACTERIAL METABOLIC ENGINEERING

Metabolic engineering is currently used to adjust the metabolic pathways of diverse native microbes for enhanced biofuel production from a spectrum of biomass feedstocks ranging from starch-based materials to lignocellulose. Genetic engineering and metabolic engineering propose significant improvements in specific product biosynthesis by the exploitation of biosynthetic pathways of cell, including its transport systems and regulatory functions (Kern et al., 2007). High production of the targeted products with little or no bottleneck is achievable through a robust fermentation employing engineered strains that are tolerant to inhibitory substances or high concentrations of end products. These strains are developed by removing the normal regulatory genes and certain enzymes associated with their metabolic pathway. Currently, techniques such as proteomics, systems biology, bioinformatics, whole-genome sequencing and metabolomics have been coupled with bacterial metabolic engineering for engineered strains development that are novel and can accomplish high-throughput performance using renewable feedstocks such as lignocellulose, thereby drastically reducing production costs (Adegboye et al., 2021). For example, several fatty acid derivatives, short-chain alcohols, isoprenoids and gaseous biofuels like hydrogen are produced by strategies like optimization of fermentation conditions (Kim et al., 2019), overexpression of enzymes, transcription regulators and metabolic pathways (Jaroensuk et al., 2019; Leplat et al., 2018; Mehrer et al., 2018; Zhang et al., 2018b; Qiao et al., 2017). Figure 4.3 gives an overview of the current trends in bacterial metabolic engineering.

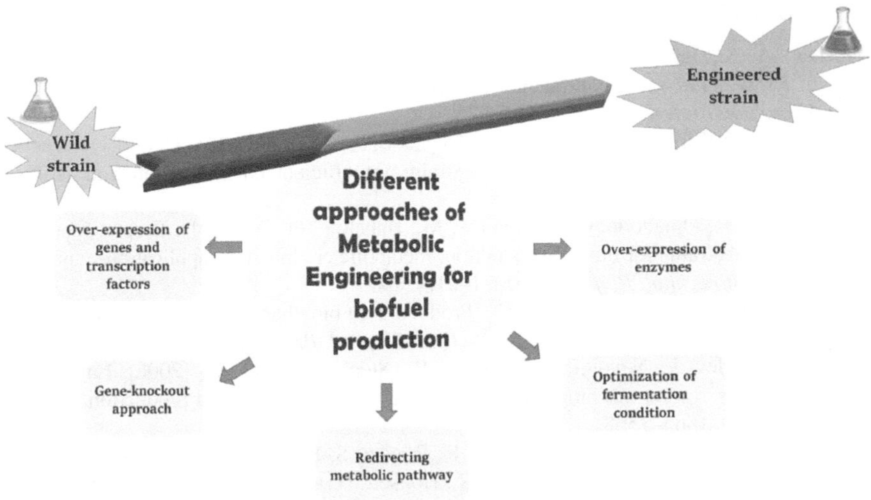

FIGURE 4.3 Current trends and aspects of bacterial metabolic engineering for biofuel production.

The CRISPR-Cas9 system and other genome engineering methods need to be improved in filamentous fungi in the near future, and these genome editing technologies will aid in the functional genomic understanding of filamentous fungi, which might enhance the efficiency of metabolic engineering. Engineered filamentous fungi were developed effectively using techniques including marker reuse, engineering of promoters, developing a library of artificial transcription factors, and editing of the genome using techniques like CRISPR-Cas9. New targets and strategies for developing hyperproducers for lignocellulosic enzymes, especially cellulose, are emerging due to the advancements in systems biology and synthetic biology (Zhao et al., 2018). Currently, scientists mostly employ strategies including the engineering of organisms at the genetic or metabolic level, optimization of processes or media components, and the use of model organisms like *Saccharomyces* to promote bioethanol production. However, the success rate of such strategies was very limited (Dakal and Dhabhai, 2019).

4.7 CONCLUSION

The current chapter generalized the recent methods employed for carrying out metabolic engineering techniques for increased second-generation biofuel production. Model organisms such as *S. cerevisiae*, *E. coli*, *K. pneumoniae*, *C. acetobytylicum*, *C. cellulolyticum* and *Z. mobilis* for bioethanol and biobutanol production have been pointed out. Further advancement in the technique could be expected in the future with the advancement of gene-editing techniques (CRISPR-Cas 9) and so on. Thus, the crisis posed by the generation of first-generation biofuels would get reduced upon the establishment of these techniques.

REFERENCES

Abdelaal, A. S., Jawed, K., Yazdani, S. S. (2019). CRISPR/Cas9-mediated engineering of *Escherichia coli* for n-butanol production from xylose in defined medium. *J. Ind. Microbiol. Biotechnol.* 46(7), 965–975.

Adebami, G. E., Kuila, A., Ajunwa, O. M., Fasiku, S. A., Asemoloye, M. D. (2021). Genetics and metabolic engineering of yeast strains for efficient ethanol production. *J. Food Process Eng.* 45(7), e13798.

Adegboye, M. F., Ojuederie, O. B., Talia, P. M., Babalola, O. O. (2021). Bioprospecting of microbial strains for biofuel production: metabolic engineering, applications, and challenges. *Biotechnol. Biofuels.* 14(5), 1–21.

Ajit, A., Sulaiman, A. Z., Chisti, Y. (2017). Production of bioethanol by *Zymomonas mobilis* in high-gravity extractive fermentations. *Food Bioprod. Process.* 102, 123–135.

Alper, H., Moxley, J., Nevoigt, E., Fink, G. R., Stephanopoulos, G. (2006). Engineering yeast transcription machinery for improved ethanol tolerance and production. *Science.* 314(5805), 1565–1568.

Argyros, D. A., Tripathi, S. A., Barrett, T. F., Rogers, S. R., Feinberg, L. F., Olson, D. G., Foden, J. M., Miller, B. B., Lynd, L. R., Hogsett, D. A. (2011). High ethanol titers from cellulose by using metabolically engineered thermophilic, anaerobic microbes. *Appl. Environ. Microbiol.* 77(23), 8288–8294.

Aristidou, A., Penttila, M. (2000). Metabolic engineering applications to renewable source utilization. *Curr. Opin. Biotechnol.* 11, 187–198.

Arora, R., Behera, S., Sharma, N. K., Kumar, S. (2017). Augmentation of ethanol production through statistically designed growth and fermentation medium using novel thermotolerant yeast isolates. *Renew. Energy.* 109, 406–421.

Atsumi, S., Cann, A. F., Connor, M. R., Shen, C. R., Smith, K. M., Brynildsen, M. P., Chou, K.J., Hanai, T., Liao, J. C. (2008). Metabolic engineering of Escherichia coli for 1-butanol production. *Metab. Eng.* 10(6), 305–311.

Banerjee, G., Scott-Craig, J. S., Walton, J. D. (2010). Improving enzymes for biomass conversion: a basic research perspective. *Bioenergy Res.* 3(1), 82–92.

Banerjee, S., Mishra, G., Roy, A. (2019). Metabolic engineering of bacteria for renewable bioethanol production from cellulosic biomass. *Biotechnol. Bioprocess Eng.* 24(5), 713–733.

Benocci, T., Aguilar-Pontes, M. V., Zhou, M., Seiboth, B., de Vries, R. P. (2017). Regulators of plant biomass degradation in ascomycetous fungi. *Biotechnol. Biofuels.* 10(1), 1–25.

Bilal, M., Iqbal, H. M., Hu, H., Wang, W., Zhang, X. (2018). Metabolic engineering and enzyme-mediated processing: a biotechnological venture towards biofuel production–a review. *Renew. Sust. Energ. Rev.* 82, 436–447.

Carneiro, S., Ferreira, E. C., Rocha, I. (2013). Metabolic responses to recombinant bioprocessing *Escherichia coli. J. Biotechnol.* 164, 396–408.

Carriquiry, M. A., Xiaodong, Du., Timilsina, G. R. (2011). Second generation biofuels: economics and policies. *Energy Policy.* 39(7), 4222–4234.

Chen, Z., Wu, Y., Huang, J., Liu, D. (2015). Metabolic engineering of *Klebsiella pneumoniae* for the de novo production of 2-butanol as a potential biofuel. *Bioresour. Technol.* 197, 260–265.

Chubukov, V., Uhr, M., Le Chat, L., Kleijn, R. J., Jules, M., Link, H., et al. (2013). Transcriptional regulation is insufficient to explain substrate-induced flux changes in *Bacillus subtilis. Mol. Syst. Biol.* 9, 709. doi: 10.1038/msb.2013.66.

Dakal, T. C., Dhabhai, B. (2019). Current status of genetic & metabolic engineering and novel QTL mapping-based strategic approach in bioethanol production. *Gene Rep.* 17, 100497.

De Bhowmick, G., Koduru, L., Sen, R. (2015). Metabolic pathway engineering towards enhancing microalgal lipid biosynthesis for biofuel application- a review. *Renew. Sustain. Energy Rev*. 50, 1239–1253.

Dexter, J., Fu, P. (2009). Metabolic engineering of cyanobacteria for ethanol production. *Energy Environ. Sci.* 2, 857–864.

Divate, N. R., Chen, G. H., Divate, R. D., Ou, B. R., Chung, Y. C. (2017). Metabolic engineering of *Saccharomyces cerevisiae* for improvement in stress tolerance. *Bioengineered.* 8(5), 524–535.

Doench, J. G., Fusi, N., Sullender, M., Hegde, M., Vaimberg, E. W., Donovan, K. F., Smith, I., Tothova, Z., Wilen, C., Orchard, R., Root, D. E. (2016). Optimized sgRNA design to maximize activity and minimize off-target effects of CRISPR-Cas9. *Nat. Biotechnol.* 34(2), 184–191.

Doudna, J. A., Charpentier, E. (2014). The new frontier of genome engineering with CRISPR-Cas9. *Science*. 346(6213), 1258096.

Dueber, J. E., Wu, G. C., Malmirchegini, G. R., Moon, T. S., Petzold, C. J., Ullal, A. V. (2009). Synthetic proteins scaffolds provide modular control over metabolic flux. *Nat. Biotechnol*. 27, 753–759.

El-Rotail, A. A., Zhang, L., Li, Y., Liu, S. P., Shi, G. Y. (2017). A novel constructed SPT15 mutagenesis library of *Saccharomyces cerevisiae* by using gTME technique for enhanced ethanol production. *AMB Expr.* 7(1), 1–12.

Fujii, T., Inoue, H., Yano, S., Sawayama, S. (2018) Strain Improvement for Industrial Production of Lignocellulolytic Enzyme by *Talaromyces cellulolyticus*. In: Fang, X., Qu, Y. (eds.) *Fungal Cellulolytic Enzymes*. Springer, Singapore. doi: 10.1007/978-981-13-0749-2_7.

Fuller, K. K., Chen, S., Loros, J. J., Dunlap, J. C. (2015). Development of the CRISPR/Cas9 system for targeted gene disruption in *Aspergillus fumigatus*. *Eukaryotic Cell*. 14(11), 1073–1080.

Gardner, J. G, Keating, D. H. (2010). Requirement of the type II secretion system for utilization of cellulosic substrates by *Cellvibrio japonicas*. *Appl. Environ. Microbiol.* 76, 5079–5087.

Guedon, E., Payot, S., Desvaux, S., Petitdemange, H. (2002). Improvement of cellulolytic properties of *Clostridium cellulolyticum* by metabolic engineering, *Appl. Environ. Microbiol*. 68, 53–58.

Gustavsson, M., Lee, S. Y. (2016). Prospects of microbial cell factories developed through systems metabolic engineering. *Microb Biotechnol*. 9, 610–617.

Gygi, S. P., Rochon, Y., Franza, B. R., Aebersold, R. (1999). Correlation between protein and mRNA abundance in yeast. *Mol. Cell. Biol*. 19, 1720–1730.

Hahn-Hägerdal, B., Karhumaa, K., Fonseca, C., Spencer-Martins, I., Gorwa-Grauslund, M. F. (2007). Towards industrial pentose- fermenting yeast strains. *Appl. Microbial. Biotechnol*. 74(5), 937–953.

Hanai, T., Atsumi, S., Liao, J. C. (2007). Engineered synthetic pathway for isopropanol production in *Escherichia coli*. *Appl. Environ. Microbiol.* 73, 7814–7818. doi: 10.1128/AEM.01140-07.

Harris, L., Desai, R., Welker, N., Papoutsakis, E. (2000). Characterization of recombinant strains of the *Clostridium acetobutylicum* butyrate kinase inactivation mutant: need for new phenomenological models for solventogenesis and butanol inhibition? *Biotechnol. Bioeng*. 67, 1–11. doi: 10.1002/(SICI)1097-0290(20000105)67:13.3.CO;2-7.

Hilton, I. B., D'ippolito, A. M., Vockley, C. M., Thakore, P. I., Crawford, G. E., Reddy, T. E., Gersbach, C. A. (2015). Epigenome editing by a CRISPR-Cas9-based acetyltransferase activates genes from promoters and enhancers. *Nat. Biotechnol.* 33(5), 510–517.

Huang, H., Liu, H., Gan, Y. R. (2010). Genetic modification of critical enzymes and involved genes in butanol biosynthesis from biomass. *Biotechnol. Adv*. 28, 651–657.

IEA (2010). Sustainable Production of second-generation Biofuels: potential and perspectives in major economies and developing countries. [Online] Available at: https://www.iea.org/publications/freepublications/publication/second_generation_biofuels.pdf [Accessed 26 July 2015].

Inderwildi, O. R., King, D. A. (2009). "Quo Vadis Biofuels". *Energy Environ. Sci.* 2(4), 343. doi: 10.1039/b822951c.

Ivanova, C., Ramoni, J., Aouam, T., Frischmann, A., Seiboth, B., Baker, S. E., Le Crom, S., Lemoine, S., Margeot, A., Bidard, F. (2017). Genome sequencing and transcriptome analysis of *Trichoderma reesei* QM9978 strain reveals a distal chromosome translocation to be responsible for loss of vib1 expression and loss of cellulase induction. *Biotechnol. Biofuels*. 10(1), 1–15.

Jaroensuk, J., Intasian, P., Kiattisewee, C., Munkajohnpon, P., Chunthaboon, P., Buttranon, S., Trisrivirat, D., Wongnate, T., Maenpuen, S., Tinikul, R., Chaiyen, P. (2019). Addition of formate dehydrogenase increases the production of renewable alkane from an engineered metabolic pathway. *J. Biol. Chem.* 294(30), 11536–11548. doi: 10.1074/jbc.RA119.008246.

Jinang, W., Bikard, D., Cox, D., Zhang, F., Marraffini, L. A. (2013). RNA guided editing of bacterial genomes using CRISPR-Cas systems. *Nat. Biotechnol.* 31, 233–239.

Kern, A., Tilley, E., Hunter, I. S., Legiša, M., Glieder, A. (2007). Engineering primary metabolic pathways of industrial micro-organisms. *J. Biotechnol.* 129(1), 6–29.

Kim, G. B., Kim, W. J., Kim, H. U., Lee, S. Y. (2020). Machine learning applications in systems metabolic engineering. *Curr. Opin. Biotechnol.* 64, 1–9.

Kim, H. M., Chae, T. U., Choi, S. Y., Kim, W. J., Lee, S. Y. (2019). Engineering of an oleaginous bacterium for the production of fatty acids and fuels. *Nat. Chem. Biol.* 15(7), 721–729. doi: 10.1038/s41589-019-0295-5.

Kim, J. H., Block, D. E., Mills, D. A. (2010). Simultaneous consumption of pentose and hexose sugars: an optimal microbial phenotype for efficient fermentation of lignocellulosic biomass. *Appl. Microbiol. Biotechnol.* 88(5), 1077–1085.

Kleinstiver, B. P., Pattanayak, V., Prew, M. S., Tsai, S. Q., Nguyen, N. T., Zheng, Z., Joung, J. K. (2016). High-fidelity CRISPR–Cas9 nucleases with no detectable genome-wide off-target effects. *Nature*. 529(7587), 490–495.

Konermann, S., Brigham, M. D., Trevino, A. E., Joung, J., Abudayyeh, O. O., Barcena, C., Hsu, P. D., Habib, N., Gootenberg, J. S., Nishimasu, H., Zhang, F. (2015). Genome-scale transcriptional activation by an engineered CRISPR-Cas9 complex. *Nature*. 517(7536), 583–588.

Kubicek, C. P., Mikus, M., Schuster, A., Schmoll, M., Seiboth, B. (2009). Metabolic engineering strategies for the improvement of cellulase production by *Hypocrea jecorina*. *Biotechnol. Biofuels*. 2(1), 1–14.

Kumar, S. J., Kumar, N. S., Chintagunta, A. D. (2020). Bioethanol production from cereal crops and lignocelluloses rich agro-residues: prospects and challenges. *SN Appl. Sci.* 2(10), 1–11.

Leplat, C., Nicaud, JM, Rossignol T. (2018). Overexpression screen reveals transcription factors involved in lipid accumulation in *Yarrowia lipolytica*. *FEMS Yeast Res*. 18(5). doi: 10.1093/femsyr/foy037. PMID: 29617806.

Lewin, G. R., Carlos, C., Chevrette, M. G., Horn, H. A., McDonald, B. R., Stankey, R. J., Fox, B. G., Currie, C. R. (2016). Evolution and ecology of Actinobacteria and their bioenergy applications. *Annu Rev Microbiol*. 70, 235–254.

Li, J., Zhang, Y., Li, J., Sun, T., Tian, C. (2020). Metabolic engineering of the cellulolytic thermophilic fungus *Myceliophthora thermophila* to produce ethanol from cellobiose. *Biotechnol. Biofuels*. 13(1), 1–15.

Li, Y., Zhang, X., Xiong, L., Mehmood, M. A., Zhao, X., Bai, F. (2017). On-site cellulase production and efficient saccharification of corn stover employing cbh2 overexpressing *Trichoderma reesei* with novel induction system. *Bioresour. Technol. 238*, 643–649.

Liang, L., Liu, R., Freed, E. F., Eckert, C. A. (2020). Synthetic biology and metabolic engineering employing *Escherichia coli* for C2–C6 bioalcohol production. *Front. Bioeng. Biotechnol.* 8, 710.

Ling, H., Teo, W., Chen, B., Leong, S. S. J., Chang, M. W. (2014). Microbial tolerance engineering toward biochemical production: from lignocellulose to products. *Curr. Opin. Biotechnol.* 29, 99–106

Liu, Q., Gao, R., Li, J., Lin, L., Zhao, J., Sun, W., Tian, C. (2017). Development of a genome-editing CRISPR/Cas9 system in thermophilic fungal *Myceliophthora* species and its application to hyper-cellulase production strain engineering. *Biotechnol. Biofuels*. 10(1), 1–14.

Liu, R., Chen, L., Jiang, Y., Zhou, Z., Zou, G. (2015). Efficient genome editing in filamentous fungus *Trichoderma reesei* using the CRISPR/Cas9 system. *Cell Discov.* 1(1), 1–11.

Liu, R., Zhu, F., Lu, L., Fu, A., Lu, J., Deng, Z., Liu, T. (2014). Metabolic engineering of fatty acyl-ACP reductase-dependent pathway to improve fatty alcohol production in *Escherichia coli. Metab. Eng*. 22, 10–21.

Lo, J., Olson, D. G., Murphy, S. J. L., Tian, L., Hon, S., Lanahan, A., Guss, A. M., Lynd, L. R. (2017). Engineering electron metabolism to increase ethanol production in *Clostridium thermocellum. Metab. Eng*. 39, 71–79.

Lönn, A., Träff-Bjerre, K. L., Otero, R. C., Van Zyl, W. H., Hahn-Hägerdal, B. (2003). Xylose isomerase activity influences xylose fermentation with recombinant *Saccharomyces cerevisiae* strains expressing mutated xylA from T*hermus thermophilus. Enzyme Microb. Technol.* 32(5), 567–573.

Lugani, Y., Sooch, B. S. (2018). Insights into fungal xylose reductases and its application in xylitol production. In: Kumar, S., Dheeran, P., Taherzadeh, M., Khanal, S. (eds.) *Fungal Biorefineries. Fungal Biology*. Springer, Cham. doi: 10.1007/978-3-319-90379-8_7.

Ma, L., Zhang, J., Zou, G., Wang, C., Zhou, Z. (2011). Improvement of cellulase activity in *Trichoderma reesei* by heterologous expression of a beta-glucosidase gene from *Penicillium decumbens. Enzyme Microb. Technol.* 49(4), 366–371.

Matsu-Ura, T., Baek, M., Kwon, J., Hong, C. (2015). Efficient gene editing in Neurospora crassa with CRISPR technology. *Fungal Biol. Biotechnol.* 2(1), 1–7.

Mehrer, C. R., Incha, M. R., Politz, M. C., Pfleger, B. F. (2018). Anaerobic production of medium-chain fatty alcohols via a β-reduction pathway. *Metab. Eng*. 48, 63–71. doi: 10.1016/j.ymben.2018.05.011.

Meyer, V. (2021). Metabolic engineering of filamentous fungi. In: Nielsen, J., Stephanopoulos, G., Lee, S. Y. (eds.) *Metabolic Engineering*. doi: 10.1002/9783527823468.ch20.

Misra, P., Shukla PK, Rao K.P, Ramteke P. (2019). Genetic engineering applications to improve cellulose production and efficiency: part II. In: Srivastava. N., Srivastava, M., Mishra, P.K., Ramteke, P.W., Ram Lakhan Singh (Eds.), *New and Future Developments in Microbial Biotechnology and Bioengineering* (pp. 227–260). Amsterdam, Netherlands: Elsevier.

Mohagheghi, A., Dowe, N., Schell, D., Chou, Y. C., Eddy, C., Zhang M. (2004). Performance of a newly developed integrant of *Zymomonas mobilis* for ethanol production on corn stover hydrolysate, *Biotechnol. Lett*. 26, 321–325.

Mukhopadhyay, A. (2015). Tolerance engineering in bacteria for the production of advanced biofuels and chemicals. *Trends Microbiol.* 23(8), 498–508.

Nakari-Setala, T., Paloheimo, M., Kallio, J., Vehmaanpera, J., Penttila, M., Saloheimo, M. (2009). Genetic modification of carbon catabolite repression in *Trichoderma reesei* for improved protein production. *Appl. Environ. Microbiol.* 75(14), 4853–4860.

Nielsen, J., Keasling, J. D. (2011). Synergies between synthetic biology and metabolic engineering. *Nat. Biotechnol.* 29, 693–695.

Nødvig, C. S., Nielsen, J. B., Kogle, M. E., Mortensen, U. H. (2015). A CRISPR-Cas9 system for genetic engineering of filamentous fungi. *PloS One*. 10(7), e0133085.

Novy, V., Schmid, M., Eibinger, M., et al. (2016). The micromorphology of *Trichoderma reesei* analyzed in cultivations on lactose and solid lignocellulosic substrate, and its relationship with cellulase production. *Biotechnol. Biofuels*. 9, 169.

Nowroozi, F., Baidoo, E. K., Ermakov, S., Redding-Johanson, A., Batth, T., Petzold, C. (2014). Metabolic pathway optimization using ribosome binding site variants and combinatorial gene assembly. *Appl. Microbiol. Biotechnol.* 98, 1567–1581.

Ohta, K., Beall, D. S., Mejia, J. P., Shanmugam, K. T., Ingram, L. O. (1991). Metabolic engineering of *Klebsiella oxytoca* M5A1 for ethanol-production from xylose and glucose. *Appl. Environ. Microbiol.* 57(10), 2810–2815.

Park, J. H., Lee, S. Y., Kim, T. Y., and Kim, H. U. (2008). Application of systems biology for bioprocess development. *Trends Biotechnol.* 26, 404–412. doi: 10.1016/j.tibtech.2008.05.001.

Pohl, C., Kiel, J. A., Driessen, A. J., Bovenberg, R. A., Nygard, Y. (2016). CRISPR/Cas9 based genome editing of *Penicillium chrysogenum. ACS Synth. Biol.* 5(7), 754–764.

Pratt, A. J., MacRae, I. J., (2009). The RNA- induced silencing complex: a versatile gene silencing machine. *J. Biol. Chem.* 284, 17897–17901.

Qiao, K., Wasylenko, T. M., Zhou, K., Xu, P., Stephanopoulos, G. (2017). Lipid production in *Yarrowia lipolytica* is maximized by engineering cytosolic redox metabolism. *Nat. Biotechnol.* 35(2), 173–177. doi: 10.1038/nbt.3763.

Roy, L., Chakraborty, S., Berab, D., Adak, S. (2020). Application of metabolic engineering for elimination of undesirable fermentation products during biofuel production from lignocellulosics. In *Genetic and Metabolic Engineering for Improved Biofuel Production from Lignocellulosic Biomass*. Elsevier Inc.

Ruffing, A., Chen, R. R. (2005). Metabolic engineering of microbes for polysaccharide and oligosaccharide synthesis. *Microb. Cell Fact.* 5, 1–9.

Rühl, M., Hardt, W. D., Sauer, U. (2011). Subpopulation-specific metabolic pathway usage in mixed cultures as revealed by reporter protein-based 13C analysis. *Appl. Environ. Microbiol.* 77, 1816–1821. doi: 10.1128/AEM.02696-10.

Rühl, M., Zamboni, N., & Sauer, U. (2010). Dynamic flux responses in riboflavin overproducing *Bacillus subtilis* to increasing glucose limitation in fed-batch culture. *Biotechnol. Bioeng.* 105(4), 795–804.

Salwan, R., Sharma, V. (2018). The role of actinobacteria in the production of industrial enzymes. In: Bhim Pratap Singh, Vijai Kumar Gupta and Ajit Kumar Passari (Eds.), *New and Future Developments in Microbial Biotechnology and Bioengineering* (pp. 165–177). Amsterdam, Netherlands: Elsevier.

Selim, K.A., El-Ghwas, D.E., Easa, S.M., Hassan, A., Mohamed, I. (2018). Bioethanol a microbial biofuel metabolite; new insights of yeasts metabolic engineering. *Fermentation*. 4(1), 16.

Senthilkumar, V., Gunasekaran, P. (2005). Bioethanol production from cellulosic substrates: engineered bacteria and process integration challenges. *J. Sci. Ind. Res.* 64, 845–853.

Shaikh, A. S., Tang, Y. J., Mukhopadhyay, A., and Keasling, J. D. (2008). Isotopomer distributions in amino acids from a highly expressed protein as a proxy for those from total protein. *Anal. Chem.* 80, 886–890. doi: 10.1021/ac071445+.

Singhal, G., Verma, V., Bhagyawant, S. S., Srivastava, N. (2020). Production of biofuel through metabolic engineering: processing, types, and applications. In *Genetic and Metabolic Engineering for Improved Biofuel Production from Lignocellulosic Biomass* (pp. 155–169). Elsevier.

Song, X., Li, Y., Wu, Y., Cai, M., Liu, Q., Gao, K., Zhang, X., Bai, Y., Xu, H. & Qiao, M. (2018). Metabolic engineering strategies for improvement of ethanol production in cellulolytic *Saccharomyces cerevisiae*. *FEMS Yeast Res.*, 18(8), foy090.

Sun, N., Zhao, H. (2013). Transcription activator-like effect or nucleases (TALENs), a highly efficient and versatile tool for genome editing. *Biotechnol. Bioeng.* 110, 1811–1821.

Techapun, C., Poosaran, N., Watanabe, M., Sasaki, K. (2003). Thermostable and alkaline-tolerant microbial cellulase-free xylanases produced from agricultural wastes and the properties required for use in pulp bleaching bioprocesses: a review. *Process Biochem.* 38(9), 1327–1340.

Wang, H., Cao, S., Wang, W. T., Wang, K. T., Jia, X. (2016). Very high gravity ethanol and fatty acid production of *Zymomonas mobilis* without amino acid and vitamin. *J. Ind. Microbiol. Biotechnol.* 43, 861–871.

Wang, S., Liu, G., Wang, J., Yu, J., Huang, B., Xing, M. (2013). Enhancing cellulase production in *Trichoderma reesei* RUT C30 through combined manipulation of activating and repressing genes. *J. Ind. Microbiol. Biotechnol.* 40(6), 633–641.

Wilson, B. J, Kovacs Wilding-Steele Markus, R., Winzer Minton, N. P. (2016). Production of a functional cell wall – anchored minicellulosome by recombinant *Clostridium acetobutylicum* ATCC 824. *Biotechnol. Biofuels.* 9, 109.

Xu, Z., Liu, Y., Williams, I., Li, Y., Qian, F., Wang, L., Lei, Y., Li, B. (2017). Flat enzyme-based lactate biofuel cell integrated with power management system: towards long term in situ power supply for wearable sensors. *Appl. Energy.* 194, 71–80.

Yang, X., Xu, M., Yang, S-T. (2015). Metabolic and process engineering of *Clostridium cellulovorans* for biofuel production from cellulose. *Metab. Eng.* 32, 39–48.

Zetsche, B., Gootenberg, J. S., Abudayyeh, O. O., Slaymaker, I. M., Makarova, K. S., Essletzbichler, P., Volz, S. E., Joung, J., Van Der Oost, J., Regev, A., Zhang, F. (2015). Cpf1 is a single RNA-guided endonuclease of a class 2 CRISPR-Cas system. *Cell.* 163(3), 759–771.

Zhang, F., Zhao, X., Bai, F. (2018a). Improvement of cellulase production in *Trichoderma reesei* Rut-C30 by overexpression of a novel regulatory gene Trvib-1. *Bioresour. Technol.* 247, 676–683.

Zhang, J., Zong, W., Hong, W., Zhang, Z. T., Wang, Y. (2018b). Exploiting endogenous CRISPR-Cas system for multiplex genome editing in *Clostridium tyrobutyricum* and engineer the strain for high-level butanol production. *Metab Eng.* 47, 49–59. doi: 10.1016/j.ymben.2018.03.007.

Zhang, K., Lu, X., Li, Y., Jiang, X., Liu, L., & Wang, H. (2019). New technologies provide more metabolic engineering strategies for bioethanol production in *Zymomonas mobilis*. *Appl. Microbiol. Biotechnol.* 103(5), 2087–2099.

Zhao, X. Q., Zhang, X. Y., Zhang, F., Zhang, R., Jiang, B. J., Bai, F. W. (2018). Metabolic engineering of fungal strains for efficient production of cellulolytic enzymes. In: *Fungal Cellulolytic Enzymes* (pp. 27–41). Springer, Singapore.

5 Fermentation of Hydrolysate Derived from Lignocellulose Biomass toward Biofuels Production

Fareeha Nadeem and Tahir Mehmood
University of Veterinary and Animal Sciences
and University of the Punjab

Bisma Meer
Quaid-i-Azam University

Muhammad Bilal
Poznan University of Technology

Kushif Meer
University of the Punjab

Ayesha Butt
Quaid-i-Azam University

CONTENTS

DOI: 10.1201/9781003203452-5

5.1 INTRODUCTION

Our society grows at a rapid rate; the present economic, social, and environmental alarms about sustainable products and energy have triggered the researchers to recover innovative, cleaner, and renewable raw material assets for materials and energy. Fossil fuels based on frequently used raw materials are nonrenewable, negatively influence the fragile environment, and pose severe threats to human beings. Therefore, fossil fuels do not fulfill the demand for chemicals and energy for a longer period of time [1–4]. Lignocellulosic biomass (LB) is one of the sustainable and renewable bioresources produced globally, amounting to up to 1.3 billion tons annually, and is considered as a substitute for fossil fuels [5,6]. The LB composition differs according to their source (e.g., softwoods, hardwoods, energy crops, and agricultural residues). It is also affected by age, origin, climate conditions, harvesting, and storage methods [7]. LB is composed of three main linked polymeric constituents (lignin, hemicellulose, and cellulose) and is found as a natural resistant biocomposite [8].

The resistant nature of LB presents a technical task for releasing fermentable sugars from the LB and is a foremost obstacle to its use in biorefineries [9]. Pretreatment is an initial step used for destroying the resistance nature of LB by increasing the cellulose surface area while decreasing the crystallinity of cellulose and lignin content [3]. Numerous pretreatment methods have been developed for the treatment of LB, such as physical, chemical, and biological [10–12]. Pretreatment techniques such as supercritical fluid-based, ionic liquids, cosolvent-enhanced lignocellulosic fractionation, and low-temperature steep delignification are considered as more advanced practices, resulting in the production of a higher sugar yield with the least amount of byproducts released. The released sugars such as pentoses and hexoses can be utilized for making organic acids, fatty acids, alcohols, polyols, and bioplastics using multiple microorganisms [13,14].

Various byproducts such as phenolics (vanillin), weak acids (acetic or formic), and furan aldehydes (5-OH methyl furfural and furfural) with fermentable sugars are released by pretreatment of LB. They are responsible for inhibition of microbial growth by energy drainage, intracellular acidification, and accumulation of reactive oxygen species [15,16]. LB hydrolysate needs detoxification before use in microbial fermentation to escape negative effects in the process of fermentation [17]. Researchers are continuously working on several strategies such as improving plant structure to reduce lignin content and engineering microbial strains to bear innumerable inhibitors [18,19].

5.2 STRUCTURAL ORGANIZATION OF LIGNOCELLULOSIC BIOMASS

LB is considered a crucial constituent of renewable resources present on the earth and is composed of three key components, including cellulose (40%–50%), hemicellulose (25%–30%), and lignin (15%–20%). Cellulose is the main polysaccharide constituent of LB. It comprises hundred to more than ten thousand d-glucose monomers linked by β-1,4-glycosidic bonds that form unbranched straight chains known as microfibrils with 3–5 nm width and some micrometers length [20].

Hemicellulose is one of the chief abundant polysaccharides and contributes 15–30 wt% of the weight of LB, which can be utilized as a low-cost and largely accessible processed and raw material for value-added alkanes and chemicals for the chemical industry [21]. It is heterogeneous in its nature and composed of different sugars (d-galactose, d-glucose, d-mannose, d-xylose, and l-arabinose) and some other organic acids. The arabinose-to-xylose ratio controls the branching rate in hemicellulose. Because of the low ratio, there is a higher rate of polymerization and a shorter chain of polymer, and vice versa [22,23].

Lignin is synthesized by the oxidation of *p*-hydroxycinnamyl alcohols: coniferyl, sinapyl, and *p*-coumaryl. Lignin formation includes three pathways of biosynthesis: the phenylpropanoid pathway, the shikimate pathway, and the synthesis of monolignols. The lignin formation units are syringyl (S), guaiacyl (G), and *p*-hydroxyphenyl (H). The three-dimensional and amorphous structure formed by lignin heteropolymers assists the cell in the case of stress, which may be due to metabolic and structural damage or provide protection to the cell wall against several pathogens [24,25]. Figure 5.1 presents the structural composition of LB and its polymer.

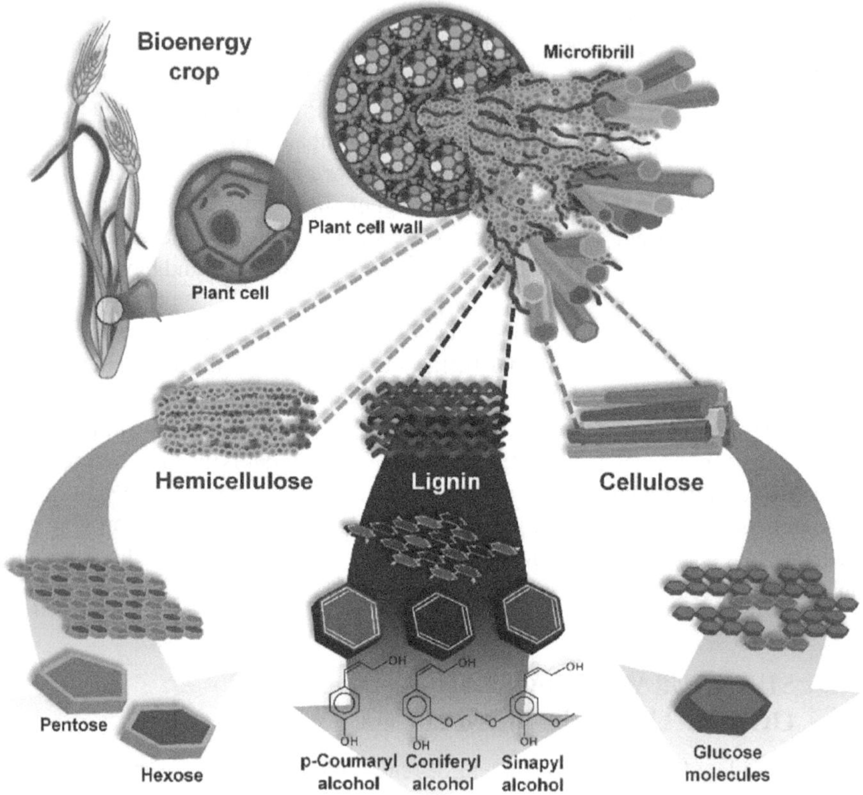

FIGURE 5.1 Structure composition of LB and its polymers [24]; this is an open access article distributed under the Creative Commons Attribution License.

5.3 PRETREATMENT OF LIGNOCELLULOSIC FEEDSTOCK

Pretreatment is an initial step because it aids in removing hemicelluloses and lignin, and enhances the material's porosity [26]. The goals of an effective pretreatment are as follows: direct or subsequent hydrolysis is responsible for the formation of sugar, to avoid degradation or loss of sugar that is formed, to restrict the formation of inhibitory products, to lessen the need for energy, and to lower the cost [27].

5.3.1 Classification of Pretreatment

Pretreatment can be categorized into biochemical, physical, and chemical pretreatment processes (Figure 5.2). Lignocellulose-based biorefinery technology is facing a major challenge to derive a cost-effective pretreatment of LB [28–30].

5.3.1.1 Chemical Pretreatment

Chemical pretreatment comprises alkaline pretreatment [31], ionic liquids [32], ozonation, acidic pretreatment, oxidation, and treatment with organic solvents [13].

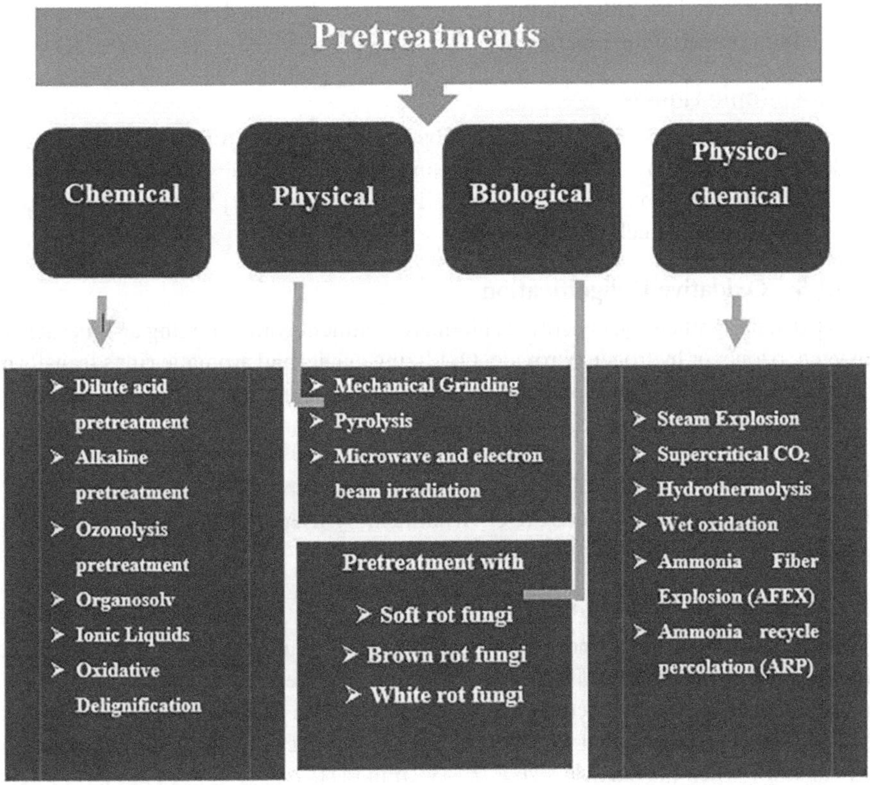

FIGURE 5.2 Classification of pretreatment methods.

5.3.1.1.1 Acid Pretreatment

Acid pretreatment is conducted by soaking the material in a dilute acid solution followed by heating to a temperature between 140°C and 200°C for a particular time [33].

5.3.1.1.2 Alkaline Pretreatment

Alkaline pretreatment used different alkaline reagents such as Na_2CO_3, $Ca(OH)_2$, KOH, and NaOH for different lignocellulosic contents [34]. The pros of this method include ambient operating conditions, low pressure, and temperature [35].

5.3.1.2 Ozonolysis

Ozonolysis refers to the process of treating material with ozone [36]. Ozone acts as a strong oxidative agent [37], thus the benefit of this method is the low generation of inhibitory compounds and the lack of need for chemical additives [38].

5.3.1.3 Organosolv

In this method, organic solvents such as methanol, ethylene glycol, ethanol, acetone, or their mixtures with water are utilized [39]. In this process, the formation of inhibitors is very low, although this technique is implemented on various biomasses with a high potential for sugar yield [40].

5.3.1.4 Ionic Liquids

A variety of biomass can be dissolved by ionic liquids of different hardness and added as selective solvents for cellulose and lignin [39]. The structure of hemicellulose and lignin is not altered after ionic liquid treatment [41] because lignin shows high solubility while cellulose shows less solubility [42].

5.3.1.5 Oxidative Delignification

Oxidative delignification constitutes biomass treatment with oxidizing agents such as oxygen, ozone, or hydrogen peroxide. Oxidizing agents and aromatic rings transform polymer of lignin into carboxylic acids. However, biomass that constitutes the hemicellulose fraction may be degraded [43].

5.3.1.6 Physical Pretreatment

Physical pretreatment methods are environmentally friendly, as they do not generate any toxic material [44].

5.3.1.6.1 Mechanical Grinding

With the help of cutting, grinding, and milling, lignocellulose residues can be crushed to lower the cellulose crystallinity; the aim is to reduce the particle size [45]. High mechanical energy is needed to grind residues, which makes it less cost-effective [46]. It is also reported that the specific energy requirement is 24.7 kW h/ton for a wheat straw when the particle size is 7.7–3.2 mm [47].

Various types of milling processes have been proposed, such as colloid, hammer, two-roll, ball, vibro-energy, and szego mill, which was also exploited. Feeding of biomass is done by gravity, or from the top; it is pumped into the mill, followed by discharge from the bottom of the mill [48–50].

5.3.1.6.2 Pyrolysis

In pyrolysis, the cellulosic content quickly decomposes to gaseous products such as CO, H_2, and residual char with a temperature of more than 300°C [27]. To improve cellulosic content alteration from ground material to glucose pyrolysis, pretreatment was effective [51].

5.3.1.6.3 Microwave and Electron Beam Irradiation

Microwave and electron beam irradiation is also feasible because it is too easy to operate them [52]. Electron beam irradiation is aligned with a continuously altering magnetic field, while microwave pretreatment exploits high heating efficiency [35].

5.3.1.7 Biological Pretreatment

Biological and enzymatic pretreatment is an energy-saving and environmentally friendly method, as it is conducted at low temperature and requires no chemicals [45,53,54]. Degradation of lignocellulose content is done with the aid of microbes such as soft rot fungi, brown rot fungi, and white rot fungi in mild conditions [55].

Enzymes such as manganese peroxidases, laccase, and lignin peroxidase can play a significant role in this type of pretreatment [56].

5.3.1.8 Physicochemical Pretreatment

Physicochemical pretreatment permits enhancing pore volume, modifying enzyme accessibility, lessening the size of the particle, and removing hemicellulose [13,57].

5.3.1.8.1 Steam Explosion

Steam explosion is a combination of the thermal, mechanical, and chemical impacts that act on the biomass [58,59]. Other benefits related to the raw material are that it applies to all types of biomass [60] and that this technique utilizes less energy than mechanical pretreatments, avoids recycling, and reduces the environmental costs [61].

5.3.1.8.2 Supercritical CO_2

This pretreatment is called gaseous compressed CO_2 fluid with a temperature above the critical point of liquid density [62], a critical pressure of 7.38 MPa, and a critical temperature of 31°C [46]. Supercritical CO_2 has various benefits, and the hydrolyzed neutralization is not needed because, in a gaseous state, CO_2 is released by decompression after pretreatment; due to mild temperature, less degraded products act as inhibitors [63].

5.3.1.8.3 Hydrothermolysis

Hydrothermolysis involves treatment in H_2O at high temperatures [54]. In contrast to steam pretreatment, the water content is much higher in hydrothermolysis, which results in the dilution of more sugar solution and thus makes the recovery process more energy-demanding [64].

5.3.1.8.4 Wet Oxidation

Wet oxidation is to release more cellulose from lignocellulose content by heating to reduce acid usage in the bioethanol process [65]. One of the significant steps to achieve a higher level of fermentable sugars for ethanol fermentation is the conversion of sugar oligomers into simple sugars by microbes [66].

5.3.1.8.5 Ammonia Fiber Explosion

Ammonia fiber explosion can be done at elevated temperature and pressure; LB residues are exposed to liquid ammonia for a particular period; and the dosage of ammonia ranges from 1 to 2 kg [45]. A reduction in the lignin content was observed due to depolymerization, resulting in swelling of the cellulose structure, and a large amount of sugar could be achieved [67].

5.4 ENZYMATIC HYDROLYSIS

Enzymatic hydrolysis of the pretreated lignocellulose content constitutes biochemical reactions that transform cellulose into glucose and hemicellulose into hexoses (mannose, glucose, and galactose) and pentoses (arabinose and xylose), catalyzed by hemicellulose and cellulose enzymes, respectively. In the production of bioethanol by technologies constituting enzymatic hydrolysis, low hydrolysis caused by sugar (product) inhibition, enzyme cost, and low microorganism productivity have been predicted as a limiting factor of downstream techniques [68].

5.5 FERMENTATION

In lignocellulosic hydrolysates, there are two significant sugars, that is, glucose and xylose. After the process of saccharification, both sugars resulted in an efficient fermentation of sugars into ethanol at a high yield [69] done by various microbes, for example, yeasts and bacteria. Some anaerobic thermophilic bacteria are significant microorganisms for ethanol production because of their ability to metabolize sugar in the wide spectrum found in lignocellulose [70]. In alcoholic fermentation, the most commonly exploited microorganisms are *Zymomonas mobilis* and *Saccharomyces cerevisiae*. The most dominant ethanol-producing yeast, that is, *S. cerevisiae*, is more robust than bacteria and capable of tolerating inhibitors or ethanol found in hydrolysates of lignocellulosic residues in a better way [71]. However, rapid and simultaneous exploitation of sugar mixtures is considered important for biofuel production that is economically feasible and from biomass hydrolysate commodity chemicals [72].

5.5.1 Fermentative Techniques

The transformation of biomass into biofuels can be done using three different processes.

5.5.1.1 Consolidated Bioprocessing Approach

To improve microbial performance for significant hydrolysis of cellulose and enhance energy efficiency during the process of fermentation, production of enzyme, saccharification, and fermentation can be further consolidated into a single process. This method has been predicted to be the most significant and promising strategy to lower the cost and environmental effects of the bioethanol production process [73]. Biological transformation of the lignocellulosic substrate into ethanol by one or more efficient organisms while conducting enzymatic hydrolysis of biomass that forms on its own and fermentation of that material possesses a large number of benefits [74]. The significant benefits of the consolidated bioprocessing (CBP) process include the removal of the need for the production of enzymes, lowering the capital investment, modifying the efficiency of hydrolysis, lowering the risk of contamination, and simplifying total reactions [35]. The major issue with CBP is that at high temperatures, saccharification and fermentation are usually carried out, and such issues could be dodged by exploiting thermophilic CBP microbes. Various fungi, such as *Fusarium*, *Trichoderma*, *Mucor*, *Rhizopus*, *Aspergillus*, *Neurospora*, *Trichoderma*, *Phlebia*, and *Monilia*, have the capability to improve the conversion efficiency of cellulose to ethanol [73].

5.5.1.2 Separate Hydrolysis and Fermentation

It is basically a two-stage process. In the first step, by exploiting suitable enzymes, hemicellulose and cellulose contents are hydrolyzed, resulting in the reduction of sugar. In the next step, this reducing sugar is converted into ethanol by fermentation utilizing specific microorganisms. It is considered as the significant benefit of this configuration as both processes possess considerable different operating conditions. Enzymatic hydrolysis possesses an optimum temperature of about 50°C that varies from the temperature of fermentation, that is, 28°C–37°C [35]. The cons of this process are that it inhibits enzymes activity by enhancing the concentration of simple sugar that is released in the vessel. This kind of inhibition process slows the cellulosic hydrolysis rate. For separate hydrolysis and fermentation (SHF), there is more economic investment because the exploitation of more than one vessel occurs at different times. Another disadvantage is that when it is conducted under high solid loading conditions, there is a reduction of cellulolytic enzymes due to inhibition. To modify the cellulose hydrolysis rate, the exploitation of various surfactants such as polyethylene glycol, ionic liquids, or Tween can perform a supreme rile while lowering the attachment of enzymes to lignin [75].

5.5.1.3 Simultaneous Saccharification and Fermentation

In contrast to SHF, simultaneous saccharification and fermentation (SSF) process is considered more favorable because, in SSF, the release of glucose through cellulase is transformed rapidly into ethanol by the microorganisms that undergo fermentation, thus lowering the inhibition of the end product to cellulase caused by the accumulation of cellobiose and glucose [76]. The combination of fermentation and saccharification processes in one vessel is cost-effective and a good alternative to SHF because of its low consumption of energy, high yield of bioethanol at high solid loading, and less time-consuming [77]. This method is suitable as it shows several benefits for converting LB to ethanol [78]. Despite several advantages, the main problem associated with SSF of cellulose is the variation in the optimum temperatures of fermentation (25°C–35°C) and saccharification (45°C–50°C). *Saccharomyces* strains are a good source of ethanol-producing microorganisms; however, they need an operating temperature of 35°C. In cellulose hydrolysis, the most frequently applied fungal cellulases have an optimum temperature of 50°C. When the temperature is low, there is a substantial reduction in the rate of hydrolysis, which is not favorable given the enhancement in processing time. *S. cerevisiae* also possesses low fermentation efficiency at elevated temperatures because of the enhanced fluidity of membranes, to which yeast responds by altering its fatty acid composition [79]. This issue is resolved by utilizing thermotolerant yeast strains instead of *Saccharomyces*, which allow for high temperature processing and thus increase the hydrolysis rate [80].

5.6 INHIBITION AND DETOXIFICATION OF LIGNOCELLULOSIC HYDROLYSATES

5.6.1 Inhibition of Lignocellulosic Hydrolysates

The main goal for the pretreatment of lignocellulose is to remove the hemicellulose and lignin in order to enhance the access of enzymes to the cellulose for a better yield [81]. But these physiochemical pretreatments also result in the production of different kinds of inhibitory products, for instance, by the degradation of lignin or

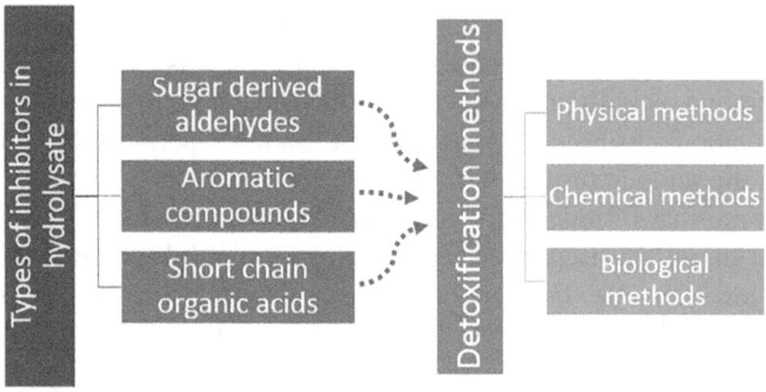

FIGURE 5.3 Types of inhibitors in the hydrolysate and their detoxification methods.

partial sugars. These inhibitors can be toxic to the microbes used in fermentation and also hamper the usage of sugars from the hydrolysate [26]. The formation of these inhibitors depends on the raw material used and the type and harshness (pH, pressure, and temperature) of the pretreatment method [81]. Types of inhibitors and LB are described in Figure 5.3.

5.6.2 Types of Inhibitors and Their Inhibitory Effects

The inhibitors are mainly produced from lignin, hemicellulose, and cellulose and can be classified into aromatic compounds, short-chain organic acids, and sugar-derived aldehydes. The feedstock composition mainly decides the type and quantity of inhibitor produced, which depends on the LB as the ratio of cellulose, hemicellulose, and lignin varies in different types of biomass [82]. In addition to these three constituents, the inhibitors can also be derived from the corrosion of pretreatment equipment as well as from the chemical additives used during the pretreatment [83].

5.6.2.1 Sugar-Derived Aldehydes

These aldehydes are mainly the products of the sugars present in the cellulose and hemicellulose constituents of the LB. The hemicellulose in the lignocellulose contains both hexoses (d-glucose, d-mannose, and d-galactose) as well as pentoses (d-arabinose and d-xylose), whereas the cellulose is made up of hexose sugar, d-glucose [82]. Furfural and hydroxymethylfurfuryl (HMF) are the two common inhibitors present in the lignocellulosic hydrolysate. Both of them are degradation products of hexose and pentose sugars. These aldehydes directly influence the enzymes, for instance, pyruvate dehydrogenase, alcohol dehydrogenase, acetaldehyde dehydrogenase, and alcohol dehydrogenase, which eventually cause a reduction in the growth rate of the microbe and the production of the ethanol. Furthermore, these aldehydes also cause oxidative stress, which has a damaging effect on the microbe. Although HMF is less toxic than furfural, its effects on microbes are long-lasting than those of furfural [84]. Glycolaldehyde is another common sugar-derived aldehyde, and it is also the degradation product of both hexoses and pentoses. The aldehyde can link covalently with macromolecules such as DNA and proteins, resulting in the perturbation of important processes in microorganisms and affecting their viability and reproduction [85].

5.6.2.2 Aromatic Compounds

The aromatic compound-based inhibitors can be subcategorized as phenolics with at least one hydroxyl group linked to the aromatic ring, such as vanillin and syringaldehyde, and nonphenolics that cannot be considered as phenolics but originated from them, such as benzoic acid and cinnamaldehyde [82]. Although some of these kinds of inhibitors can also be derived from extracts of the biomass, they mostly result from the degradation of the lignin [86]. The inhibitory effect of the aromatic inhibitors against microorganisms comes partially from their hydrophobic properties, which cause them to enter the membranes, resulting in membrane disintegration and damaging the microbes' barrier [84]. The aromatic compounds form a complex around the cellulase enzyme, causing enzyme deactivation and precipitation, resulting in enzyme inhibition [87].

5.6.2.3 Short-Chain Organic Acids

Some of the common short-chain organic acids include lactic acid, acetic acid, formic acid, and levulinic acid. Acetylation of hemicellulose and lignin results in the formation of acetic acid, while the degradation of furan and HMF produces formic acid and levulinic acid, respectively [83]. These acids mainly inhibit the microbes; for instance, acetic acid has been found to hamper the nutrient uptake by *S. cerevisiae*. In addition, acetic acid also causes acidification of the microbe's intracellular environment but doesn't inhibit the cellulases. These acids also disturb the structure and function of the cellular membranes. Moreover, the stress produced by these acids reduced the synthesis of DNA and RNA in the cell [82].

5.7 DETOXIFICATION OF INHIBITORS

To improve yield from the LB, inhibitors must be removed from the hydrolysate before the hydrolysis and fermentation. Different methods have been used to remove these inhibitors [88], each of which is briefly discussed below.

5.7.1 Physical Methods

The production of inhibitory compounds strongly depends on the composition of the biomass, its structure, and especially the lignin content. So, the selection of feedstock must be done carefully so that it contains a less recalcitrant noncellulosic portion. For instance, *Populus trichocarpa*, with its low recalcitrance, has been considered a suitable candidate for the production of bioethanol [81]. A physical approach for the removal of inhibitors is the use of activated charcoal. The use of activated charcoal has been found useful for the removal of furans, phenolic compounds, and HMF. It is believed that the removal of inhibitors is due to the hydrophobic nature of activated charcoal and components in the hydrolysate [86].

Membrane filtration is an appropriate choice for a variety of biorefineries due to the flexibility of its operation and low consumption of energy. The technique involves the use of nanofiltration (NF), microfiltration (MF), and reverse osmosis (RO). The method is also used for sugar concentration and detoxification of hydrolysates [89]. Evaporation is another technique for reducing volatile inhibitors from the hydrolysate such as acetic acid, furfural, and vanillin. But the evaporation does not remove volatile inhibitors completely; for instance, furfural is frequently removed completely from the hydrolysate as compared to other inhibitors [84].

5.7.2 Chemical Methods

Besides physical methods, there are also a variety of chemical methods for the detoxification of hydrolysate. One of them is overliming of hydrolysate. In this technique, a base such as CaOH is used to increase the pH of the hydrolysate to 10 or even higher. The increased pH degrades the furfural and HMF inhibitors almost completely. A drawback of this treatment is that most sugars can be lost from the hydrolysate at pH 12 [90]. Liquid–liquid extraction (LLE) is another approach for the removal of inhibitors using a solvent. For instance, in LLE, ethyl acetate removed the acetic acid up to 90% and increased the yield by 11%. But the practical usage of this method is challenging because it requires additional additives and an extraction process, increasing the overall cost [81].

5.7.3 Biological Methods

For the detoxification of hydrolysates, biological methods can be categorized into microbial or enzymatic treatments and biotechnological modifications. The enzymatic treatment involves the use of peroxidase and laccase in order to remove phenolics selectively. Microbial treatment is also exploited to remove toxic compounds, including furfural, HMF, acetic acid, and phenolic compounds, but the treatment also slightly consumes sugars [84]. A soil bacterium, *Bordetella* sp. BTIITR, was found to detoxify the sugarcane bagasse hydrolysate by degrading acetic acid (82%), HMF (94%), and furfuran (100%) [91].

A widely accepted approach is the production of recombinant strains using genetic engineering. For example, in a study, the engineering of *S. cerevisae* strain caused the overexpression of particular enzymes such as transaldolase and alcohol dehydrogenase in hydrolysate with a high level of furfural, which eventually not only caused the furfural inhibition but also increased the yield of ethanol [81].

5.8 EXTRACTION OF BIOBUTANOL

Biobutanol production has increased tremendously in recent years by using advanced technology. Various techniques have been used to get an enormous amount of biobutanol [92].

5.8.1 Immobilized and Cell Recycle Continuous Bioreactors

Several factors inhibit biobutanol production in batch processes. Productivity achieved in a batch reactor is not more than $0.50\,g^{-1}L^{-1}h^{-1}$. The factors that limit the productivity of biobutanol are as follows:

- Product inhibition
- Low concentration of cell
- Down time [93]

One can get more than 4 $g^{-1}L^{-1}h^{-1}$ of cell concentration productivity by using the batch fermentation process. However, two techniques proved themselves very helpful in the case of increasing the concentration of cells inside bioreactors. These techniques are 'immobilization' or 'cell recycling' [94]. Rector productivity for *Clostridium beijerinckii* increased up to 15.8 $g^{-1}L^{-1}h^{-1}$ using other cell reactors such as clay bricks. Using the immobilized cell technique, the reactivity product of *C. acetobutylicum* was obtained at 4.6 $g^{-1}L^{-1}h^{-1}$ in a continuous reactor with fibrous support [95]. In the cell recycling process, return of cells to bioreactors occurs with the help of a filter. By this process, such liquid is removed that is transparent. Cell recycling is used to increase the concentration of cells and their reactivity in a bioreactor. Fermentation of biobutanol increases up to 6.5 $g^{-1}L^{-1}h^{-1}$ using this approach. However, without this technique, 0.5 $g^{-1}L^{-1}h^{-1}$ of productivity was obtained [93].

5.8.2 Gas Stripping

In situ recovery of biobutanol is done efficiently by the gas stripping technique from simultaneous fermentation broth (Figure 5.4). In ABE (acetone-butanol-ethanol) fermentation, CO_2 and H_2 gases are generated in fermented liquid, and then the solvent is recovered by passing it through a condenser. After passing through the condenser, the stripped gas is moved into the fermenter for recycling. The recycling process is not completed until all sugar is consumed by liquid [93,96]. Butanol is inhibited through reduction and a concentrated sugar solution [97]. However, sometimes, there is a need for another stripper. This stripper is used separately for solvent equipping. After equipping off the solvent, this stripper is returned back into the reactor vessel [98]. This technique is used for solvent recovery in a batch reactor from the fermented

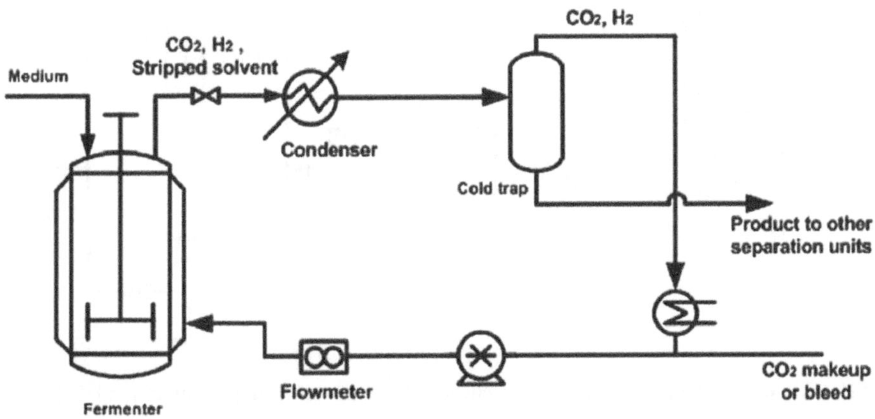

FIGURE 5.4 Removal of biobutanol from fermentation broth by gas stripping [101].

broth of *C. beijerinckii* BA101 [99]. In this process, 161.7 gl^{-1} of sugar solution was fermented with the utilization of 75.9 gl^{-1} of solvent [100].

5.8.3 Pervaporation

With the help of a membrane, selective volatile compounds are removed from fermentation broth by the pervaporation technique (Figure 5.5). As in this process, a membrane is placed between the fermentation broth and the volatile compound. Then, by diffusion, they moved out in the form of vapor from the membrane [92,102]. One can use the liquid as well as the solid pervaporation membrane [103]. Pervaporation involves a change of phase strategy, where the liquid phase is converted into the vapor phase. At a particular heat of vaporization, the selective component is removed. This is also referred to as solution diffusion, as in it, one gets rid of volatile components by pervaporation [94, 104]. Solvent productivity is increased from 0.35 to 0.98 $g^{-1}L^{-1}h^{-1}$ as a reduction in product inhibition occurs [105]. By using an ultrafiltration membrane of ionic liquid polydimethylsiloxane solvent, productivity increases up to 2.34 $g^{-1}L^{-1}h^{-1}$ by pervaporation technique of *C. acetobutylicum* fermentation [106].

5.8.4 Liquid–Liquid Extraction

The LLE technique involves a method of mixing a water-insoluble organic extractant with fermentation broth. However, in this case, both extractant and fermentation broth are immiscible, so the separation of extractant takes place very easily from fermentation broth after successful recovery of butanol [107,108].

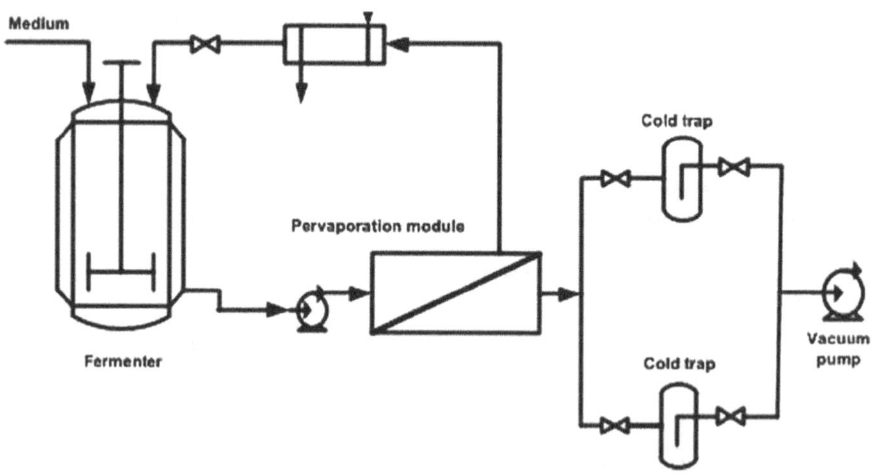

FIGURE 5.5 Schematic diagram of recovery of biobutanol from pervaporation [101].

5.8.5 Perstraction

Recovery of biobutanol by LLE technique faces some major issues such as emulsification, cell toxicity, emulsion formation, solvent extraction loss, extractant toxicity, and microbial cell accumulation at interphase. To deal with all these problems, researchers developed a new method named 'perstraction' [93,109]. With the help of a membrane, both fermentation broth and extractant are separated in a perstractive technique. Thus, diffusion of butanol takes place easily across the membrane in such an environment, while other reagents are retained in the aqueous phase. These reagents are fermentation intermediates such as butyric acids and acetic acids [92,108]. The recovery of butanol between two immiscible phases entirely depends on the rate of diffusion of butanol across the membrane. The membrane acts as a physical barrier as it hinders the butanol extraction rate [110].

5.8.6 Reverse Osmosis

In ABE fermentation, the formation of dilute product concentrations can be minimized by reverse osmosis. To terminate fermentation, one can use reverse osmosis to dewater the fermenting broth [111]. The solvent can be removed from fermentation broth with the help of a specific membrane by using reverse osmosis technique. As in reverse osmosis, instead of separating suspended solids, their removal is given priority by using a hollow ultrafiltration membrane [112]. Dealing with an ultrafiltration membrane is very convenient. However, its regular washing and replacement are necessary. Through reverse osmosis, 90% of butanol is removed, but flux is very low as other components are present. Recovery of butanol increased up to 85%, and flux decreased from 0.6 to 0.15 L/min [101].

5.8.7 Adsorption

One technique for the separation of butanol from fermentation broth is adsorption. A suitable adsorbent is used for the successful adsorption of butanol. The selection of adsorbents is done very carefully. Several factors are considered important during selection, such as

- Rate and capacity of adsorption
- Desorption ease
- Desired product selectivity
- Adsorbent cost [113,114]

The solid adsorbent is generally used for solutions and fermenting broth, as it is made up of zeolites, polymeric resins, activated carbon, and polyvinyl pyridine [115, 116]. As fermenting broth contains substrates and nutrients, adsorbent selectivity determines the extraction of butanol [117]. Adsorption and desorption both determine the fate of successful butanol recovery [115,118].

5.9 CONCLUSION AND FUTURE PERSPECTIVES

Quick and efficient fermentation of different sugars found in lignocellulosic hydrolysates is vital for the cost-effective bioconversion of biomass into biofuels. In the past several years, much research work and progress have been made in the genetic modifications of microbial strains, but in the future, to overcome the challenges, more innovative ideas will be needed to optimize and enhance the fermentation methodology and the genetic improvement of microbial strains for industrial appliances.

REFERENCES

1. Mahmood, H., et al., Recent advances in the pretreatment of lignocellulosic biomass for biofuels and value-added products. *Current Opinion in Green and Sustainable Chemistry*, 2019. 20: pp. 18–24.
2. Clark, J.H., Green biorefinery technologies based on waste biomass. *Green Chemistry*, 2019. 21: pp. 1168–1170.
3. Chen, J., et al., A review on recycling techniques for bioethanol production from lignocellulosic biomass. *Renewable and Sustainable Energy Reviews*, 2021. 149: p. 111370.
4. Mehmood, T., et al., Recent trends on the food wastes valorization to value-added commodities, in Anish Khan, Mohammad Jawaid, Antonia Pizzi, Naved Azum, Abdulla M.Asiri, Illyas Md Isa (Eds.) *Advanced Technology for the Conversion of Waste into Fuels and Chemicals*. 2021, Elsevier, Amsterdam. pp. 171–196.
5. Tu, W.C. and J.P. Hallett, Recent advances in the pretreatment of lignocellulosic biomass. *Current Opinion in Green and Sustainable Chemistry*, 2019. 20: pp. 11–17.
6. Baruah, J., et al., Recent trends in the pretreatment of lignocellulosic biomass for value-added products. *Frontiers in Energy Research*, 2018. 6: p. 141.
7. Bhatia, S.K., H.-S. Joo, and Y.H. Yang, Biowaste-to-bioenergy using biological methods–a mini-review. *Energy Conversion and Management*, 2018. 177: pp. 640–660.
8. Pattanaik, L., et al., Biofuels from agricultural wastes, in Angelo Basile and Francesco Dale (Eds.) *Second and Third Generation of Feedstocks*. 2019, Elsevier, Amsterdam. pp. 103–142.
9. Bhatia, S.K., et al., Bioconversion of barley straw lignin into biodiesel using *Rhodococcus sp*. YHY01. *Bioresource Technology*, 2019. 289: p. 121704.
10. Cai, Y., et al., A review about pretreatment of lignocellulosic biomass in anaerobic digestion: achievement and challenge in Germany and China. *Journal of Cleaner Production*, 2021: p. 126885.
11. Sankaran, R., et al., Recent advances in the pretreatment of microalgal and lignocellulosic biomass: a comprehensive review. *Bioresource Technology*, 2020. 298: p. 122476.
12. Hoang, A.T., et al., Insight into the recent advances of microwave pretreatment technologies for the conversion of lignocellulosic biomass into sustainable biofuel. *Chemosphere*, 2021. 281: p. 130878.
13. Bhatia, S.K., et al., Recent developments in pretreatment technologies on lignocellulosic biomass: effect of key parameters, technological improvements, and challenges. *Bioresource Technology*, 2020. 300: p. 122724.
14. Jagtap, S.S., et al., Production of galactitol from galactose by the oleaginous yeast *Rhodosporidium toruloides* IFO0880. *Biotechnology for Biofuels*, 2019. 12: pp. 1–13.
15. Moreno, A.D., et al., Evolutionary engineered *Candida intermedia* exhibits improved xylose utilization and robustness to lignocellulose-derived inhibitors and ethanol. *Applied Microbiology and Biotechnology*, 2019. 103: pp. 1405–1416.

16. Jonsson, L.J. and C. Martin, Pretreatment of lignocellulose: formation of inhibitory by-products and strategies for minimizing their effects. *Bioresource Technology*, 2016. 199: pp. 103–112.
17. Santoso, S.P., et al., Atmospheric cold plasma-assisted pineapple peel waste hydrolysate detoxification for the production of bacterial cellulose. *International Journal of Biological Macromolecules*, 2021. 175: pp. 526–534.
18. Bhatia, S.K., et al., Microbial biodiesel production from oil palm biomass hydrolysate using marine *Rhodococcus sp.* YHY01. *Bioresource Technology*, 2017. 233: pp. 99–109.
19. Shafrin, F., et al., Modification of monolignol biosynthetic pathway in jute: different gene, different consequence. *Scientific Reports*, 2017. 7: pp. 1–12.
20. Bilal, M. and H.M. Iqbal, Recent advancements in the life cycle analysis of lignocellulosic biomass. *Current Sustainable/Renewable Energy Reports*, 2020. 7: pp. 1–8.
21. Han, X., et al., Catalytic conversion of lignocellulosic biomass into hydrocarbons: a mini review. *Catalysis Today*, 2019. 319: pp. 2–13.
22. Mehmood, T., et al., Bioconversion of agro-industrial waste into value-added compounds in Inamuddin and Anish Khan (Eds.) *Sustainable Bioconversion of Waste to Value Added Products.* 2021, Springer, Cham. pp. 349–368.
23. Wang, F., et al., Lignocellulosic biomass as sustainable feedstock and materials for power generation and energy storage. *Journal of Energy Chemistry*, 2021. 57: pp. 247–280.
24. Hernandez-Beltran, J.U., et al., Insight into pretreatment methods of lignocellulosic biomass to increase biogas yield: current state, challenges, and opportunities. *Applied Sciences*, 2019. 9: p. 3721.
25. Bilal, M., et al., Biotransformation of lignocellulosic materials into value-added products-A review. *International Journal of Biological Macromolecules*, 2017. 98: pp. 447–458.
26. Suhag, M. and J. Singh, Recent advances in fermentation of lignocellulosic biomass hydrolysate to ethanol. *Education*, 2012. 3: pp. 517–526.
27. Sarkar, N., et al., Bioethanol production from agricultural wastes: an overview. *Renewable Energy*, 2012. 37: pp. 19–27.
28. Lu, X., et al., Enzymatic hydrolysis of corn stover after pretreatment with dilute sulfuric acid. *Chemical Engineering & Technology: Industrial Chemistry-Plant Equipment-Process Engineering-Biotechnology*, 2007. 30: pp. 938–944.
29. Krishnan, C., et al., Alkali-based AFEX pretreatment for the conversion of sugarcane bagasse and cane leaf residues to ethanol. *Biotechnology and Bioengineering*, 2010. 107: pp. 441–450.
30. Sasmal, S. and K. Mohanty, Pretreatment of lignocellulosic biomass toward biofuel production, in Sachin Kumar, Rajesh K. Sani (Eds.) *Biorefining of Biomass to Biofuels.* 2018, Springer, Cham. pp. 203–221.
31. Mosier, N., et al., Optimization of pH controlled liquid hot water pretreatment of corn stover. *Bioresource Technology*, 2005. 96: pp. 1986–1993.
32. Klein-Marcuschamer, D., B.A. Simmons, and H.W. Blanch, Techno-economic analysis of a lignocellulosic ethanol biorefinery with ionic liquid pre-treatment. *Biofuels, Bioproducts and Biorefining*, 2011. 5: pp. 562–569.
33. Palmqvist, E. and B. Hahn-Hagerdal, Fermentation of lignocellulosic hydrolysates. II: inhibitors and mechanisms of inhibition. *Bioresource Technology*, 2000. 74: pp. 25–33.
34. Sun, S., et al., The role of pretreatment in improving the enzymatic hydrolysis of lignocellulosic materials. *Bioresource Technology*, 2016. 199: pp. 49–58.
35. Dey, P., et al., Lignocellulosic bioethanol production: prospects of emerging membrane technologies to improve the process–a critical review. *Reviews in Chemical Engineering*, 2020. 36: pp. 333–367.

36. Omar, W.N.N.W. and N.A.S. Amin, Multi response optimization of oil palm frond pretreatment by ozonolysis. *Industrial Crops and Products*, 2016. 85: pp. 389–402.
37. Travaini, R., et al., Ozonolysis: an advantageous pretreatment for lignocellulosic biomass revisited. *Bioresource Technology*, 2016. 199: pp. 2–12.
38. Ab Rasid, N.S., M.M. Zainol, and N.A.S. Amin, Pretreatment of agroindustry waste by ozonolysis for synthesis of biorefinery products, in R. Praveen Kumar, Edgard Gnansounou, Jegannathan Kenthorai Raman, and Gurunathan Baskar (Eds.) *Refining Biomass Residues for Sustainable Energy and Bioproducts*. 2020, Elsevier, Amsterdam. pp. 303–336.
39. Badiei, M., et al., Comparison of chemical pretreatment methods for cellulosic biomass. *APCBEE Procedia*, 2014. 9: pp. 170–174.
40. Harmsen, P.F., et al., *Literature Review of Physical and Chemical Pretreatment Processes for Lignocellulosic Biomass*. 2010: Wageningen UR-Food & Biobased Research.
41. Wyman, C.E., et al., Comparative sugar recovery and fermentation data following pretreatment of poplar wood by leading technologies. *Biotechnology Progress*, 2009. 25: pp. 333–339.
42. Zhi, S., et al., Enzymatic hydrolysis of cellulose after pretreated by ionic liquids: focus on one-pot process. *Energy Procedia*, 2012. 14: pp. 1741–1747.
43. Sun, F. and H. Chen, Evaluation of enzymatic hydrolysis of wheat straw pretreated by atmospheric glycerol autocatalysis. *Journal of Chemical Technology & Biotechnology: International Research in Process, Environmental & Clean Technology*, 2007. 82: pp. 1039–1044.
44. Raud, M., et al., Biomass pretreatment with the Szego Mill™ for bioethanol and biogas production. *Processes*, 2020. 8: p. 1327.
45. Sun, Y. and J. Cheng, Hydrolysis of lignocellulosic materials for ethanol production: a review. *Bioresource Technology*, 2002. 83: pp. 1–11.
46. Arenas-Cardenas, P., et al., Current pretreatments of lignocellulosic residues in the production of bioethanol. *Waste and Biomass Valorization*, 2017. 8: pp. 161–181.
47. Chiaramonti, D., et al., Review of pretreatment processes for lignocellulosic ethanol production, and development of an innovative method. *Biomass and Bioenergy*, 2012. 46: pp. 25–35.
48. Aftab, M.N., et al., *Different Pretreatment Methods of Lignocellulosic Biomass for Use in Biofuel Production in Biomass for Bioenergy-Recent Trends and Future Challenges*. 2019. IntechOpen, United Kingdom.
49. Chen, X., et al., Comparison of different mechanical refining technologies on the enzymatic digestibility of low severity acid pretreated corn stover. *Bioresource Technology*, 2013. 147: pp. 401–408.
50. Song, S. and O. Trass, Floc flotation of Prince coal with simultaneous grinding and hydrophobic flocculation in a Szego mill. *Fuel*, 1997. 76: pp. 839–844.
51. Leustean, I., Bioethanol from lignocellulosic materials. *Journal of Agroalimentary Processes and Technology*, 2009. 15: pp. 94–101.
52. Bjerre, A.B., et al., Pretreatment of wheat straw using combined wet oxidation and alkaline hydrolysis resulting in convertible cellulose and hemicellulose. *Biotechnology and Bioengineering*, 1996. 49: pp. 568–577.
53. Lee, J., Biological conversion of lignocellulosic biomass to ethanol. *Journal of Biotechnology*, 1997. 56: pp. 1–24.
54. Galbe, M. and G. Zacchi, *Pretreatment of Lignocellulosic Materials for Efficient Bioethanol Production in Biofuels*. 2007, Springer Verlag, Berlin Heidelberg. pp. 41–65.
55. Prasad, S., A. Singh, and H. Joshi, Ethanol as an alternative fuel from agricultural, industrial and urban residues. *Resources, Conservation and Recycling*, 2007. 50: pp. 1–39.

56. Floudas, D., et al., Spatafora Joseph Botany Plant Pathology Paleozoic Origin Enzymatic.pdf. 2017.
57. Carneiro, T., et al., Biomass pretreatment with carbon dioxide. In Solange Ines Mussatto (ed.), *Biomass Fractionation Technologies for a Lignocellulosic Feedstock Based Biorefinery*, 2016, Elsevier, Amsterdam. pp. 385–407.
58. Duque, A., et al., Steam explosion as lignocellulosic biomass pretreatment. In Solange Ines Mussatto (ed.), *Biomass Fractionation Technologies for a Lignocellulosic Feedstock Based Biorefinery*, 2016, Elsevier, Amsterdam. pp. 349–368.
59. Cara, C., et al., Enhanced enzymatic hydrolysis of olive tree wood by steam explosion and alkaline peroxide delignification. *Process Biochemistry*, 2006. 41: pp. 423–429.
60. Alvira, P., et al., Pretreatment technologies for an efficient bioethanol production process based on enzymatic hydrolysis: a review. *Bioresource Technology*, 2010. 101: pp. 4851–4861.
61. Duff, S.J. and W.D. Murray, Bioconversion of forest products industry waste cellulosics to fuel ethanol: a review. *Bioresource Technology*, 1996. 55: pp. 1–33.
62. Kumar, P., et al., Methods for pretreatment of lignocellulosic biomass for efficient hydrolysis and biofuel production. *Industrial & Engineering Chemistry Research*, 2009. 48: pp. 3713–3729.
63. Pienkos, P.T. and M. Zhang, Role of pretreatment and conditioning processes on toxicity of lignocellulosic biomass hydrolysates. *Cellulose*, 2009. 16: pp. 743–762.
64. Van, G.P., Conversion of lignocellulosics pretreated with liquid hot water to ethanol. *Applied Biochemistry and Biotechnology*, 1996. 57: p. 157.
65. Ahou, Y.S., et al. Wet oxidation pretreatment effect for enhancing bioethanol production from cassava peels, water hyacinth, and green algae (Ulva). in *AIP Conference Proceedings*. 2020. AIP Publishing LLC.
66. Amore, A. and V. Faraco, Potential of fungi as category I Consolidated BioProcessing organisms for cellulosic ethanol production. *Renewable and Sustainable Energy Reviews*, 2012. 16: pp. 3286–3301.
67. Hayes, D.J., An examination of biorefining processes, catalysts and challenges. *Catalysis Today*, 2009. 145: pp. 138–151.
68. Gonzalez, R., et al., Converting Eucalyptus biomass into ethanol: financial and sensitivity analysis in a co-current dilute acid process. Part II. *Biomass and Bioenergy*, 2011. 35: pp. 767–772.
69. Singh, A. and N.R. Bishnoi, Enzymatic hydrolysis optimization of microwave alkali pretreated wheat straw and ethanol production by yeast. *Bioresource Technology*, 2012. 108: pp. 94–101.
70. Crespo, C.F., et al., Ethanol production by continuous fermentation of d-(+)-cellobiose, d-(+)-xylose and sugarcane bagasse hydrolysate using the thermoanaerobe Caloramator boliviensis. *Bioresource Technology*, 2012. 103: pp. 186–191.
71. Olsson, L. and B. Hahn-Hägerdal, Fermentation of lignocellulosic hydrolysates for ethanol production. *Enzyme and Microbial Technology*, 1996. 18: pp. 312–331.
72. Kim, J.-H., D.E. Block, and D.A. Mills, Simultaneous consumption of pentose and hexose sugars: an optimal microbial phenotype for efficient fermentation of lignocellulosic biomass. *Applied Microbiology and Biotechnology*, 2010. 88: pp. 1077–1085.
73. Hasunuma, T., et al., A review of enzymes and microbes for lignocellulosic biorefinery and the possibility of their application to consolidated bioprocessing technology. *Bioresource Technology*, 2013. 135: pp. 513–522.
74. Olson, D.G., J.E. McBride, A.J. Shaw, and L.R. Lynd, Recent progress in consolidated bioprocessing. *Current Opinion in Biotechnology*, 2012. 23: pp. 396–405.
75. Eriksson, T., J. Börjesson, and F. Tjerneld, Mechanism of surfactant effect in enzymatic hydrolysis of lignocellulose. *Enzyme and Microbial Technology*, 2002. 31: pp. 353–364.

76. Jang, J.S., et al., Optimization of saccharification and ethanol production by simultaneous saccharification and fermentation (SSF) from seaweed, *Saccharina japonica. Bioprocess and Biosystems Engineering*, 2012. 35: pp. 11–18.
77. Ye Lee, J., et al., Ethanol production from *Saccharina japonica* using an optimized extremely low acid pretreatment followed by simultaneous saccharification and fermentation. *Bioresource Technology*, 2013. 127: pp. 119–125.
78. Olofsson, K., M. Bertilsson, and G. Liden, A short review on SSF–an interesting process option for ethanol production from lignocellulosic feedstocks. *Biotechnology for Biofuels*, 2008. 1: pp. 1–14.
79. Suutari, M., K. Liukkonen, and S. Laakso, Temperature adaptation in yeasts: the role of fatty acids. *Microbiology*, 1990. 136: pp. 1469–1474.
80. Kadar, Z., Z. Szengyel, and K. Reczey, Simultaneous saccharification and fermentation (SSF) of industrial wastes for the production of ethanol. *Industrial Crops and Products*, 2004. 20: pp. 103–110.
81. Kim, D., Physico-chemical conversion of lignocellulose: inhibitor effects and detoxification strategies: a mini review. *Molecules*, 2018. 23: p. 309.
82. Sjulander, N. and T. Kikas, Origin, impact and control of lignocellulosic inhibitors in bioethanol production—a review. *Energies*, 2020. 13: p. 4751.
83. Yang, Y., et al., Progress and perspective on lignocellulosic hydrolysate inhibitor tolerance improvement in *Zymomonas mobilis. Bioresources and Bioprocessing*, 2018. 5: pp. 1–12.
84. Jung, Y.H. and K.H. Kim, *Acidic pretreatment*, in Ashok Pandey, Sangeeta Negi, Parameswaran Binod, and Christian Larroche (Eds.) *Pretreatment of Biomass*. 2015, Elsevier, Amsterdam. pp. 27–50.
85. Jayakody, L.N., et al., Identification and detoxification of glycolaldehyde, an unattended bioethanol fermentation inhibitor. *Critical Reviews in Biotechnology*, 2017. 37: pp. 177–189.
86. Jönsson, L.J., B. Alriksson, and N.-O. Nilvebrant, Bioconversion of lignocellulose: inhibitors and detoxification. *Biotechnology for Biofuels*, 2013. 6: pp. 1–10.
87. Kim, Y., et al., Soluble inhibitors/deactivators of cellulase enzymes from lignocellulosic biomass. *Enzyme and Microbial Technology*, 2011. 48: pp. 408–415.
88. Li, Y., B. Qi, and Y. Wan, Separation of monosaccharides from pretreatment inhibitors by nanofiltration in lignocellulosic hydrolysate: fouling mitigation by activated carbon adsorption. *Biomass and Bioenergy*, 2020. 136: p. 105527.
89. Kumar, V., et al., A critical review on current strategies and trends employed for removal of inhibitors and toxic materials generated during biomass pretreatment. *Bioresource Technology*, 2020. 299: p. 122633.
90. Millati, R., C. Niklasson, and M.J. Taherzadeh, Effect of pH, time and temperature of overliming on detoxification of dilute-acid hydrolyzates for fermentation by *Saccharomyces cerevisiae. Process Biochemistry*, 2002. 38: pp. 515–522.
91. Singh, B., et al., A biotechnological approach for degradation of inhibitory compounds present in lignocellulosic biomass hydrolysate liquor using *Bordetella sp.* BTIITR. *Chemical Engineering Journal*, 2017. 328: pp. 519–526.
92. Ezeji, T.C., N. Qureshi, and H.P. Blaschek, Bioproduction of butanol from biomass: from genes to bioreactors. *Current Opinion in Biotechnology*, 2007. 18: pp. 220–227.
93. Ezeji, Q. and K. Blaschek, Butanol production from corn, in *Alcoholic Fuels: Fuels for Today and Tomorrow*. 2006, Taylor and Francis. pp. 99–122.
94. Qureshi, N. and H.P. Blaschek, 20 butanol production from agricultural biomass, in Anthony Pometto, Kalidas Shetty, Gopinadhan Paliyath, and Robert E. Levin (Eds.) *Food Biotechnology*. 2005, Boca Raton: CRC Press. pp. 552–579.

95. Huang, W.C., D.E. Ramey, and S.T. Yang. Continuous production of butanol by *Clostridium acetobutylicum* immobilized in a fibrous bed bioreactor, in *Proceedings of the Twenty-Fifth Symposium on Biotechnology for Fuels and Chemicals* Held May 4–7, 2003, in Breckenridge, CO. 2004. Springer.
96. Qureshi, N., A. Lolas, and H. Blaschek, Soy molasses as fermentation substrate for production of butanol using *Clostridium beijerinckii* BA101. *Journal of Industrial Microbiology and Biotechnology*, 2001. 26: pp. 290–295.
97. Maddox, I., N. Qureshi, and K. Roberts-Thomson, Production of acetone-butanol-ethanol from concentrated substrate using clostridium acetobutylicum in an integrated fermentation-product removal process. *Process Biochemistry*, 1995. 30: pp. 209–215.
98. Ezeji, T.C., N. Qureshi, and H.P. Blaschek, Industrially relevant fermentations, in Peter Duerre (Ed.) *Handbook on Clostridia*. 2005, Boca Raton: CRC Press. pp. 797–812.
99. Ezeji, T.C., N. Qureshi, and H.P. Blaschek, Butanol fermentation research: upstream and downstream manipulations. *The Chemical Record*, 2004. 4: pp. 305–314.
100. Ezeji, T., N. Qureshi, and H. Blaschek, Acetone butanol ethanol (ABE) production from concentrated substrate: reduction in substrate inhibition by fed-batch technique and product inhibition by gas stripping. *Applied Microbiology and Biotechnology*, 2004. 63: pp. 653–658.
101. Abdehagh, N., F.H. Tezel, and J. Thibault, Separation techniques in butanol production: challenges and developments. *Biomass and Bioenergy*, 2014. 60: pp. 222–246.
102. Gholizadeh, L., *Enhanced Butanol Production by Free and Immobilized* Clostridium sp. *Cells Using Butyric Acid as Co-Substrate*. 2010, University of Boras/School of Engineering.
103. Matsumura, M., et al., Energy saving effect of pervaporation using oleyl alcohol liquid membrane in butanol purification. *Bioprocess Engineering*, 1988. 3: pp. 93–100.
104. Qureshi, N. and T.C. Ezeji, Butanol, 'a superior biofuel' production from agricultural residues (renewable biomass): recent progress in technology. *Biofuels, Bioproducts and Biorefining: Innovation for a Sustainable Economy*, 2008. 2: pp. 319–330.
105. Qureshi, N., et al., Continuous solvent production by *Clostridium beijerinckii* BA101 immobilized by adsorption onto brick. *World Journal of Microbiology and Biotechnology*, 2000. 16: pp. 377–382.
106. Izak, P., et al., Increased productivity of *Clostridium acetobutylicum* fermentation of acetone, butanol, and ethanol by pervaporation through supported ionic liquid membrane. *Applied Microbiology and Biotechnology*, 2008. 78: pp. 597–602.
107. Evans, P.J. and H.Y. Wang, Enhancement of butanol formation by *Clostridium acetobutylicum* in the presence of decanol-oleyl alcohol mixed extractants. *Applied and Environmental Microbiology*, 1988. 54: pp. 1662–1667.
108. Karcher, P., et al., Microbial production of butanol: product recovery by extraction, in T. Satyanarayann and B.N. Johri (Eds.) *Microbial Diversity: Current Prospectives and Potential Applications*, 2005, IK International, New Delhi, India.
109. Baroghi, L.G., Enhanced butanol production by free and immobilized *Clostridium sp.* Cells using Butyric Acid as Co-Substrate. PhD in biology, 2009.
110. Eckert, G. and K. Schuegerl, Continuous acetone/butanol fermentation with direct product removal. *Chem.Ing.Tech (Germany, Federal Republic of)*, 1987. 59.
111. Zheng, Y.-N., et al., Problems with the microbial production of butanol. *Journal of Industrial Microbiology and Biotechnology*, 2009. 36: pp. 1127–1138.
112. Garcia III, A., E.L. Iannotti, and J.L. Fischer, Butanol fermentation liquor production and separation by reverse osmosis. *Biotechnology and Bioengineering*, 1986. 28: pp. 785–791.
113. Oudshoorn, A., L.A. Van Der Wielen, and A.J. Straathof, Assessment of options for selective 1-butanol recovery from aqueous solution. *Industrial & Engineering Chemistry Research*, 2009. 48: pp. 7325–7336.

114. Groot, W., R. Van der Lans, and K.C.A. Luyben, Technologies for butanol recovery integrated with fermentations. *Process Biochemistry*, 1992. 27: pp. 61–75.
115. Abdehagh, N., F. Tezel, and J. Thibault, Adsorbent screening for biobutanol separation by adsorption: kinetics, isotherms and competitive effect of other compounds. *Adsorption*, 2013. 19: pp. 1263–1272.
116. Takeuchi, Y., et al., Adsorption of l-butanol and p-xylene vapour with high silica zeolites and their mixtures. *Separations Technology*, 1995. 5: pp. 23–34.
117. Dürre, P., Biobutanol: an attractive biofuel. *Biotechnology Journal: Healthcare Nutrition Technology*, 2007. 2: pp. 1525–1534.
118. Saravanan, V., et al., Recovery of 1-butanol from aqueous solutions using zeolite ZSM-5 with a high Si/Al ratio; suitability of a column process for industrial applications. *Biochemical Engineering Journal*, 2010. 49: pp. 33–39.

6 Rector Configurations for Thermochemical Conversion of Lignocellulosic Biomass

Subramaniyasharma Sivaraman, Bhuvaneshwari Veerapandian, Dayavathi Madhavan, Saravanan Ramiah Shanmugam, and Ponnusami V.
SASTRA Deemed University

CONTENTS

DOI: 10.1201/9781003203452-6

6.1 INTRODUCTION

This particular chapter is written with the aim of providing basic information on lignocellulosic biomass thermochemical conversion for applications in energy, fuels, and fine chemicals. The first section of this chapter explains to the reader the fundamental ideas and definitions involved in lignocellulosic biomass conversion. These are prepared with the intention of assisting those who are unfamiliar with biomass conversion principles in comprehending the fundamental concepts. The following are the topics covered in the second portion of this chapter: Current reactor configurations are discussed, as well as critical characteristics involved in reactor design and how these variables impact reactor performance. The reader will get an understanding of the key properties of reactors used for lignocellulosic biomass conversion, as well as their strengths and weaknesses in relation to the planned end-use applications, after reading this chapter.

Basic Definitions – Biofuels: The term "lignocellulosic biomass conversion" is mainly used for the production of "biofuels." Hence, it becomes important to clearly define certain definitions before we further start to describe different reactor configurations used for such conversion. Those who are very familiar with such concepts can skip this section and proceed to the next section. The majority of the definitions in this part will aid in understanding the subsequent section's discussion topics. Biofuels are fuels that are created from renewable biomass and are available in gaseous, liquid, or solid form. Going by this definition, production of ethanol from sugarcane or corn starch, biodiesel, syngas from agricultural crop/forest residues, or any other fuel originating from lignocellulosic biomass or algae can be termed as "Biofuels." Compared to conventional fossil fuels, which are non-renewable and polluting in nature, biofuels, on the other hand, are renewable and provide a sustainable source of energy. From the pollution perspective, biofuel production can be termed as zero or less polluting because the CO_2 produced during the process of producing biofuels can be offset by the amount of CO_2 used by the plants. Hence, the lignocellulosic biomass conversion process can also be termed as "carbon neutral process." It's also thought that when biofuels are utilized to generate electricity, they produce lower NOx and SOx emissions than when fossil fuels are burned. In addition to the environmental benefits mentioned above, biofuel production has the potential to lessen the country's reliance on foreign oil supplies and is seen as strategically vital in terms of energy security. Due to these reasons, research on biofuel generation has increased several folds during the last few years. The major issue in the development of biofuels stems from their economic viability. Researchers around the globe are working toward addressing these concerns.

Biofuels are divided into three categories: first-generation, second-generation, and third-generation biofuels, depending on the order in which different types of biofuels became popular. The creation of biodiesel from vegetable oils via the trans-esterification process, as well as ethanol from maize grain or sugarcane, is an example of first-generation biofuels that are commercially accessible. The generation of biodiesel from triglycerides is a practical and well-established technology (Xiao, Mathew, and Obbard 2009). The limited availability of feedstock and the "food vs. fuel" debate prevent large-scale biodiesel production

(Muscat et al. 2020). Currently, alternative feedstocks such as used cooking oil, other non-edible oils, and lipids from animal wastes have been evaluated for biodiesel production (Malode et al. 2021). Ethanol is the only first-generation biofuel that has been generated in substantial quantities around the world. The ethanol produced is combined at a 10% volume ratio with gasoline. Recent increases in engine efficiency, on the other hand, have allowed the vehicles (flex cars) to operate at any ethanol-to-gasoline ratio (Campos and Viglio 2021).

Biofuels that are produced using lignocellulosic biomass are commonly referred to as second-generation biofuels. Utilization of lignocellulosic biomass for biofuel production addresses some of the issues involved with the first-generation biofuels, such as the "food vs. fuel debate," as these feedstocks is readily available as waste and do not require separate land for production. Furthermore, using such leftovers will aid in the elimination of residues that would otherwise be deemed garbage. Because these feedstocks mostly consist of agricultural crop leftovers, forest residues, specialized energy crops, and other natural biopolymers such as cellulose, hemicellulose, and lignin, they are high in natural biopolymers such as cellulose, hemicellulose, and lignin. Cellulose is a linear biopolymer made up of glucose units connected by ether bonds, with a polymerization degree ranging from 7,000 to 10,000. Cellulose can be either crystalline or amorphous in nature. The chemical structure of lignocellulosic biomass is shown in Figures 6.1–6.3. Hexoses, pentoses, and glucuronic acid make up hemicellulose (Figure 6.2), which is made up of ramified and amorphous polymers (Romaní et al. 2020). Degree of polymerization of hemicellulose varies between 50 and 300. Glucose, arabinose, mannose, xylose, galactose, and glucuronic acid are the major components of hemicellulose.

The third important polymer that is present in lignocellulosic biomass is lignin (Figure 6.3). They are amorphous and cross-linked phenolic polymers. The presence of an aromatic structure in the lignin macromolecular network confers high mechanical and chemical stability. The "glue" that holds the components together is known as lignin (Peral 2016). Although second-generation biofuels have significant environmental benefits, the conversion process is often more expensive in nature, which is a key barrier to the deployment of successful, commercial-scale facilities.

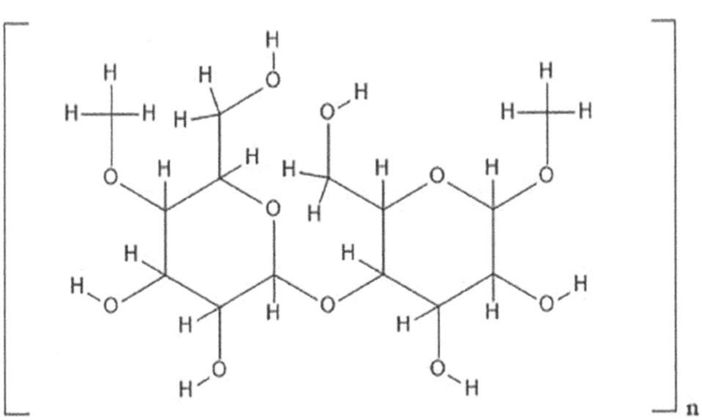

FIGURE 6.1 Structure of cellulose.

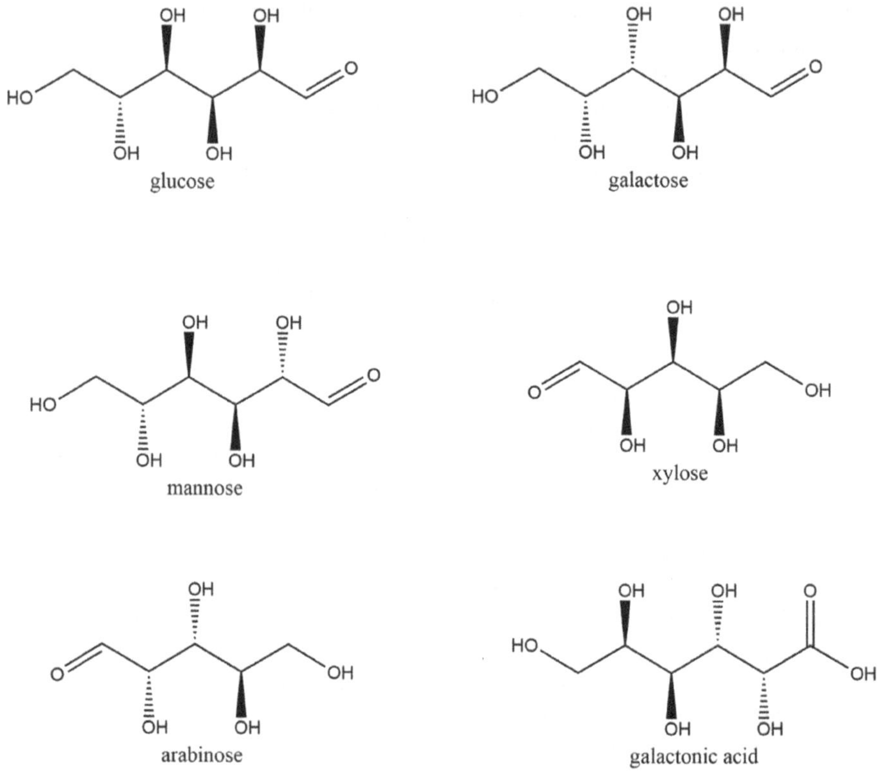

FIGURE 6.2 Hemicellulose monomers.

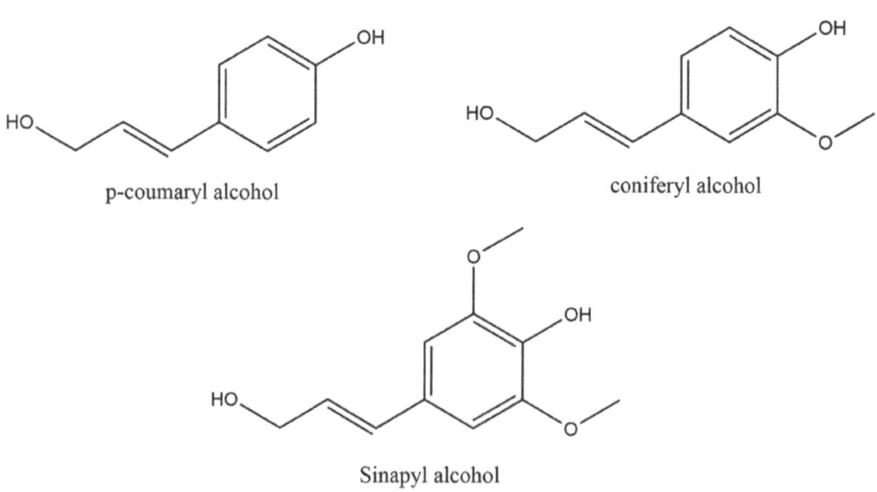

FIGURE 6.3 Monomers of lignin.

Biofuels that originate from algae are commonly termed as third-generation biofuels. Algae is an excellent feedstock owing to its high growth rates and its ability to grow in ponds, thereby eliminating the need for separate arable land. Currently, the major research interest is in the development of algae-based biofuels. The growth requirements of algae, including reactor configurations, open vs. closed pond algal cultivation, and increasing CO_2 solubility, are some of the key areas in which the current research is focused.

The technology and reactor layouts used to produce second-generation biofuels are discussed in this chapter. It is likely that similar platforms may also be utilized to make additional high-value chemicals and products, which could be a different way to boost the economic viability of operations. Any plant that uses biomass to produce "biofuels" and "fine chemicals" is referred to as a "biorefinery."

6.2 LIGNOCELLULOSIC BIOMASS CONVERSION TECHNOLOGIES

It should be remembered that lignocellulosic biomass in its solid form is by itself a fuel used for energy generation via the process of combustion. However, because of their decreased density and the presence of moisture, they are considered low-value fuels, resulting in an inefficient process with restricted applicability. These biomass leftovers, on the other hand, can be easily transformed into gas and liquid forms, which may then be used in a variety of energy applications ranging from power to transportation fuels if appropriately managed. Lignocellulosic biomass may be used to make biofuels in two ways: Biochemical conversion and thermochemical conversion are the two types of conversion. Since the first generation, biochemical conversion, which includes both hydrolysis and fermentation, has been utilized to create a range of biofuels such as ethanol, butanol, and others. The biomass must be processed to remove the lignin barrier and fractionate the cellulose and hemicellulose components in order to convert lignocellulosic biomass utilizing a biochemical conversion method (dos Santos et al. 2019). Hydrolysis of cellulose and hemicellulose components results in formation of sugar monomers, which can be then fermented using microbes or biocatalysts (enzymes) to produce biofuels. In comparison, the thermochemical conversion process utilizes high temperatures for the production of gases, liquids, and solids. Depending on the nature of the product desired, the temperatures can vary between 250°C and 1,200°C. The reactions occur in the presence of a limited supply of oxygen under strictly anoxic conditions. Several hundred chemical reactions occur simultaneously during the breakdown of complex structure of lignocellulosic biomass. At high working temperatures, these complex polymers degrade into monomers, oligomers, and other aromatic components. As a result, the fundamental restriction in this subject is typically considered as a clear knowledge of these processes utilizing a specific set of chemical reactions. Consequently, researchers in this field frequently employ "lumped models," which divide chemical species into distinct categories based on a set of criteria such as "volatiles," "secondary char," "non-condensable gases," and so on (Teh et al. 2021).

On the other hand, a well-defined collection of reactions is used in biochemical technologies for the conversion of lignocellulosic biomass. As a result, the question of whether to convert lignocellulosic biomass via a biochemical or thermochemical method is frequently debated. Major limitations with biochemical conversion are associated with the efficient fractionation of holocellulose from the lignin fractions, the high cost of enzymes, and the low yields of biochemical systems (Concepts 1992). On the other hand, the limitations of thermochemical conversion include high operational temperatures (which increase the energy costs associated with the process) and selectivity of desired products. A successful operation of a biorefinery requires both biochemical and thermochemical conversion to operate synergistically for complete valorization of lignocellulosic biomass. For example, cellulose and hemicellulose fractions of the lignocellulosic biomass can be converted using biochemical conversion, whereas lignin can be efficiently converted using thermochemical conversion for the production of value-added bioproducts. As a result, we will concentrate on both of these processes in this chapter in order to maximize the usage of lignocellulosic biomass.

Based on the operating temperature and amount of oxygen present in the reactor, thermochemical conversion processes may be divided into three categories: combustion, pyrolysis, and gasification. To complete the combustion of biomass, high operating temperatures (more than 900°C) are required, as are stoichiometric levels of oxygen. In reality, however, an excessive amount of oxygen is given to achieve full biomass burning. Combustion is an exothermic reaction that produces a huge quantity of heat as well as carbon dioxide (CO_2) and steam (H_2O) (van Loo and Koppejan 2012). This heat is recovered and utilized for operating the reactor to continue this process. In the case of paper and pulp industry, the lignin that is separated from the cellulose undergoes combustion to supply the heat that is required for operation of the plant. Pyrolysis, on the other hand, requires an anoxic environment where the biomass is the only reactant involved in the process. At high temperatures between 400°C and 800°C, the solid biomass breaks its polymeric constituents into monomers or smaller units. Pyrolysis is usually done in a gaseous environment such as nitrogen, helium, or argon. Liquids (bio-oil), solids (biochar), and non-condensable gases are the three types of pyrolysis products. Bio-oil is a complex combination of products created in vapor or aerosol form during pyrolysis and recovered as liquid when condensable aerosols are cooled to ambient temperature. Biochar is a solid carbonaceous substance produced directly from lignocellulosic biomass or by secondary processes involving char and the organic vapors released during pyrolysis. The majority of non-condensable gases are hydrogen and carbon monoxide, with minor quantities of carbon dioxide, methane, and other short-chain hydrocarbons with a boiling point lower than the ambient temperature. As a result, even under pyrolysis conditions and at room temperature, they remain in the form of gas (Lam et al. 2016).

Pyrolysis conditions affect the ratio of bio-oil, biochar, and non-condensable gases. Pyrolysis is typically divided into two categories: slow pyrolysis and quick pyrolysis. In these two types of pyrolysis, the difference is in the operating temperature, heating rate, and residence period of the organic vapors. Slow pyrolysis, for example, occurs at low temperatures of 300°C–400°C, and the residence time of organic vapors is often measured in minutes. As a result, primary products may react with one another or with the char, resulting in significant volumes of char being produced while consuming condensable goods. Biochar is the desired end product of the slow pyrolysis process, with average yields ranging from 30% to 40% biochar,

30% to 40% liquid, and 30% to 40% permanent gases. The slow pyrolysis process heats at a rate of 5°C–50°C per minute. Fast pyrolysis, on the other hand, can reach temperatures of up to 20,000°C every second (in some lab-scale pyro probe equipment). This is done to heat the biomass as rapidly as possible, reducing the chances of secondary reactions involving volatile organic chemicals. The dwell time of such a process is usually only a few seconds. The created char is separated in a cyclone separator, and the organic volatiles are quickly cooled down in condensers, resulting in liquid bio-oil. The fast pyrolysis method produces 70%–75% bio-oil, 10%–15% biochar, and 10%–15% non-condensable gases by weight as a result of the process. Biochar from the slow pyrolysis method has been utilized in agriculture, but bio-oil from the fast pyrolysis process can be used for heating, power generation, and fine chemical synthesis. The evaluation of bio-oil as a transportation fuel has sparked a lot of interest, but there are a lot of roadblocks in the way (Adhikari, Nam, and Chakraborty 2018). Pyrolysis, in general, is a high-energy process due to the requirement of high operating temperatures.

Gasification is a thermochemical technique for converting solid lignocellulosic biomass to gaseous fuel (Syngas). The temperature required for the gasification process is often greater than 800°C. Gasification occurs in the presence of a partial oxidant environment, resulting in the desired end products of carbon monoxide and hydrogen. When compared to pyrolysis, the gasification process produces less char and volatiles (Damartzis and Zabaniotou 2011). The volatiles produced during this process do not condense fast and are transported away with the gas product, resulting in a sticky material known as tar. For gasification applications, the production of tar and char is undesirable. One of the most difficult technological issues in the gasification process is tar separation. The presence of oxygen during the gasification process promotes biomass combustion, supplying energy to the system and lowering heat requirements. The syngas produced by gasification has a wide range of uses, from energy generation in turbines to the manufacture of fine compounds like methanol (Plácido and Capareda 2015).

Apart from pyrolysis and gasification, there are other thermochemical conversion processes that involve the presence of additional substances as reaction media. For example, hydropyrolysis involves the presence of hydrogen gas during the pyrolysis process, which affects the product distribution (Straw 2020). Similarly, the use of liquid water as a reactant medium occurs in the hydrothermal liquefaction process (e.g., hydrothermal liquefaction of algae). Other hot fluids or supercritical fluids, such as supercritical water gasification of biomass, are being explored as reactant media in thermochemical conversion processes such as supercritical water gasification (Leng et al. 2021). In most cases, hydrothermal processes need a high operating pressure in addition to a high temperature. When compared to pyrolysis and gasification, the energy needs and building costs for these processes are often greater. The reactor setups for thermochemical conversion of biomass will be discussed in the next section.

6.3 PYROLYSIS REACTOR CONFIGURATIONS

Research on fast pyrolysis has gained momentum over the last decade because of its ability to produce liquid fuel (bio-oil) in a single step as a major end product from lignocellulosic biomass. Currently, the bio-oil produced from this process is mainly used for heating applications, and considerable research efforts are under way to use it as a transportation fuel, which largely reduces the import dependency of a country and improves the energy security of a country. The reactor configurations commonly used in a fast pyrolysis process are shown below.

6.3.1 Fluidized Bed Reactor with Internal Gas Bubbling

The name of the reactor alludes to the fact that the reactor bed is kept under permanent fluidization by a continuous upward gas flow. This reactor system comprises a vertical cylindrical tube with a perforated gas distribution plate at the bottom that is well insulated to avoid heat loss (Figure 6.4). This reactor's bed is normally made out of sand, and the upward gas flow at fluidization velocity keeps the sand particles suspended (fluidization). A fixed bed and a bubbling fluidized bed can both be used in the same reactor. When the gas velocity is low, the gas simply flows through the holes in the sand particles and out the top of the reactor. Fixed-bed reactor operation is the name for this sort of operation. The particle is kept suspended by increasing the velocity of gas to the point where the force applied upwardly on each particle matches

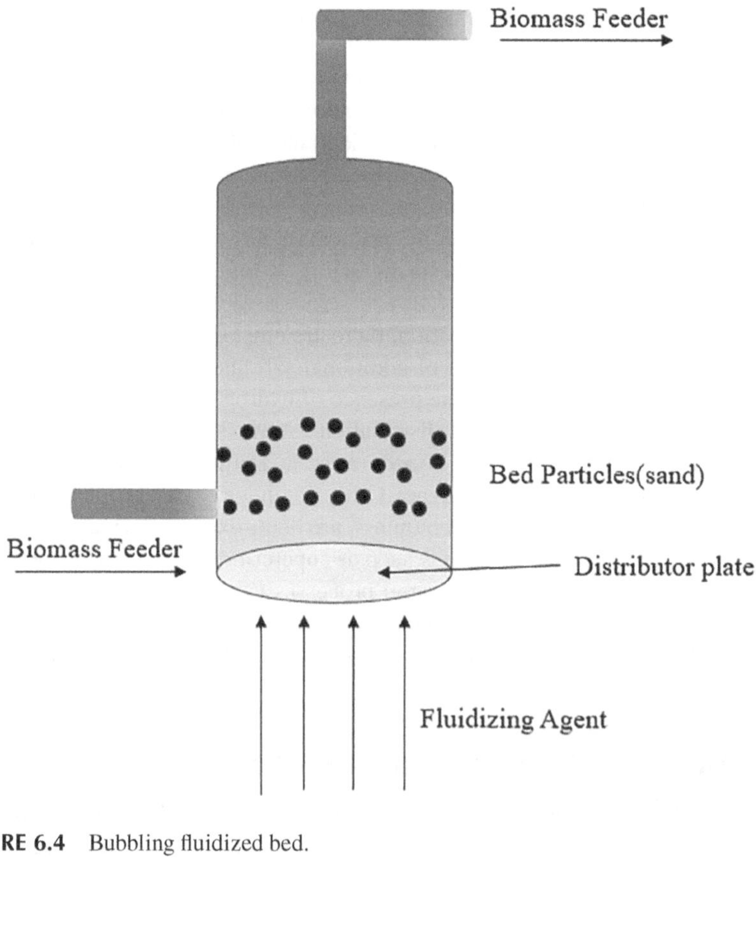

FIGURE 6.4 Bubbling fluidized bed.

the particle weight. This is referred to as the minimal fluidization velocity. In this system, an increase in gas velocity causes gas bubbles to rapidly move upward. The bubbling fluidized bed reactor is the name for this type of reactor (Sricharoenchaikul 2009).

Fluidized bed bioreactor for fast pyrolysis is used commercially due to its ease of scale up. The important characteristics of this system include uniform heat distribution within the reactor due to efficient mixing throughout the reactor bed. This aids in good control of process temperature. One of the important properties of the fast pyrolysis process is that the residence period of vapors may be regulated by altering the gas velocity. The main disadvantage of quick pyrolysis processes is that they frequently demand particle sizes of 2–3 mm to reduce heat/mass transfer loss, which necessitates costly lignocellulosic biomass size-reduction procedures prior to the start of this process (Liu 2014).

6.3.2 Circulating Fluidized Bed Reactor

This circulating fluidized bed reactor works in the same way as a bubbling fluidized bed reactor. The gas velocity in this sort of reactor is extremely high, allowing the biomass particles to be transported outside of the bed by the force imposed on them (Figure 6.5). Elutriation is the term for this phenomenon. Both the sand and the char particles leave the bed and enter the cyclone separator. The sand is removed from the cyclone, and the char is burned to provide more energy to the sand particles, which heat up and return to the reactor. The heat carrier in this sort of reactor is sand, which delivers the significant quantity of heat needed to keep the reactor at the desired

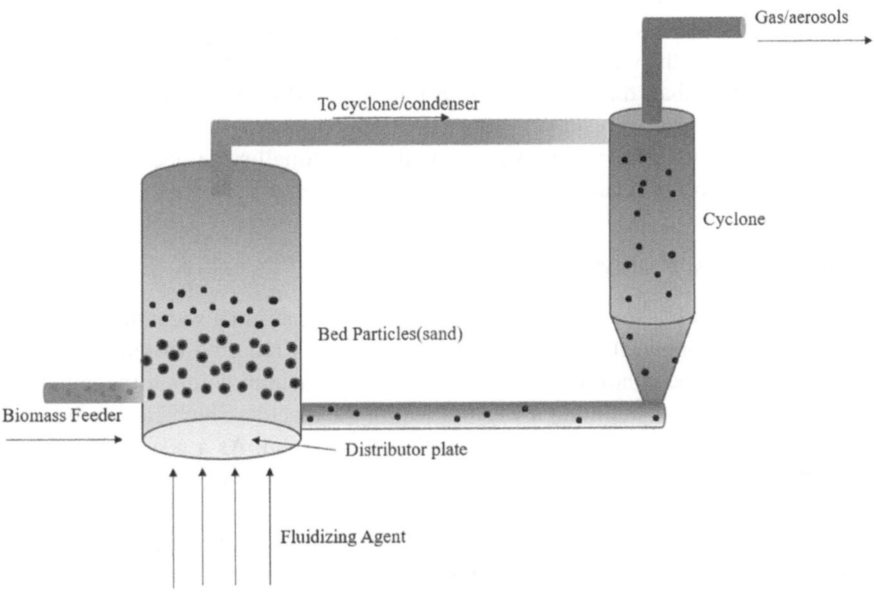

FIGURE 6.5 Circulating fluidized bed reactor.

temperature. When compared to a bubbling fluidized bed reactor, this type of reactor uses less energy. This reactor is most commonly employed in fluid catalytic cracking, which converts low-value, highly viscous crude oil into value-added transportation fuels such as gasoline and diesel. Zeolite in a fluidized condition is the solid catalyst employed in this process. During the fluid catalytic cracking process, a solid carbonaceous residue known as "coke" will deposit on the surface of the zeolite. The solid residues obtained in this type of thermochemical conversion process are known as "coke" and "biochar." The term "coke," on the other hand, is most widely used in the petrochemical industry. This coke is burned off to provide heat to the zeolite catalyst, which is then re-circulated back into the reactor, allowing the catalyst surface to be re-activated by eliminating the coke-blocking pores. Circulating fluidized bed reactors, like bubbling fluidized bed reactors, offer uniform temperature control and are simple to scale up. Furthermore, this system's energy requirements are lower (Grace and Lim 2013). However, compared to a bubbling fluidized bed reactor, this reactor requires more careful management of reactor operation and design. Due to the precise control requirements of such a reactor, they are only used in commercial-scale operations and very limited to lab scale.

6.3.3 Auger Pyrolysis Reactor

The lignocellulosic biomass is continually fed to a screw feeder (auger) that is normally filled with sand in this type of pyrolysis reactor. In this form of reactor, the sand is warmed and serves as a carrier system. With the spinning of the auger, the sand and biomass are combined, and the mixture is moved down the auger's axis until it reaches the end of the reactor. The char and hot sand are gathered at the bottom of the reactor, while the gas and volatile vapors exit at the top. In these reactors, both sand and biomass must be supplied on a continual basis. Although this reactor is simple to operate, the yield of bio-oil obtained from this type of reactor is often more than 50% lower than that obtained from fluidized bed reactors (Nam et al. 2015). This is due to the longer residence period of organic vapors within the reactor, which causes secondary reactions and biochar production. Small-scale operations are best served by this sort of reactor.

6.3.4 Vacuum Pyrolysis

The name of this reactor alludes to the fact that it operates in a vacuum. As a result, no continuous supply of nitrogen, argon, or helium is required to remove air and maintain an anoxic condition within the reactor. A vacuum pump at the end of the reactor is required to maintain a vacuum within the reactor, which aids in the faster removal of organic volatiles produced within this system. As a result, this type of reactor produces bio-oil at a yield of 60%–65%. Poor heating rates are one of the technical limits of this type of reactor due to the lack of convection currents within the reactor. This reactor is commonly used in laboratories. Commercial operation of this type of reactor is difficult due to the requirements of a large-scale vacuum pump (Wentrup 2017).

6.3.5 Ablative Pyrolysis Reactors

In this sort of reactor, heat is transferred mostly through solid–solid contact (Figure 6.6). Direct contact with a hot metallic surface heats either the biomass particles or the complete log of woody biomass. Due to conduction heat transmission via solid–solid contact, heat transfer in this type of reactor is higher than in vacuum pyrolysis reactors. The small surface area of contact with this type of reactor is a key drawback, making scale-up difficult. As a result, this sort of reactor is best suited for pyrolysis in a lab setting. The vortex reactor and the cone reactor are the two most common forms of ablative pyrolysis reactors. The cortical reactor, sometimes known as the cyclone reactor, was created at the National Renewable Energy Laboratory (NREL) in the United States. In tangential motion, the biomass and inert gas are supplied to a cylindrical reactor. The biomass particles are pressed against the interior walls of the reactor, which are heated by wall heaters on the outside. Due to centrifugal forces, the biomass particles clash with the wall. Because of the high temperature in this reactor, the biomass is heated and pyrolyzed through solid–solid contact. The rapid velocity of the entering biomass separates the biochar from the walls. Because the reaction takes place mostly on the surface of the particles due to the solid–solid contact, big particles can be employed in this process as long as they are less dense than the entering gas. Heat transfer from the heater to the reactor walls, rather than from the walls to the particles, controls the overall heat transfer rate. Because this sort of reactor is surface-area restricted, commercial scale-up of such systems is problematic (Peacocke and Bridgwater 1994). As a result, they are restricted to laboratory-scale operations.

At the University of Twente in the Netherlands, a rotating cone-type ablative reactor was invented. This type of reactor is shaped like an inverted cone that rotates continually. The biomass is fed into the wide base (top) along with the sand particles,

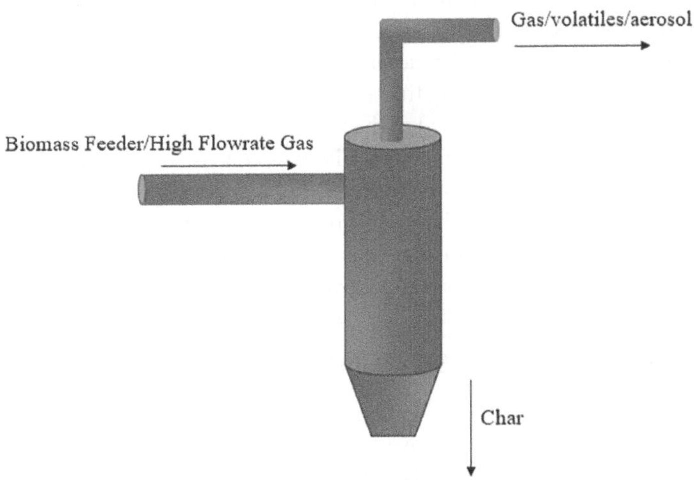

FIGURE 6.6 Vortex/cyclone reactor.

and the biomass is transported down by the centrifugal rotation of the cone. The biochar, as well as the sand, is gathered at the cone's narrow end (base). The biochar is burned off, and the heated sand is recycled into the process. Organic vapors and non-condensable gases exit the reactor through the top. The spinning cone pyrolyzer, a riser for sand particle recycling, and a bubbling bed char combustor are the three primary components of this system. In addition, gas is necessary for sand delivery and biochar burn-off. Similar to vortex pyrolysis reactors, this type of reactor is extremely difficult to scale up.

6.4 FACTORS INFLUENCING PYROLYSIS REACTOR SELECTION

The factors influencing pyrolysis reactor system include the following: (1) End-product application, (2) Desired Heat Transfer Characteristics, and (3) Operational Scale. Because ablative pyrolysis reactors do not require feedstock particle size, they may be a good alternative if the feedstock has a high particle size and grinding is a problem. For the generation of bio-oil, the bubbling and fluidized bed types of pyrolysis systems are well-proven setups. When compared to newer, less-proven types of reactors, this type of reactor may be used if a popular, well-proven configuration with low difficulties is required. Because of their simple design and straightforward operation, auger-type pyrolysis reactors are the optimum choice for small-scale unit operation. One of the greatest impediments to the thermochemical conversion of agricultural crop leftovers in India is the supply chain for lignocellulosic biomass feedstock. Due to the lack of size-reduction procedures and suitability for small-scale operations, a mobile pyrolysis ablative type reactor may be the best option for resolving this problem.

Table 6.1 summarizes different types of pyrolysis reactors based on their operational parameters on a pilot scale. Different types of LB were studied with these reactors. Comparisons were made on the following parameters: operational complexity, particle size, biomass variability, scale-up difficulty, and flow rate. The auger reactor is the least complex type of reactor when compared on the basis of operational complexity, followed by both rotating cone reactors and fixed-bed reactors. Higher particle sizes of LB can be handled with ease in fixed-bed reactors, ablative pyrolysis reactors, and vacuum pyrolysis reactors. Fluidized bed reactor has low biomass variability compared to other types of reactors. Except the auger reactor and, to some extent, rotating cone reactor, all other types of pyrolysis reactors have difficulties with scale-up. High flow rate of inert gas is required in the case of a fluidized bed reactor, which will add to the operational cost in large-scale applications.

6.5 GASIFICATION – BASIC TERMINOLOGIES AND CONCEPTS

In comparison to pyrolysis, gasification is the most studied thermochemical conversion process. Gasification involves the partial oxidation of solid fuels at high temperatures, resulting in syngas (H_2 and CO) as a by-product. The gasification process takes roughly 25%–30% of the oxygen required for complete biomass burning. By combusting a portion of the lignocellulosic biomass, oxygen is employed to produce energy to keep the reactor temperature constant. Due to a shortage of

TABLE 6.1
Biomass Pyrolysis Pilot-Scale Reactor Classification and Comparison of Operating Parameters

Pyrolysis Reactors	Feed Rate(kg/h)	Material	Yield (%) Liquid	Yield (%) Solid	Yield (%) Gas	Bio-Oil Heating Value (MJ/kg)	Comparison of Various Reactor Parameters (*Low, **Medium, ***High) Operational Complexity	Particle Size	Biomass Variability	Scale-Up Difficulty	Inert Gas Flow Rate	Ref
Auger reactor	5	Waste tire	42.6	40.5	16.9	42.5	*	**	***	*	*	Martínez et al. (2013)
Rotating cone reactor	26	Sawdust	65	15.4	18.8	32	**	**	***	**	*	Tanoh et al. (2020)
Fluidized bed reactor	45	Sugarcane bagasse	78.07	9.25	12.68	18.4	***	**	*	***	***	Treedet and Suntivarakorn (2018)
Circulating fluidized bed reactor	42	Sawdust	60	15	25	20	***	**	*	***	***	Park et al. (2019)
Vacuum pyrolysis reactor	25	Açaí seeds	38	35	27	N/A	***	***	**	***	*	Rocha de Castro et al. (2021)
Ablative pyrolysis reactor	5	Beech wood	60	N/A	N/A	18.3	***	***	***	***	*	Auersvald et al. (2020)
Fixed-bed reactor	20	Palm kernel shell	30	35	40	28	**	***	***	***	*	Haryati et al. (2018)

liquid transportation fuel in Germany during the Second World War, gasification was used commercially. They immediately discovered that combining gasification and the Fischer-Tropsch method could produce liquid fuels. The syngas produced during gasification reacts with liquid hydrocarbons in a Fischer-Tropsch system. Another key use of a gasification system is the production of hydrogen, which can be used in fertilizer (ammonia/urea) synthesis (Tijmensen 2002).

When compared to complete biomass combustion, the gasification process requires lower operating temperatures. Furthermore, process pollutants such as nitrogen oxides are not present in the product gas (NO_x). Nitrogen oxides are one of the most prevalent pollutants produced during combustion, when nitrogen from the environment reacts with oxygen from the air or fuel at high temperatures. It's worth noting that while nitrogen is commonly referred to as an "inert" gas, it really serves as a reactant in the combustion process. Syngas contains trace quantities of sulfur, which may be easily eliminated, making it a cleaner process than direct biomass combustion (Susmozas, Iribarren, and Dufour 2013). The primary by-products of gasification are gases (CO and H_2), char, and tar. Tar is comparable to pyrolysis bio-oil, but it is much more viscous. Steam and pure oxygen, in addition to air, can be employed as an oxidant environment during the gasification process. The type of gasification reagent to use is mostly determined by the end-product use.

The heating value, which is the amount of energy produced by full combustion of a reactant in the presence of oxygen, is a crucial concept to grasp in all thermochemical conversion processes. It is divided into two categories: lower heating value (LHV) and higher heating value (HHV) (Figure 6.7). The key distinction is the physical condition of the created product, which is water. The heating value of water created in a gaseous condition is known as LHV. If the created water is liquid, however, we

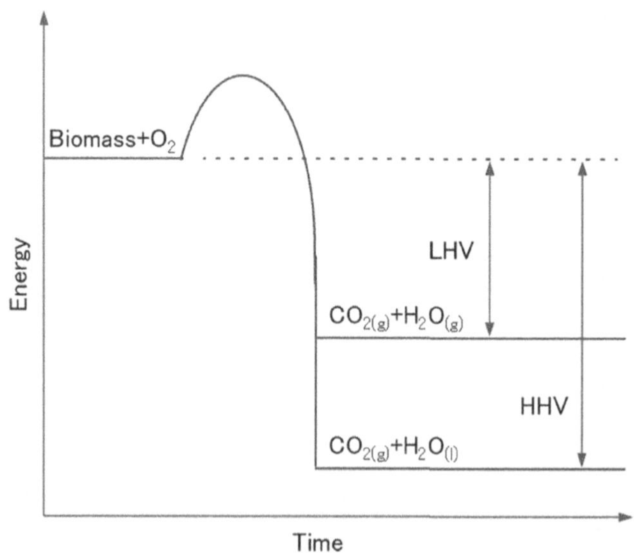

FIGURE 6.7 Energy diagram of HHV and LHV.

refer to it as having a greater heating value. The phrases "lower" and "higher" refer to the heating value's magnitude. To condense into a liquid, heat must be removed from a gas (water vapor). As a result, when compared to water in its vapor phase, the system must lose more energy. In simple terms, when water is in a liquid form, its heating value is higher, and when water is in a vapor state, its heating value is lower.

The comprehension of the gasification process necessitates the use of two more terms: the carbon conversion efficiency (CCE) (Equation 6.1) and the cold gas efficiency (CGE) (Equation 6.2).

$$\text{CCE}(\%) = \frac{\text{Carbon content in gas}}{\text{Carbon content in biomass}} \times 100 \tag{6.1}$$

$$\text{CGE}(\%) = \frac{\text{Lower heating value of gas}}{\text{Lower heating value of biomass}} \times 100 \tag{6.2}$$

CCE is linked to the mass balance of the gasification process, which indicates how much of the initial carbon in biomass is transformed into carbon in the final gas. CCE is an important feature of a gasification system. In some gasification reactors, it can reach 98%–99%. On the other side, CGE is determined by the energy balance. It's the ratio of a gas's LHV to that of a solid. It provides a decent indicator of the amount of energy that can be obtained by combusting the gasification product compared to the amount of energy that can be obtained by directly combusting the solid. The CGE for gasification systems ranges from 75% to 90%.

6.5.1 Steps Involved in the Gasification Process

Drying, pyrolysis, combustion, and reduction are the four basic phases in the gasification process (Figure 6.8). It's worth noting that pyrolysis is a process in and of itself as well as one of the processes in the gasification process. To remove the water from the biomass, the drying process uses a lot of energy. The moisture content of lignocellulosic biomass wastes can be as high as 50% by weight. As a result, during the initial drying stage of the gasification process, this moisture is removed. In the absence of oxygen, the biomass is pyrolyzed, resulting in gasification char and tar. The oxygen in the air reacts with the char and tar from the pyrolysis step during

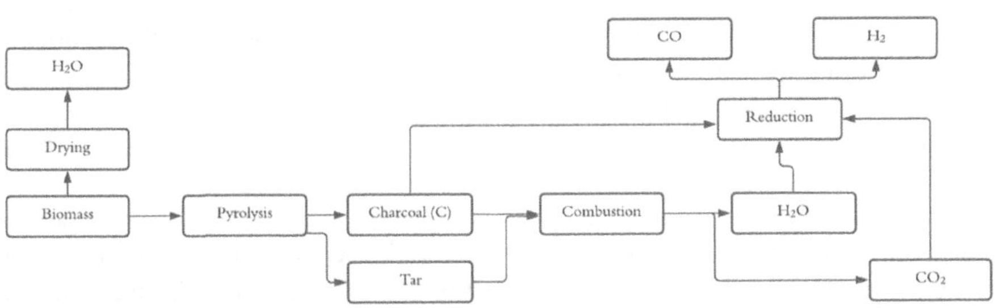

FIGURE 6.8 Gasification steps.

the combustion process. Carbon dioxide and water from the combustion step react with gasification char in the reduction step to create syngas (CO and H_2), the desired gasification end products. The most critical phase in the gasification process is reduction, which converts zero-heat-of-combustion molecules (CO_2 and H_2O) into calorific end products (CO and H_2). As indicated in the diagram below, these reduction reactions in the gasification process are known as "Boudouard reaction" (Equation 6.3) and "Water–gas reaction" (Equation 6.4), respectively.

$$CO_2 + C \leftrightarrow 2CO \tag{6.3}$$

$$H_2O + C \leftrightarrow H_2 + CO \tag{6.4}$$

The "C" in the above equations represents the carbon present in the char generated during the pyrolysis step. The other reactions involved in the gasification process include gasification with oxygen (Equation 6.5), combustion with oxygen (Equation 6.6), hydrogasification (Equation 6.7), and so on.

$$C + 0.5O_2 \leftrightarrow CO \tag{6.5}$$

$$C + O_2 \leftrightarrow CO_2 \tag{6.6}$$

$$C + 2H_2 \leftrightarrow CH_4 \tag{6.7}$$

When air or pure oxygen is utilized as the oxidant medium, the foregoing processes predominate. The reactions with molecular oxygen are not regarded as relevant when steam is utilized as the oxidant medium. Instead, in these conditions, the water–gas shift reaction (Equation 6.8) and methanation (Equation 6.9) become the most significant stages.

$$CO + H_2O \leftrightarrow H_2 + CO_2 \tag{6.8}$$

$$CO + 3H_2 \leftrightarrow CH_4 + H_2O \tag{6.9}$$

Increase in availability of water from steam promotes the water–gas shift reaction. If the ratio of H_2:CO increases to 3:1, then methanation reaction takes place, thereby producing methane. This indicates that when steam is used as a gasification agent, a large amount of H_2, CO_2, and CH_4 along with low amounts of CO, will be formed. The two most important parameters that affect gasification include the heating rate and the gasification oxidant. Heating rate affects the nature of gasification products produced and also the sequence of gasification process. For example, at low heating rates (<1 °C/sec), the reduction reactions start only after the completion of the pyrolysis reaction, which increases the quantity of volatile organics within the reactor, and there is a chance for removal of some of these gases without undergoing reduction. On the other hand, at high heating rates (>100 °C/sec), both pyrolysis and reduction reactions occur simultaneously, thereby quickly converting the organic volatiles and leaving behind a cleaner gas product that is free from tar (Resende 2014).

The second most important parameter affecting the gasification process is the type of oxidant used. Air, oxygen, and steam are the three most commonly utilized oxidants. The most common oxidant for gasification systems is air, which provides huge amounts of free oxygen. However, the gas generated by such a method has a poor calorific value (4–7 MJ/Nm^3). This is mostly due to the enormous amount of unreacted N_2 in the air, which reduces the calorific value. For applications that demand clean syngas, gasification with pure oxygen is preferable to overcome this issue. The use of oxygen also facilitates the conversion of methane and any other volatile organics created in the product gas to solely H_2 and CO via burning. The heating value of this method ranges from 10 to 18 MJ/Nm^3.

Steam is another oxidant used in gasification reactors that reduces the yield of H_2 and CO compared to pure oxygen-based gasification systems. When steam is used, methane is produced, which serves to increase the calorific value of the product gas since methane adds to the calorific value of the product gas. Therefore, the heating value of such a process varies between 15 and 20 MJ/Nm^3.

Recently, supercritical water, which is akin to steam, has been employed. In its supercritical state (374°C and 22 MPa), water alternates between being a gas and a liquid. As solvents, supercritical fluids combine the properties of a liquid and a gas: they are denser than a gas and have much higher diffusivity rates than liquids, which aid in dissolving lignin and cellulose, changing reaction pathways during the gasification process, and lowering the amount of tar and char produced (Singh and Harvey 2010).

The syngas generated by the gasification process is mostly utilized for heating and energy generation. Syngas is co-fired with natural gas for heating without requiring any alterations to the current combustion facility's infrastructure. The presence of contaminants in syngas is not a big problem for this application. However, the presence of tar in syngas makes the combustion of gas in gas turbines unsuitable for power production. Chemical conversion of syngas into vital chemicals and fuels is another key use. For example, the following equation may be used to synthesize methanol from syngas (assuming the CO and H_2 contents are greater than 70%) (Equation 6.10).

$$CO + 2H_2 \leftrightarrow CH_3OH \tag{6.10}$$

The use of O_2 as a gasifying agent is appropriate for this application since it produces high-purity H_2 and CO with a minimum production of CH_4, hydrocarbons, and tar. The previously described Fischer-Tropsch method is another chemical use of syngas.

This method mixes CO and H_2 in various ratios at high temperatures to produce hydrocarbons according to Equation 6.11:

$$(2n+1)H_2 + n\,CO \leftrightarrow C_nH_{(2n+2)} + n\,H_2O \tag{6.11}$$

Because of the large range of values for "n" that may be employed as coefficients in this reaction, a mixture of hydrocarbons is usually generated. The Fischer-Tropsch method is most commonly used to convert coal to liquid fuels. Similarly, lignocellulosic biomass can be used in the same way (Resende 2014).

6.6 REACTORS FOR GASIFICATION PROCESS

In the gasification process, many types of reactors are employed. Some are comparable to those used in the pyrolysis process, while others are very different since gasification does not need a shorter residence period of the lignocellulosic biomass within the reactor. The reactor configurations utilized in the gasification process are briefly discussed in the next section.

6.6.1 Updraft Gasification Reactor

The biomass is supplied into the system from the top, while the oxidant is put into the reactor at the bottom, in the opposite direction of the biomass flow (Figure 6.9). As a result, this sort of reactor is often referred to as a "counter-current gasification reactor." As the biomass goes down the reactor, it heats up. As a result, the volatile organics created in this process remain uncracked in the gaseous stream at low temperatures, causing a considerable quantity of tar to develop. Although the syngas generated by this method is unsuitable for a variety of applications, it has a significant improvement in heating value and CCE. From bottom to top, the steady heating of the biomass throughout the length of the reactor generates various reaction zones: drying zone, pyrolysis zone, combustion zone, and reduction zone. The bottom of the reactor is where the ash is removed (Lin 2007).

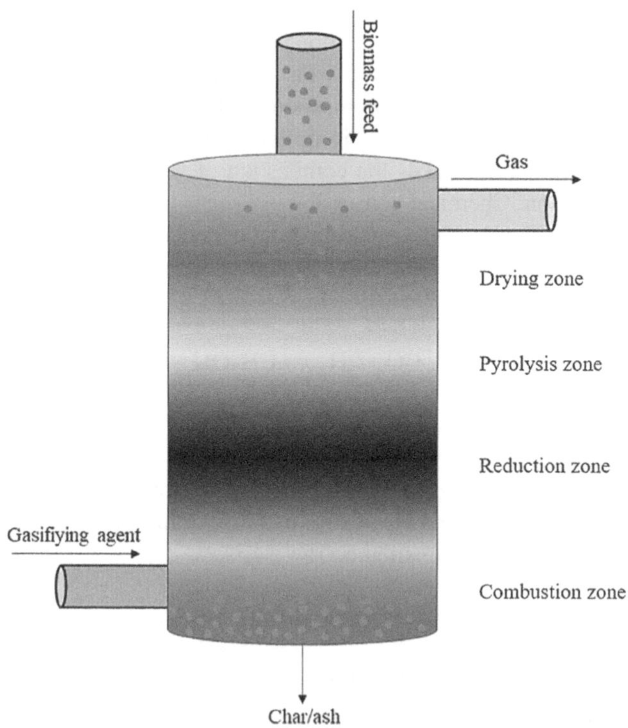

FIGURE 6.9 Updraft gasifier.

6.6.2 Downdraft Gasification Reactor

The biomass moves along with the oxidant in a co-current fashion in this type of reactor from top to bottom (Figure 6.10). The gasifying agent is supplied along the length of the reactor at a certain height but not at the reactor's bottom. The gas generated by the gasification process heats up the incoming oxidant stream, thereby minimizing the need for additional oxygen. Therefore, this system operates at high thermal efficiency. Furthermore, the product gas exits the reactor at the bottom from the hottest zone, assisting in the cracking the tar components and producing a large amounts of CO. More than 75% of commercial gasifiers are of this reactor type, with processing capacities up to 4 dry metric tons per hour (Patra and Sheth 2015).

6.6.3 Bubbling Fluidized Bed Reactor

This sort of reactor, as previously mentioned in the pyrolysis section, has great mixing qualities, good temperature distribution, and high reaction rates. Because the concentration of reactants/products is uniform throughout the reactor, mixing can be a problem. As a result, the tar content of the gas leaving the reactor is identical to that seen in other regions of the reactor system, together with a high level of

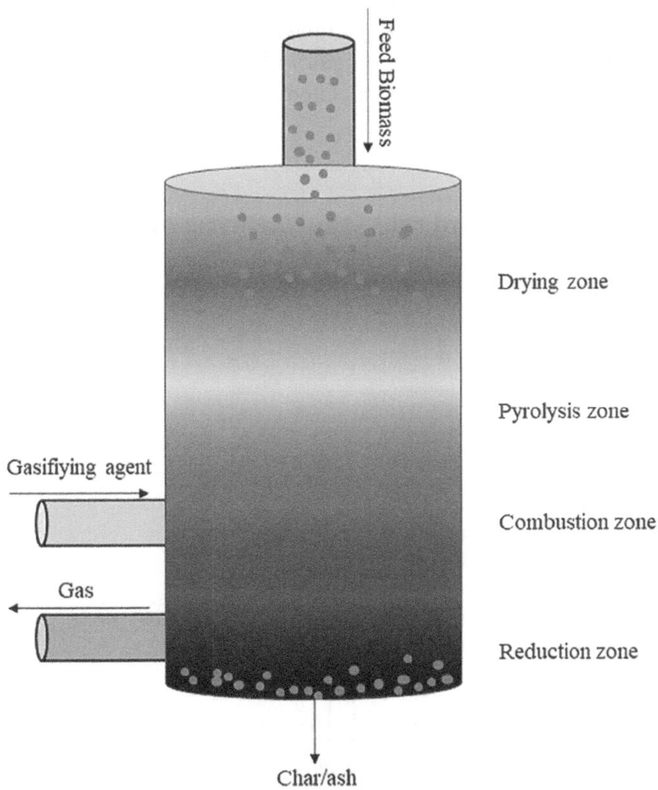

FIGURE 6.10 Downdraft gasifier.

particulate matter. As a result, a tar cracking catalyst should be used in the bed to reduce the amount of tar in the product gas. Some fuel may remain unreacted in the system. This reduces the process' CCE. Low quantities of SO_x and NO_x emissions are produced, and this sort of reactor is simple to scale up (Cordiner, De Simone, and Mulone 2012).

6.6.4 Circulating Fluidized Bed Gasification Reactor

The reactor's architecture is fairly similar to what was covered earlier in the pyrolysis section. CCE is improved by recirculating product contents, and gasification char can be burned to supply some process energy. The circulating fluidized bed gasification reactor has several primary benefits. This reactor layout may be desirable at larger scales, but it would be more expensive at smaller scales. It is said to be more dependable in terms of performance and easier to scale up (Chen et al. 2019).

6.6.5 Entrained Flow Gasification Reactor

The lignocellulosic biomass is present in this reactor as very tiny particles (Figure 6.11). An oxidant (gasifying agent) propels the particles through the system at extremely high speeds. Lower residence periods result, and high operating temperatures (>1,200°C) are required to guarantee full gasification. The high operating temperatures provide a highly clean syngas with minimal tar and condensable volatile concentrations. The main issue associated with this type of reactor configuration

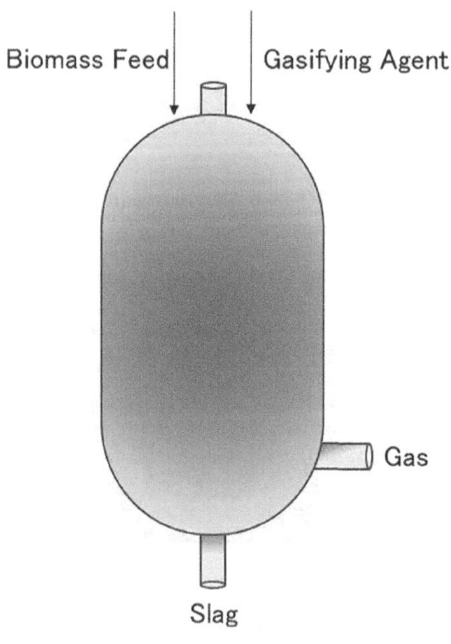

FIGURE 6.11 Entrained flow gasifier.

TABLE 6.2
Examples of Gasification Reactors and Their Characteristics

Reactor Scale	Biomass Feed Stock	Thermal Capacity (kWe)	Reactor Type	Gasification Agent	LHV (MJ/Nm^3)	Ref
Pilot	Almond shells	1,000	Fluidized bed reactor	$Steam/O_2$	10.9–11.7	González-Vázquez et al. (2018)
Pilot	Pine chips	80	Bubbling fluidized bed	Air	6.5	Pio et al. (2020)
Pilot	Palm kernel shells	35	Updraft gasifier	Air	7.94	Jeong et al. (2020)
Pilot	Empty fruit bunch/ mesocarp fiber	75	Downdraft gasifier	Air	12	Anyaoha et al. (2020)
Bench	Tomato peels	N/A	Entrained flow gasifier	Steam	5.4	Brachi et al. (2018)

is the removal of ash. Ash formed will be melted at these high temperatures and solidify at the bottom of the reactor. Another major disadvantage with this system is the requirement for high amounts of oxygen (Yang et al. 2018).

Table 6.2 summarizes the results from the pilot-scale operation of different types of gasification reactors with different types of gasification agents ($steam/O_2$/ Air). Different types of LB waste have been studied with these reactors. The LHV of the syngas is highest (12 MJ/Nm^3) in the case of a downdraft gasifier (air as oxidant) using empty fruit punch/mesocarp fiber. The thermal capacity of this unit is 75 kWe. Fluidized bed reactor using almond shells and steam as oxidants has shown LHV around 11.5 MJ/Nm^3. The scale of operation of this plant is also high at 1,000 kWe. The syngas composition varies with the use of $steam/O_2$ compared to air as the gasification agent because CH_4 formation is lower in the case of gasification using $steam/O_2$.

6.7 CONCLUSION

The conversion of lignocellulosic biomass into biofuels and bioproducts using a thermochemical conversion process has great potential in the coming years. In this chapter, an overview of basic concepts, reactor configurations, operational parameters, and scalability of two types of well-established technologies, namely pyrolysis and gasification, were discussed. Currently, research on the generation of liquid fuels using gas-to-liquid technology such as gasification and pyrolysis has gained momentum to meet the growing demands for liquid fuels. The major challenge with these technologies is bringing their costs down to par with those of conventional sources of liquid fuels.

REFERENCES

Adhikari, S., H. Nam, and J. P. Chakraborty. 2018. "Conversion of Solid Wastes to Fuels and Chemicals Through Pyrolysis." *Waste Biorefinery*. Elsevier B.V. doi:10.1016/b978-0-444-63992-9.00008-2.

Anyaoha, K. E., R. Sakrabani, K. Patchigolla, and A. M. Mouazen. 2020. "Co-Gasification of Oil Palm Biomass in a Pilot Scale Downdraft Gasifier." *Energy Reports* 6. Elsevier Ltd: 1888–1896. doi:10.1016/j.egyr.2020.07.009.

Auersvald, M., T. Macek, T. Schulzke, M. Staš, and P. Šimáček. 2020. "Influence of Biomass Type on the Composition of Bio-Oils from Ablative Fast Pyrolysis." *Journal of Analytical and Applied Pyrolysis* 150 (September): 104838. doi:10.1016/j.jaap.2020.104838.

Brachi, P., R. Chirone, F. Miccio, M. Miccio, and G. Ruoppolo. 2018. "Entrained-Flow Gasification of Torrefied Tomato Peels: Combining Torrefaction Experiments with Chemical Equilibrium Modeling for Gasification." *Fuel* 220 (February): 744–753. doi:10.1016/j.fuel.2018.02.027.

Campos, J. N., and J. E. Viglio. 2021. "Drivers of Ethanol Fuel Development in Brazil: A Sociotechnical Review." *MRS Energy & Sustainability*, November. doi:10.1557/s43581-021-00016-6.

Chen, X., J. Ma, X. Tian, J. Wan, and H. Zhao. 2019. "CPFD Simulation and Optimization of a 50 KWth Dual Circulating Fluidized Bed Reactor for Chemical Looping Combustion of Coal." *International Journal of Greenhouse Gas Control* 90 (November): 102800. doi:10.1016/j.ijggc.2019.102800.

Concepts, B. 1992. "Bioprocess Engineering: Basic Concepts." *Journal of Controlled Release* 22 (3): 293. doi:10.1016/0168-3659(92)90106-2.

Cordiner, S., G. De Simone, and V. Mulone. 2012. "Experimental–Numerical Design of a Biomass Bubbling Fluidized Bed Gasifier for Paper Sludge Energy Recovery." *Applied Energy* 97 (September): 532–542. doi:10.1016/j.apenergy.2011.11.024.

Damartzis, T., and A. Zabaniotou. 2011. "Thermochemical Conversion of Biomass to Second Generation Biofuels through Integrated Process Design-A Review." *Renewable and Sustainable Energy Reviews* 15 (1). Elsevier Ltd: 366–378. doi:10.1016/j.rser.2010.08.003.

dos Santos, A. C., E. Ximenes, Y. Kim, and M. R. Ladisch. 2019. "Lignin–Enzyme Interactions in the Hydrolysis of Lignocellulosic Biomass." *Trends in Biotechnology* 37 (5): 518–531. doi:10.1016/j.tibtech.2018.10.010.

González-Vázquez, M. P., R. García, M. V. Gil, C. Pevida, and F. Rubiera. 2018. "Comparison of the Gasification Performance of Multiple Biomass Types in a Bubbling Fluidized Bed." *Energy Conversion and Management* 176 (November): 309–323. doi:10.1016/j.enconman.2018.09.020.

Grace, J. R., and C. J. Lim. 2013. "Properties of Circulating Fluidized Beds (CFB) Relevant to Combustion and Gasification Systems." In *Fluidized Bed Technologies for Near-Zero Emission Combustion and Gasification*, pp. 147–176. Elsevier. doi:10.1533/9780857098801.1.147.

Haryati, Z., S. K. Loh, S. H. Kong, and R. T. Bachmann. 2018. "Pilot Scale Biochar Production from Palm Kernel Shell (PKS) in a Fixed Bed Allothermal Reactor." *Journal of Oil Palm Research* 30 (3): 485–494. doi:10.21894/jopr.2018.0043.

Jeong, Y-S., Y-K. Choi, B-S. Kang, J-H. Ryu, H-S. Kim, M.-S. Kang, L-H. Ryu, and J-S. Kim. 2020. "Lab-Scale and Pilot-Scale Two-Stage Gasification of Biomass Using Active Carbon for Production of Hydrogen-Rich and Low-Tar Producer Gas." *Fuel Processing Technology* 198: 106240. doi:10.1016/j.fuproc.2019.106240.

Lam, S. S., R. K. Liew, A. Jusoh, C. T. Chong, F. N. Ani, and H. A. Chase. 2016. "Progress in waste oil to sustainable energy, with emphasis on pyrolysis techniques." *Renewable and Sustainable Energy Reviews* 53. Elsevier: 741–753. doi:10.1016/j.rser.2015.09.005.

Leng, L., L. Yang, J. Chen, Y. Hu, H. Li, H. Li, S. Jiang, H. Peng, X. Yuan, and H. Huang. 2021. "Valorization of the Aqueous Phase Produced from Wet and Dry Thermochemical Processing Biomass: A Review." *Journal of Cleaner Production* 294. Elsevier Ltd: 126238. doi:10.1016/j.jclepro.2021.126238.

Lin, J-C. M. 2007. "Combination of a Biomass Fired Updraft Gasifier and a Stirling Engine for Power Production." *Journal of Energy Resources Technology* 129 (1): 66–70. doi:10.1115/1.2424963.

Liu, H. 2014. "CFD Modeling of Biomass Gasification Using a Circulating Fluidized Bed Reactor," [Doctoral Thesis, University of Waterloo, Ontario, Canada], https://core.ac.uk/download/pdf/144147219.pdf (Accessed Feb 23, 2023).

Malode, S. J., K. Keerthi Prabhu, R. J. Mascarenhas, N. P. Shetti, and T. M. Aminabhavi. 2021. "Recent Advances and Viability in Biofuel Production." *Energy Conversion and Management: X* 10 (June): 100070. doi:10.1016/j.ecmx.2020.100070.

Martínez, J. D., R. Murillo, T. García, and A. Veses. 2013. "Demonstration of the Waste Tire Pyrolysis Process on Pilot Scale in a Continuous Auger Reactor." *Journal of Hazardous Materials* 261. Elsevier B.V.: 637–645. doi:10.1016/j.jhazmat.2013.07.077.

Muscat, A., E. M. de Olde, I. J.M. de Boer, and R. Ripoll-Bosch. 2020. "The Battle for Biomass: A Systematic Review of Food-Feed-Fuel Competition." *Global Food Security* 25. doi:10.1016/j.gfs.2019.100330.

Naidu, V. S., P. Kumar, N. Karri, P. K.Singh, S. Gandham, and B. R. Rao. 2020. "Effect of pressure on in-situ catalytic hydropyrolysis of rice straw." *Austin Chemical Engineering* 7 (2): 1076.

Nam, H., S. C. Capareda, N. Ashwath, and J. Kongkasawan. 2015. "Experimental Investigation of Pyrolysis of Rice Straw Using Bench-Scale Auger, Batch and Fluidized Bed Reactors." *Energy* 93. Elsevier Ltd: 2384–2394. doi:10.1016/j.energy.2015.10.028.

Park, J. Y., J. K. Kim, C. H. Oh, J. W. Park, and E. E. Kwon. 2019. "Production of Bio-Oil from Fast Pyrolysis of Biomass Using a Pilot-Scale Circulating Fluidized Bed Reactor and Its Characterization." *Journal of Environmental Management* 234 (August 2018). Elsevier: 138–144. doi:10.1016/j.jenvman.2018.12.104.

Patra, T. K., and P. N. Sheth. 2015. "Biomass Gasification Models for Downdraft Gasifier: A State-of-the-Art Review." *Renewable and Sustainable Energy Reviews* 50 (October): 583–593. doi:10.1016/j.rser.2015.05.012.

Peacocke, G. V. C., and A. V. Bridgwater. 1994. "Ablative Plate Pyrolysis of Biomass for Liquids." *Biomass and Bioenergy* 7 (1–6): 147–154. doi:10.1016/0961-9534(94)00054-W.

Peral, C. 2016. "Biomass Pretreatment Strategies (Technologies, Environmental Performance, Economic Considerations, Industrial Implementation)." In *Biotransformation of Agricultural Waste and By-Products*, pp. 125–160. Elsevier. doi:10.1016/B978-0-12-803622-8.00005-7.

Pio, D. T, L. A. C. Tarelho, A. M. A. Tavares, M. A. A. Matos, and V. Silva. 2020. "Co-Gasification of Refused Derived Fuel and Biomass in a Pilot-Scale Bubbling Fluidized Bed Reactor." *Energy Conversion and Management* 206: 112476. doi:10.1016/j.enconman.2020.112476.

Plácido, J., and S. Capareda. 2015. "Production of Silicon Compounds and Fulvic Acids from Cotton Wastes Biochar Using Chemical Depolymerization." *Industrial Crops and Products* 67. Elsevier B.V.: 270–280. doi:10.1016/j.indcrop.2015.01.027.

Resende, F. L. P. 2014. "Reactor Configurations and Design Parameters for Thermochemical Conversion of Biomass into Fuels, Energy, and Chemicals." *Reactor and Process Design in Sustainable Energy Technology*, 1–25. doi:10.1016/B978-0-444-59566-9.00001-6.

Rocha de Castro, D., H. da Silva Ribeiro, L. H. Guerreiro, L. P. Bernar, S. J. Bremer, M. C. Santo, H. da Silva Almeida, S. Duvoisin, L. P. Borges, and N. T. Machado. 2021. "Production of Fuel-Like Fractions by Fractional Distillation of Bio-Oil from Açaí (Euterpe Oleracea Mart.) Seeds Pyrolysis." *Energies* 14 (13): 3713. doi:10.3390/en14133713.

Romaní, A., C. M. R. Rocha, M. Michelin, L. Domingues, and J. A. Teixeira. 2020. "Valorization of Lignocellulosic-Based Wastes." In *Current Developments in Biotechnology and Bioengineering*, pp. 383–410. Elsevier. doi:10.1016/B978-0-444-64321-6.00020-3.

Singh, O. V., and S. P. Harvey. 2010. "Sustainable Biotechnology: Sources of Renewable Energy." *Sustainable Biotechnology: Sources of Renewable Energy*, 1–323. doi:10.1007/978-90-481-3295-9.

Sricharoenchaikul, V. 2009. "Assessment of Black Liquor Gasification in Supercritical Water." *Bioresource Technology* 100 (2). Elsevier Ltd: 638–643. doi:10.1016/j.biortech.2008.07.011.

Susmozas, A., D. Iribarren, and J. Dufour. 2013. "Life-Cycle Performance of Indirect Biomass Gasification as a Green Alternative to Steam Methane Reforming for Hydrogen Production." *International Journal of Hydrogen Energy* 38 (24): 9961–9972. doi:10.1016/j.ijhydene.2013.06.012.

Tanoh, T. S., A. A. Oumeziane, J. Lemonon, F. J. E. Sanz, and S. Salvador. 2020. "Green Waste/Wood Pellet Pyrolysis in a Pilot-Scale Rotary Kiln: Effect of Temperature on Product Distribution and Characteristics." *Energy & Fuels* 34 (3). American Chemical Society: 3336–3345. doi:10.1021/acs.energyfuels.9b04365.

Teh, J. S., Y. H. Teoh, H. G. How, and F. Sher. 2021. "Thermal Analysis Technologies for Biomass Feedstocks: A State-of-the-Art Review." *Processes* 9 (9). doi:10.3390/pr9091610.

Tijmensen, M. 2002. "Exploration of the Possibilities for Production of Fischer Tropsch Liquids and Power via Biomass Gasification." *Biomass and Bioenergy* 23 (2): 129–152. doi:10.1016/S0961-9534(02)00037-5.

Treedet, W., and R. Suntivarakorn. 2018. "Design and Operation of a Low Cost Bio-Oil Fast Pyrolysis from Sugarcane Bagasse on Circulating Fluidized Bed Reactor in a Pilot Plant." *Fuel Processing Technology* 179 (May). Elsevier: 17–31. doi:10.1016/j.fuproc.2018.06.006.

van Loo, S., and J. Koppejan. 2012. "The Handbook of Biomass Combustion and Co-Firing." *The Handbook of Biomass Combustion and Co-Firing*, 1–442. doi:10.4324/9781849773041.

Wentrup, C. 2017. "Flash Vacuum Pyrolysis: Techniques and Reactions." *Angewandte Chemie International Edition* 56 (47): 14808–14835. doi:10.1002/anie.201705118.

Xiao, M., S. Mathew, and J. P. Obbard. 2009. "Biodiesel Fuel Production via Transesterification of Oils Using Lipase Biocatalyst." *GCB Bioenergy* 1 (2): 115–125. doi:10.1111/j.1757-1707.2009.01009.x.

Yang, S., B. Li, J. Zheng, and R. K. Kankala. 2018. "Biomass-to-Methanol by Dual-Stage Entrained Flow Gasification: Design and Techno-Economic Analysis Based on System Modeling." *Journal of Cleaner Production* 205: 364–374. doi:10.1016/j.jclepro.2018.09.043.

7 Advanced Pretreatment Process for Lignocellulosic Biomass

Itha Sai Kireeti, Ponnusami V., and Arumugam A.
SASTRA Deemed University

CONTENTS

DOI: 10.1201/9781003203452-7

7.1 INTRODUCTION

Renewable forms of energy became the major interest of every country attempting to shiftaway from nonrenewable sources. Coal reserves and crude reserves are the major sources of energy production. Burning crude and coal reserves produces several toxic chemicals, which are depleting due to overextraction (Khare et al. 2019). Renewable sources of energy include solar energy, tidal, wind, and hydro energy from agricultural waste. In recent times, energy production from various agricultural wastes has been discussed (Owusu and Asumadu-Sarkodie 2016). One of the most available agricultural feedstocks is "lignocellulosic biomass".

Since most of the energy produced is from crude and coal reserves, approximately 81% of the energy is produced from coal and crude resources (Marta G. Plaza 2019). According to statistics, in India, during 2019–2020, it is estimated that 730.87 MT of coal and 32.17 MT of crude oil were extracted for energy production. These sources of energy are not readily available at any time, and they are depleting as demand for fuel increases day by day. To tackle this problem, we need to focus on the alternative energy sources that act as efficient substitutes (Moriarty and Honnery 2017). Among all the renewable energy sources, "lignocellulosic biomass" has major importance because of its availability in various forms and low cost. Lignocellulosic biomass is available as plant waste, agricultural waste, and so on (de Paula et al. 2019). Reports state that approximately 550 billion tonnes of biomass are available, and around 200 billion metric tonnes of lignocellulosic biomass reserves are also available. India is the leading producer of cereals; therefore, India has 605 million tonnes of agricultural residue annually. Similarly, China is left with 900 million tonnes of straw annually, and the United States is left with 105 million tonnes of corn stover feedstock per annum (Arevalo-Gallegos et al. 2017).

Lignocellulosic biomass is an inexpensive feedstock that is available in the form of agricultural waste, wood chips, and so on. These can be processed to obtain various chemicals because lignocellulose is mainly composed of lignin, cellulose, and hemicellulose (Karimi 2015). Green fuels such as biohydrogen and bioethanol are the major products that can be produced from the conversion of cellulose and hemicellulose. Lignin can be converted into phenols, dibasic acids, olefins, and so on. Various alcohols, hydrocarbons, and furfurals can be produced from the biomass. C_5 and C_6 sugars are also one of the major products obtained when the biomass is synthesized. First-generation and second-generation fuels are major products of lignocellulosic material. Genetic modifications are being made to increase the efficiency of the conversion (Chen et al. 2017).

Cellulose, hemicellulose, and lignin are the main components of lignocellulose, whose composition varies from crop to crop. Cellulose is made up of glucose molecules linked together by a **β**-(1-4)-glycosidic bond (Yang 2007). Due to its strong hydrogen bonds, cellulose naturally resists degradation. Hemicellulose is the second major component of lignocellulose, which is composed of pentose and hexose sugars (Saini et al. 2015). Due to the amorphous nature of hemicellulose, these are easily degradable when compared to cellulose. Lignin usually fills in the gaps between cellulose and hemicellulose. Lignin is composed of *p*-hydroxyphenyl, guaiacyl, and syringyl with chemical bonds. Due to strong interactions between compounds of lignin and heterogeneity, degradation of lignin is difficult. Therefore, pretreatment techniques have a crucial role in obtaining the maximum conversion of lignocellulosic biomass (Romano 2018).

7.2 CHEMISTRY OF LIGNOCELLULOSE

Lignocellulose is the most abundant biomass available. As of now, we know that we can convert the available biomass into biofuel and other valuable products. To study how one can convert biomass into valuable products, we should know the properties of lignocellulose.

7.2.1 Structure of Lignocellulose

Lignocellulose specifically contains three important components. They are:

1. Cellulose
2. Hemicellulose
3. Lignin

Cellulose is the main component of lignocellulose, consisting of 1-4 β links between glucose molecules and hydrogen bonds between the different layers (Reading 2020).

Hemicellulose is the second component of lignocellulose, consisting of various 5- and 6-carbon sugars, such as arabinose, galactose, glucose, and mannose. Lignin is the third component of lignocellulose, which is composed of three phenolic components: *p*-coumaryl alcohol (H), coniferyl alcohol (G), and sinapyl alcohol (S) (Jonge and Elliott 2002). Cellulose and hemicellulose mainly consist of hydrogen bonds, and some chemical bonds are found between hemicellulose and lignin, which makes the structure of the lignocellulose compact and stronger. The chemical bonds between hemicellulose and lignin are mainly due to the bonds between galactose and arabinose molecules. Based upon the type of biomass, the proportion of the components varies (Table 7.1). The cell wall is composed of cellulose, hemicellulose, and lignin in a proportion of 4:3:3.

7.2.1.1 Chemical Structure of Cellulose

Cellulose is a linear homopolymer consisting of d-glucopyranose units linked to each other with β-1,4-glycosidic bonds. The structural formula of cellulose is $(C_6H_{10}O_5)_n$. The primary components of cellulose are 44.44% carbon, 6.17% hydrogen, and 49.39% oxygen (Wu et al. 2022).

7.2.1.1.1 Chemical Properties of Cellulose

Three active hydroxyls are present in every glucosyl ring. These 3-OH molecules are involved in chemical reactions related to the substitution of hydrogen or oxidation reactions. The reactivity of the cellulose mainly depends on the reactivity of the primary and secondary hydroxyl groups (Knott et al. 2014). Due to steric hindrance, the secondary hydroxyl is less reactive when compared to the primary. For example, esterification of "toluenesulfonyl chloride" generally takes place at the primary

TABLE 7.1
Structural Components of Lignocellulose (Chen 2014)

	Lignin	Hemicellulose	Cellulose
Subunits	Guaiacylpropane (G), syringylpropane (S), *p*-hydroxyphenylpropane(H)	d-Xylose, mannose, l-arabinose, galactose, glucuronic acid	d-Pyran glucose units
Bonds between the subunits	Ether bonds and carbon–carbon bonds, mainly β-O-4 ether bond	β-1,4-Glycosidic bonds in main chains; β-1,2-, β -1,3-, β-1,6-glycosidic bonds inside chains	β-1,4-Glycosidic bonds
Polymerization	4,000	Less than 200	Several hundred to tens of thousands
Polymer	G lignin, GS lignin, GSH lignin	Polyxylose, galactoglucomannan (Gal-Glu-Man), glucomannan (Glu-Man)	β-Glucan
Composition	Amorphous, inhomogeneous, nonlinear three-dimensional polymer	Three-dimensional in homogenous molecular with a small crystalline region	Three-dimensional linear molecular composed of the crystalline regions and the amorphous region
Bonds between three components	Contain chemical bond with hemicellulose	Contains chemical bond with lignin	Without chemical bond

hydroxyl group. Etherification is also another reaction that takes place at hydroxyl groups (Jia et al. 2017).

7.2.1.1.2 Physical Properties of Cellulose

Cellulose is an odorless, white powdery substance. Solubility of the crystalline form of cellulose is very low, whereas the amorphous form dissolves easily. The solubility of amorphous cellulose is due to the formation of additional hydrogen bonds with water molecules. Cellulose is found to dissolve in "aq. cupric ammonium hydroxide ($Cu(NH_3)_4(OH)_2$)." Cellulose starts decomposing when it is heated in the range of 300°C–375°C. The heating of cellulose in the range of 200°C–280°C causes dehydration, thus forming charcoal and other gas products. The heating of cellulose at very high temperatures makes it a volatile material known as "tar." The important product obtained at high temperatures is "levoglucosan" (Heinze 2016).

7.2.1.2 Chemical Structure of Hemicellulose

The main chain of hemicellulose consists of more types of glycols with various bonds. Due to this, the hemicellulose content varies from plant to plant. The main components of hemicellulose are "xylan, xyloglucan, glucomannan, galactomannan, callose, etc." (Isikgor and Becer 2015).

Xylan hemicellulose is one of the glucans, which has 1,4-β-d-xylopyranose as the backbone and branch chains of 4-O-methylglucuronic acid. The hemicellulose present in hardwood and gramineous forbs is mainly composed of l-arabinofuranose, where the hemicellulose in gramineous consists of β-d-xylopyranosyl with β-1,4-glucosidic bonds (Bajpai 2014). Similarly, the hemicellulose of softwood contains xylan hemicellulose with 4-O-methyl-glucuronic acid and arabinose-xylan with no acetyl, while O-acetyl-l-O-methyl-glucuronic acid xylan is an important compound hemicellulose in hardwood (Jönsson and Martín 2016).

7.2.1.3 Chemical Structure of Lignin

Lignin has the exclusive composition of "gymnosperm and angiosperm." The structure of lignin is complex, and it is composed of phenyl propane units that are nonlinearly and randomly linked. "Coumaryl Alcohol, Coniferyl Alcohol, and Sinapyl Alcohol" are the main components of lignin. Due to different contents in lignin, it is classified into three types: "syringyl lignin polymerized by syringyl propane, guaiacyl lignin polymerized by guaiacyl propane, and hydroxy-phenyl lignin polymerized by hydroxy-phenyl propane." Similar to other main components of lignocellulose, lignin composition also varies from plant to plant, as do its origin and conditions of growth (Kögel-Knabner and Amelung 2013).

7.2.2 Various Pretreatment Methods

Pretreatment methods have a significant role in the conversion. These methods make the conversion of biomass into biofuels easier. Generally, pretreatment methods are categorized into two types as traditional and advanced emerging pretreatment techniques. Traditional methods include "physical pretreatments, chemical pretreatments, and biological pretreatments" (Haldar and Purkait 2021) (Figure 7.1).

7.2.2.1 Physical Pretreatments

7.2.2.1.1 Mechanical Pretreatment

In this method, shear force is applied to the biomass, which improves the accessibility of chemicals to the cellulose and hemicellulose. Commonly used mechanical operations such as grinding, shearing, stirring, and so on are performed. Mechanical pretreatment is mainly preferable for the biomass, which has low lignin content. Biomass materials such as wheat straw and meadow are examples of low lignin biomass that are subjected to extrusion grinding and cutting, which reduce the degree of polymerization and improve the reaction activity (Millati et al. 2020).

7.2.2.1.2 Ultrasonic Pretreatment

Ultrasonic vibrations can disrupt the lignocellulose crystal structure with their high energy and penetrability. In general, ultrasonic treatment is performed at 20–8 Hz for about 20–150 min, which reduces the degree of polymerization and crystallinity of cellulose. Sometimes, ultrasonic pretreatment is employed along with chemical pretreatment to obtain better results (Zhao et al. 2022).

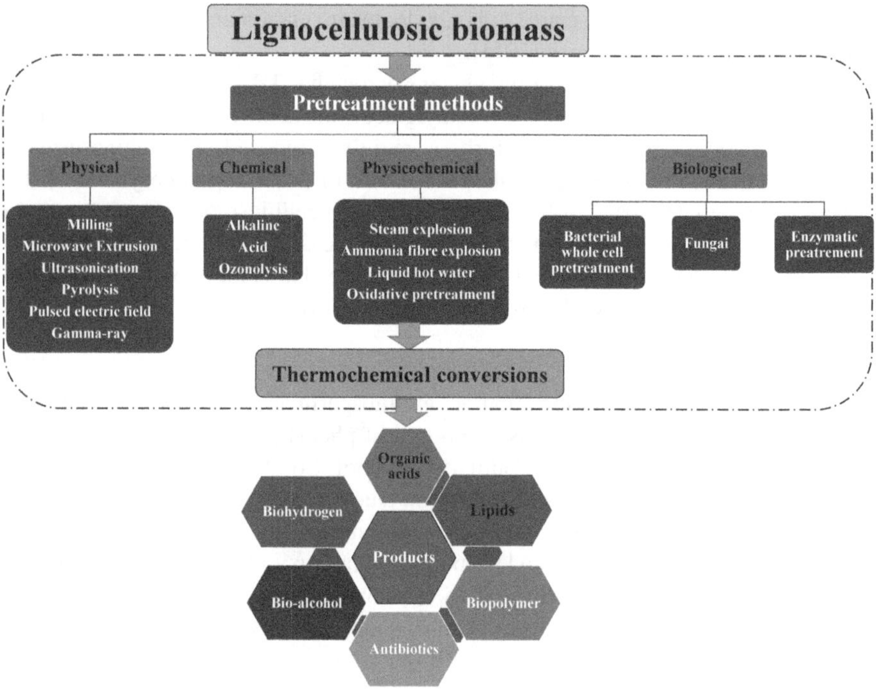

FIGURE 7.1 Pretreatment methods for lignocellulosic biomass and various products derived from plant biomass.

7.2.2.1.3 Thermal Pretreatment

Dissolution of xylan, mannan, and araban takes place when the biomass is subjected to thermal treatment. Thermal pretreatment improves accessibility for enzymatic hydrolysis. It is proved that performing this treatment at 200°C for 90 min aids in the extraction of lignin content and the dissolution of oligomers (Zhao et al. 2022).

7.2.2.2 Chemical Pretreatments

7.2.2.2.1 Acid Pretreatment

It is known that acids can destroy the glycosidic bonds in lignocellulosic biomass and also dissolve cellulose, hemicellulose, and a small amount of lignin, which promote the fermentable sugars (de Araújo Padilha et al. 2020). The most commonly used acids are sulfuric acid, acetic acid, and phosphoric acid. The use of high concentrations leads to excessive corrosion and the formation of more inhibitors. To avoid the formation of inhibitors, acids of 0%–5% concentration (wt/wt) are commonly used at 120°C–215°C for 10–120 min. Optimal conditions are to be followed to obtain the maximum amount of desired products with low inhibitors (Rocha et al. 2014).

7.2.2.2.2 Alkali Pretreatment

High-concentration alkali can dissolve more lignin and hemicellulose and also be able to make changes in cellulose crystallinity (Arumugam et al. 2016). Alkali is also capable of promoting enzymatic hydrolysis. This pretreatment is mainly suitable for biomass that has a high lignin content. NaOH, KOH, NH_4OH, $Ca(OH)_2$, and so on are the most commonly used alkalis. The alkali concentration is 2%–7% at 100°C–200°C for a short time. The addition of alkaline pretreatment reagents increases the dissolving capacity of alkali (Ghasemian et al. 2016).

7.2.2.2.3 Oxidative Pretreatment

In this pretreatment, hydrogen peroxide is used, which is capable of destroying alkyl propylene ether bonds and aromatic nuclei and can break the lignin bonds without inhibitor formation. A concentration of 2%–8% H_2O_2 in the presence of 10% NaOH at 90°C for 12 h can reduce the lignin content to 9%–63%. Sometimes ozone is used for pretreatment, which reacts with lignin and decomposes hemicellulose. The activity of ozone is completely dependent upon the concentration and reaction conditions (Chen et al. 2017).

7.2.2.2.4 Organosolv Pretreatment

Organic solvents such as methanol, ethanol, tetrahydrofuranol, and so on can break the various aryl and ester bonds, which leads to the degradation of lignin and hemicellulose. This also increases the activity of cellulose toward enzymatic hydrolysis. These solvents are usually applied at 150°C–200°C for efficient removal of lignin content. To increase the lignin conversion, a combination of organic solvents is used. It has been found that a combination of organic solvents produces the most desirable results; 82% higher lignin conversion is observed when a combination of solvents is used (Mahmood et al. 2019).

7.2.2.3 Biological Pretreatment

7.2.2.3.1 Fungal Pretreatment

Fungal pretreatment is one of the most conventional methods in which fungi such as white rot fungi, brown rot fungi, and soft rot fungi release enzymes called lignin peroxidase, manganese peroxidase, and laccase, which are found to be effective in degrading lignin (Kainthola et al. 2021). In biofuel production, fungal pretreatment plays a significant role, which improves the lignocellulosic conversion during the hydrolysis and fermentation processes. When specific white rot fungi called "Trametes Versicolor" are used to treat cow dung, barley straw, and wheat, the hydrolysis of cellulose is increased to 80% and the methane production is increased to 10%–18% (Kainthola et al. 2021), as in a study by Ma et al. in which white rot fungi called "Peniphora incarnata" are used to treat poplar trees. The hemicellulose and lignin content are reduced by 48% and 70%, respectively, over the course of seven days (Ma et al. 2021).

7.2.2.3.2 Bacterial Pretreatment

Bacteria are capable of producing enzymes and reducing the degree of polymerization of lignocellulose. Bacillus is a specific bacterium that is widely distributed in soil and can secrete both cellulase and peroxidase, which degrade cellulose and lignin (Barati et al. 2021). Auer et al. (2017) used multiple microorganisms to significantly increase lignocellulose degradation. The use of multiple organisms increased the solubility of wheat straw up to 47% within three days, which includes 40% lignin degradation (Auer et al. 2017).

7.2.2.3.3 Termite Pretreatment

Termites are the most commonly known lignocellulose-decomposing organism. It is found that termites are capable of reducing the size of wood chips and destroying the structure of the wood at 27°C for 45 days (Verma et al. 2009). Dumond et al. (2021) used wheat straw that was exposed to different termite organisms for 20 days; the breakdown of cellulose, hemicellulose, and lignin was found to be in the range of 28%–47%, 12.5%–23.1%, and 7%–32%, respectively (Dumond et al. 2021).

7.2.2.4 Physicochemical Pretreatment

7.2.2.4.1 Steam Explosion Treatment

In this method, high-pressure steam is passed over the biomass, which breaks the β-O-4 bond of lignocellulose and destroys the cell wall structure (Rastogi and Shrivastava 2017). The steam under a pressure of 0.5–5 MPa at high temperatures of 160°C–250°C is used (Paudel et al. 2017). It is found that using this method, over 80% of the lignin is removed by passing the steam at 2 MPa at 215°C for about 5 min (Tanpichai et al. 2019). This method is associated with some disadvantages, including inhibitor compound formation, severe pretreatment conditions, and high energy consumption.

7.2.2.4.2 Alkali-Heat Pretreatment

In this process, the alkali pretreatment method is subjected to some heat applied. During this process, temperatures of around 75°C–215°C are maintained. In this temperature range with alkalis such as NaOH, Na_2CO_3, and alkaline peroxide, an increase in lignin degradation is noticed, crystallinity is reduced, and the surface area of the biomass is increased (Soares et al. 2017). The main challenges associated with this process are high treatment costs and complex subsequent treatment.

7.2.2.4.3 Extrusion Pretreatment

In this particular pretreatment process, shear force is applied to the biomass in a closed container. Due to this, the temperature of the biomass increases, causing changes in both the physical and chemical properties of the biomass. It is found that this process is effective with the little addition of NaOH, which increased hemicellulose and lignin conversion, with maximum conversions of 62.88% and 86.05%, respectively (Khatri et al. 2018).

7.2.3 Advances in Pretreatment Technologies

Pretreatment is always a major step in the conversion of biomass into biofuels. The pretreatment methods aim to bring about changes in the properties of the biomass and make it vulnerable to subsequent steps in biofuel production (Galbe et al. 2019). The type of pretreatment depends on the properties of biomass, its composition, and the type of product required. Every pretreatment is associated with certain advantages and disadvantages. It is found that single pretreatment is not completely effective, and a combination of pretreatment techniques is found to be effective, i.e., treatments such as biochemical, physicochemical, and emerging pretreatment techniques. The pretreatment step aims to achieve greater conversion of biomass to energy (Karimi and Taherzadeh 2016).

This is a combination of both chemical and biological pretreatment methods. This is well suited for the low efficiency and low duration of biological processes and also reduces chemical usage and inhibitor product formation. The overall conversion is greatly dependent on the sequential steps followed during the process. A study by Si et al. found that 53.58%–68.2% and 19.65%–33.44% of lignin and hemicellulose are removed when the biomass is subjected to alkali pretreatment in the presence of bacteria called "Acinetobacter sp. B-2, Bacillus sp. B-3, Pandoraea sp. B-6, and Comamonas sp. B-9 biodegradation", respectively (Liu et al. 2018). Ionic liquids (ILs) are the molten form of salts or electrolytes that are composed of cations and anions at room temperature. ILs are combinations of imidazole, pyridine, and choline, with the most familiar anions being chloride and acetate. They are known to be green solvents when compared with organic solvents. These are capable of destroying the structure of lignin by breaking the specific **β**-O-4 bond. It is found that treating the corn straw with $[NH_3(CH_2)_2OH][OAc]$ dissolved 68.3% of the lignin. Deep eutectic solvents (DESs) are considered green organic solvents, which also act as a substitute for ILs (Ma et al. 2018). When compared with ILs, DESs are found to be more compatible with enzymes and microorganisms (Jeong et al. 2017). DESs are composed of a hydrogen bond donor (HBD) and a hydrogen bond acceptor (HBA), which form a hydrogen bond between them. The degradation of biomass is completely dependent on the HBD and HBA, both of which have different effects. It is found that lignin removal from poplar wood is about 92.98%–95.13% when the biomass is treated with DES (Malaeke et al. 2018).

Supercritical carbon dioxide and supercritical water are the most commonly used chemicals that are highly efficient in dissolving biomass. SC-CO_2 reduces the pH, which is associated with an improvement in the solubility of the polar compounds, thus increasing the activity toward lignocellulosic hydrolysis. It has been found that using SC fluids at higher pressures increases the conversion of biomass. The little addition of water to biomass with SC-CO_2 led to the formation of carbonic acid, which increased the lignocellulose hydrolysis efficiency to 60%.

7.2.4 Ionic Liquids

ILs are nothing but liquid salts, which have a boiling point of less than 100°C. Paul Walden was the first person to identify the IL "ethyl ammonium nitrate" in 1914. ILs have a wide range of applications in the entire chemistry region. They are found to have specific applications in material sciences, chemical engineering, and environmental studies. ILs have significant importance because of their many physical and chemical properties (Liu et al. 2021). ILs can be used as "solvents, catalysts, and reagents or any combination of these." ILs are classified based on their use in a particular aspect. They are "room temperature ILs, task-specific ILs, polyionic liquids, and supported IL membranes." The physical properties of ILs are generally "non-volatile, non-flammable, and stable with air and water." Due to these desirable properties of ILs, the area of interest in ILs is increasing day by day (Isosaari et al. 2019). Generally, ILs are used as a solvent in the lignocellulosic conversion to various valuable products and biofuel and as a catalyst in the conversion of cellulose, hemicellulose, and lignin. From the basic idea of what the ILs are, let us start with how they are used in biomass conversion, their mechanism, and other aspects.

7.2.4.1 ILs in Biomass Conversion

Biomass is primarily produced from plants that include cellulose, lignin, and lignocellulose. These are abundant resources that are renewable, completely biodegradable, and completely environmentally friendly. The major problems with other solvents include incomplete conversion, solvents not being eco-friendly, and so forth. These solvents are replaced with ILs, which have a higher conversion of biomass to useful products and are more environmentally friendly. Biomass has strong hydrogen bond interactions, which can be broken using ILs more efficiently (Gares et al. 2020).

7.2.4.2 Dissolution of Biomass in ILs

Lignocellulose is a type of biomass that contains lignin, hemicellulose, and cellulose as its main components. Due to the rigid 3D and complex structure of lignocellulose, called "biomass recalcitrance," it is of major interest to obtain biofuel using ILs (Roy et al. 2020). It is found that ILs of both polar and non-polar groups have a significant impact on the conversion of lignocellulose. ILs contain "imidazolium cations," found to have good stability in lignocellulose conversion. It is found that the dissolution of lignocellulosic biomass in "AmimCl" is more efficient than that in "BmimCl." The ILs with higher hydrogen bond basicity (e.g. EmimAc) are found to dissolve biomass more efficiently (Liu et al. 2019). The dissolution of lignocellulose is due to the loss of rigid and complex structures. The dissolved lignocellulose is regenerated to form "films, fibers, and gels" as the products. Extending the dissolution time or increasing the temperature causes a higher conversion of biomass. Usually, this biomass is pretreated in a ball mill and treated with hot water for better conversion.

Cellulose is another biomass that can be regenerated into useful products (Xia et al. 2020). Usually "1-butyl-3-methylimidazolium chloride (BmimCl)" is used for dissolving cellulose. IL 1-ethyl-3-methylimidazolium diethyl phosphate (EmimDEP) and EmimDEP with dimethylsulfoxide solution can also be used, which have significant effects on the physical properties of the products. Certain mechanical operations are employed to increase the few properties of the products obtained from cellulose (Wei et al. 2012).

Lignin, another kind of biomass, is the most abundant natural material. Lignin has a complex structure that contains guaiacyl (G), syringyl (S) and *p*-hydroxyphenyl (H) as repeating units with linkages such as β–1, β–5, β–β, 5–5 carbon–carbon, and aryl ether linkages include β-O-4, α-O-4, 4-O-5, and dibenzodioxins. Among all linkages, the β-O-4 bond is the most prominent linkage. Around 45%–50% of links include this type of linkage. Hence, this is called as the "predominant linkage". The main purpose of lignin recycling is due to its harmful impact on the environment. To reduce the environmental pollution due to lignin, ILs are used to dissolve the lignin and transform it into useful products. Studies found that ILs such as

1-hexyl-3-methylimidazolium methylsulfate ([Hmim][CF_3SO_3]), 1-butyl-3-dimethyl imidazolium methylsulfate ([Bmim][$MeSO_4$]), and 1,3-dimethyl imidazolium methylsulfate ([Mmim][$MeSO_4$]) have the ability to dissolve 20 wt% of lignin (Geniselli da Silva 2021). To improve the lignin conversion, the IL containing 1-ethyl-3-methylimidazolium cation along with the mixture of alkylbenzene sulfonates and xylene sulfonate was used as an anion and showed 93% lignin conversion. In lignin, it is found that hydrogen-bonding basicity of the anions, hydrogen-bonding acidity of the cations, pi–pi interactions, and the hydrophobic interactions between the IL and lignin cause lignin to dissolve in the ILs, and the conversion depends upon the type of IL (Zhang et al. 2019).

7.2.5 Deep Eutectic Solvents

These DESs are another type of ILs and have similar properties with ILs. DESs contain a variety of anionic or cationic parts, whereas ILs contain discrete anions and cations. The physical properties of DESs and ILs are found to be similar, while the chemical properties are significantly different. Due to their large and nonsymmetric ions, DESs reportedly have low melting points. Low melting points are also due to low lattice energy. DESs are prepared by the complexation of quaternary ammonium or imidazolium salts and metal salts. It is also found that a low melting point is due to charge dislocation between the anion in metal salt and the hydrogen donor moiety (Tan et al. 2020). DES obtained from $ZnCl_2$ and quaternary ammonium salt is found to have a low melting point of 23°C–25°C. The general formula of DESs is "Cat+X-zY". Cat+ is ammonium, phosphonium, or sulfonium cation, and X is Lewis Base (A halide ion). Complex anionic species are formed between X and Lewis or Bronsted acid Y, where z refers to the number of Y molecules that are required to interact with the anion.

7.2.5.1 Deep Eutectic Solvents in Biomass Conversion

In recent years, the term "renewable energy" has become familiar due to the various adverse effects of energy production from "nonrenewable sources". Biomass has the most promising renewable feedstock, which is highly abundant. The energy production from biomass required pretreatment for effective energy conversion (Dutta et al. 2014). There are many methods for pretreating the biomass; previously, dilute acids and alkaline hydrolysis were used for the pretreatment. To achieve higher conversion, several other solutions are being investigated. Among all the solutions, DESs are found to be one of the most effective solvents for the pretreatment of biomass. DESs possess similar physiochemical properties to ILs but with less toxicity, better biodegradability, and enzyme compatibility. Previous DESs were used in electrochemical industries, and in recent years, DESs have also found their applications in biodiesel processing, sugar conversion, extraction of several compounds, and so on because of their potential applications (Smith et al. 2014).

7.2.5.2 Dissolution of Biomass in DESs

Lignocellulose is one of the most abundant materials and is made of lignin, cellulose, and hemicellulose. Plants, agricultural wastes, forest residues, and paper waste are the main sources of lignocellulosic biomass. The biomass is used as feedstock for the production of second-generation fuel and various bio-products. To produce the fuel, the biomass is pretreated with certain solvents such as ILs, DESs, and so on. The conversion of biomass discussed is based on the HBD and HBA activities. Various DESs are introduced to find the best solvent for greater conversion.

7.2.5.2.1 Carbohydrate-Based DES

Carbohydrate-based DESs consist of sugar compounds such as glucose, fructose, xylitol, and so on as HBDs. These DESs exhibit a neutral pH when HBD pairs with HBA, such as choline chloride (ChCl). Specific DES ChCl:fructose can dissolve rice straws up to 0.65 wt% (Florindo et al. 2017). When compared to other HBDs, the fructose-based DES showed higher dissolution ability.

Polyalcohol-based DESs: These types of DESs are one of the most investigated DESs due to their better enzyme stability. Glycerol and ethylene glycol are the most popular HBDs, and choline chloride is used as an HBA. It has been discovered that using ethylene glycol as the HBD in polyalcohol-based DESs rather than 1,2- and 1,3-propanediol improves lignin dissolution (Hou et al. 2018). For oil palm feedstock, ethylene glycol DES is more effective than glycerol DES.

Acid-based DESs: These DESs are one of the most investigated DESs and have been found to be effective in biomass conversion. It is also evident that lignin and xylan removal were enhanced by 30% by using acid-based DESs instead of organic acids (Pan et al. 2017). Most acid-based DESs perform well in xylan hydrolysis when pretreated at temperatures above 100°C. These DESs are effective in dissolving lignin and xylose, but cellulose was usually completely retained. Although cellulose is completely retained, these DESs can bring about certain changes in cellulose structure and properties, which have a greater effect on efficiency. Carboxylic acids such as formic acid, acetic acid, and so on are used as HBD and ChCl as HBA. The highest lignin extraction efficiency is obtained when ChCl:formic acid is used as DESs than other DESs, around 62% efficiency is achieved with ChCl:formic acid DESs. Amine- and amide-based DESs: This type of DESs showed moderate dissolution of biomass when compared to acid-based and polyalcohol-based DESs. One of the amine-based DESs is ChCl:urea, which showed the dissolution of lignin and xylan up to 30% and 20%, respectively (Pan et al. 2017). It is found that the better dissolution of biomass is attributed to its basicity and among all the amine-based DESs, monoethanolamine-DESs exhibited the highest conversion of lignin and xylan, which is around 81% and 47%, respectively (Xia et al. 2018). Phenolic compound-based DESs: There are several phenol-based DESs which has HBD as 4-hydroxybenzyl alcohol, catechol, vanillin, and *p*-coumaric acid (PCA). DES ChCl:PCA achieved the highest conversion of lignin and xylan, which is about 61% and 71%, respectively. It is found that the acidic characteristics of the DES contributed to the highest conversion of biomass. ChCl:vanillin can convert 53% and 50% of lignin and xylan, respectively (Kim et al. 2018). Apart from the mentioned HBDs phenol, alpha-naphthol and resorcinol also act as good HBDs (Malaeke et al. 2018).

7.2.6 Supercritical Fluids

Supercritical fluids are the special state of the fluid, which generally exist beyond the "critical temperature and pressure." It is not possible to distinguish between the liquid and gas phases when the fluid attains beyond the critical state. Critical fluids have very low viscosities as gases and high densities as liquids (Basu 2018). Due to this unique nature, SCF cannot be liquefied by lowering the pressure. SCFs have no surface tension, which indicates that there are no interactions between the molecules and that the properties of the fluid are completely different from those of SCFs. SCF fluids are more likely to form at the bases of volcanoes, deep beneath the ocean surface (Sun et al. 2019). SCFs have several applications, and they are used in the extraction of floral essence, the process of making decaffeinated coffee, food industry, polymer industry, nono-systems, and so on. In recent discoveries, SCFs are found to be very useful for the pretreatment of biomass in the process of preparing biofuels. SC-CO_2 is found to be the best solvent for the pretreatment of biomass due to its various properties, such as being non-toxic, non-flammable, very volatile, cheap, and recyclable (Haq et al. 2021).

7.2.6.1 Supercritical Fluids in Biomass Conversion

ILs and DESs are the most commonly used solvents for the pretreatment of biomass feedstock. It is found that SCFs also have characteristics in digesting the biomass and help in the pretreatment of the feedstock. Apart from its use as a pretreatment solvent, it is also used as an extraction solvent to yield the most-valued compounds (Xu et al. 2020). The primary aim of the SCF is to enhance the substrate's accessibility to enzymatic hydrolysis by changing the physical and chemical properties of the substrate. SCFs react with carbohydrates and lignin of biomass and turn that into liquid fuel, char, and so on. This is due to the high temperature and pressures under which the SCF tends to improve its penetration through the pores of the substrate, which dissolves biomass (Morais et al. 2015).

7.2.6.2 Supercritical Water

Supercritical water is one of the most widely used solvents for the pretreatment of lignocellulosic biomass. SCF for the hydrolysis of beet pulp yielded 71% and 61% of C5 and C6 sugars, respectively. It is also evident that the longer time of reaction reduced the sugar yield (Martínez et al. 2018). The addition of a small concentration of H_2SO_4 increased the sugar yield significantly from 19.7% to 35.3% at 380°C and 230 bar (Jeong et al. 2017). Although these are found to have many applications, they also have certain drawbacks. High operational conditions, formation of inhibitors, low residence time, and so forth are the major challenges of the SCW.

7.2.6.3 Supercritical Carbon Dioxide

SC-CO_2 is one of the most widely used pretreatment solvents in biomass conversion. The main advantage of using scCO_2 as a solvent is that it enhances the mass transfer diffusivity and lowers the viscosity of the solvent system (Narayanaswamy et al. 2011). If moisture is present in the biomass, it is converted into carbonic acid, which promotes hemicellulose hydrolysis. Another advantage of $scCO_2$ pretreatment is that no fermentation inhibitor is formed during hydrolysis. The product obtained from the pretreatment can be hydrolyzed without any detoxification or purification.

7.2.7 Cosolvent

ILs and DESs are the advanced emerging pretreatment technologies for dissolving the biomass. These pretreatment techniques have a greater advantage when compared with traditional methods. This is due to various facts such as inhibitor formation, toxicity, ecofriendliness, flammable characteristics, and so on. One of the emerging pretreatment methods that adds to the list of ILs and DESs is "cosolvent pretreatment." Cosolvent-enhanced lignocellulosic fractionation (CELF) is a recently developed advanced pretreatment method that was found to be efficient in dissolving the lignin content when compared with other pretreatment techniques. The miscible mixture of "tetrahydrofuran (THF) and water with the addition of dilute acid" is used in the CELF treatment. THF is a green solvent that is relatively non-toxic and easily miscible with water. Although THF is not easily recovered after use, it is found that it has high solubility of lignin content. In a study by Zhanhui Shen, et al., 2018 it was found that THF–water pretreatment requires high operating conditions for effective conversion. It is found that 84% of the lignin content is removed from maple wood using a THF–water–miscible mixture at temperatures between 160°C and 170°C with the addition of 0.05 M H_2SO_4, and 90% of the lignin in corncob is converted when the system temperature is about 200°C, and a low conversion of 13.5% cellulose is converted. A low concentration of acid is preferred to avoid undesirable products, inhibitor formation, and corrosion. After the CELF pretreatment, a sugar- and ash-free product called CELF lignin is obtained by heating the THF-treated biomass (Shen et al. 2018). In a study by Xianzhi Meng et al., 2019 CELF pretreatment is performed in a specific reactor called the Hastelloy Parr reactor, which is equipped with a double-stacked pitch blade impeller operated at 200 rpm. The pretreatment was operated at 180°C for 20 min with a 4:1 THF/water ratio. The CELF lignin formed is precipitated using hydrolyzate at 80°C for 6 h in a fume hood, and it is left for 15 h to evaporate the remaining THF at room temperature. The precipitated dark lignin is then washed extensively with water to remove the non-lignin impurities. CELF lignin dissolves in several solvents, such as toluene, gasoline, ethanol, diesel, and THF, at room temperature for 24 h. The undissolved lignin is separated by centrifugation, and dissolved lignin is obtained by rotary evaporation under reduced pressure (Meng et al. 2019).

7.2.8 Challenges in Energy Production from Lignocellulose

Every energy production method is associated with some challenges to be focused on. Since the structure of lignocellulose is rigid and strong due to various strong hydrogen bond interactions, it is difficult to achieve high conversion efficiency. Inhibitor formation is another important challenge during the pretreatment steps. Inhibitors are formed in every pretreatment technique that reduces the production of biofuel by restricting the activity of biomass. This leads to a low biomass energy yield. Inhibitors such as furfurals, phenolics, and other aromatic compounds are the major challenges. When the phenolic compounds are formed, it results in the leakage of intercellular components, changes the protein–lipid ratio, and reduces the selective bacterial function (Wang et al. 2017, Wang et al. 2018a).

REFERENCES

de Araújo Padilha, C.E., da Costa Nogueira, C., Oliveira Filho, M.A., de Santana Souza, D.F., de Oliveira, J.A., and dos Santos, E.S., 2020. Valorization of cashew apple bagasse using acetic acid pretreatment: Production of cellulosic ethanol and lignin for their use as sunscreen ingredients. *Process Biochemistry*, 91, 23–33.

Arevalo-Gallegos, A., Ahmad, Z., Asgher, M., Parra-Saldivar, R., and Iqbal, H.M.N., 2017. Lignocellulose: A sustainable material to produce value-added products with a zero waste approach—A review. *International Journal of Biological Macromolecules*, 99, 308–318.

Arumugam, A., Saravanan, M., and Harini, S., 2016. Bioethanol production and optimization by response surface methodology from corn cobs by alkali pretreated. *Journal of Pure and Applied Microbiology*, 10 (1), 547–552.

Auer, L., Lazuka, A., Sillam-Dussès, D., Miambi, E., O'Donohue, M., and Hernandez-Raquet, G., 2017. Uncovering the potential of termite gut microbiome for lignocellulose bioconversion in anaerobic batch bioreactors. *Frontiers in Microbiology*, 8 (DEC), 1–14.

Bajpai, P., 2014. Xylan. *Xylanolytic Enzymes*, pp. 9–18. Massachusetts, United States.

Barati, B., Zafar, F.F., Rupani, P.F., and Wang, S., 2021. Environmental Technology & Innovation Bacterial pretreatment of microalgae and the potential of novel nature hydrolytic sources. *Environmental Technology & Innovation*, 21, 101362.

Basu, P., 2018. Hydrothermal conversion of biomass. *Biomass Gasification, Pyrolysis and Torrefaction: Practical Design and Theory*. Oxford: Elsevier.

Chen, H., 2014. *Biotechnology of lignocellulose: Theory and practice* Springer, Dordrecht.

Chen, H., Liu, J., Chang, X., Chen, D., Xue, Y., Liu, P., Lin, H., and Han, S., 2017. A review on the pretreatment of lignocellulose for high-value chemicals. *Fuel Processing Technology*, 160, 196–206.

Dumond, L., Lam, P.Y., Van Erven, G., Kabel, M., Mounet, F., Grima-Pettenati, J., Tobimatsu, Y., and Hernandez-Raquet, G., 2021. Termite Gut Microbiota Contribution to Wheat Straw Delignification in Anaerobic Bioreactors. *ACS Sustainable Chemistry and Engineering*, 9 (5), 2191–2202.

Dutta, K., Daverey, A., and Lin, J.G., 2014. Evolution retrospective for alternative fuels: First to fourth generation. *Renewable Energy*, 69, 114–122.

Florindo, C., Oliveira, M.M., Branco, L.C., and Marrucho, I.M., 2017. Carbohydrates-based deep eutectic solvents: Thermophysical properties and rice straw dissolution. *Journal of Molecular Liquids*, 247, 441–447.

Galbe, M., Wallberg, O., and Zacchi, G., 2019. Techno-economic aspects of ethanol production from lignocellulosic agricultural crops and residues. In Murray Moo-Young (Ed.) *Comprehensive Biotechnology*, Third Edition. Elsevier, pp. 615–628.

Gares, M., Hiligsmann, S., and Kacem Chaouche, N., 2020. Lignocellulosic biomass and industrial bioprocesses for the production of second generation bio-ethanol, does it have a future in Algeria? *SN Applied Sciences*, 2 (10), 1–19.

Geniselli da Silva, V., 2021. Laccases and ionic liquids as an alternative method for lignin depolymerization: A review. *Bioresource Technology Reports*, 16 (July), 100824.

Ghasemian, M., Zilouei, H., and Asadinezhad, A., 2016. Enhanced biogas and biohydrogen production from cotton plant wastes using alkaline pretreatment. *Energy and Fuels*, 30 (12), 10484–10493.

Haldar, D. and Purkait, M.K., 2021. A review on the environment-friendly emerging techniques for pretreatment of lignocellulosic biomass: Mechanistic insight and advancements. *Chemosphere*, 264, 128523.

Haq, I.U., Qaisar, K., Nawaz, A., Akram, F., Mukhtar, H., Zohu, X., Xu, Y., Mumtaz, M.W., Rashid, U., Ghani, W.A.W.A.K., and Choong, T.S.Y., 2021. Advances in valorization of lignocellulosic biomass towards energy generation. *Catalysts*, 11 (3), 309.

Heinze, T., 2016. Cellulose chemistry and properties: Fibers, nanocelluloses and advanced materials. *Advances in Polymer Science* 271, 1–52.

Hou, X.D., Lin, K.P., Li, A.L., Yang, L.M., and Fu, M.H., 2018. Effect of constituents molar ratios of deep eutectic solvents on rice straw fractionation efficiency and the micro-mechanism investigation. *Industrial Crops and Products*, 120 (December 2017), 322–329.

Isikgor, F.H. and Becer, C.R., 2015. Lignocellulosic biomass: A sustainable platform for the production of bio-based chemicals and polymers. *Polymer Chemistry*, 6 (25), 4497–4559.

Isosaari, P., Srivastava, V., and Sillanpää, M., 2019. Science of the Total Environment Ionic liquid-based water treatment technologies for organic pollutants : Current status and future prospects of ionic liquid mediated technologies. *Science of the Total Environment*, 690, 604–619.

Jeong, H., Park, Y.C., Seong, Y.J., and Lee, S.M., 2017. Sugar and ethanol production from woody biomass via supercritical water hydrolysis in a continuous pilot-scale system using acid catalyst. *Bioresource Technology*, 245, 351–357.

Jia, H., Li, Q., Bayaguud, A., She, S., Huang, Y., Chen, K., and Wei, Y., 2017. Tosylation of alcohols: An effective strategy for the functional group transformation of organic derivatives of polyoxometalates. *Scientific Reports*, 7 (1), 2–10.

Jonge, V.N. De and Elliott, M., 2002. Causes, historical development, effects and future challenges of a common environmental problem: eutrophication. In Orive, E., Elliott, M., de Jonge, V.N. (Eds.) *Nutrients and eutrophication in estuaries and coastal waters*. Sciences New York, Springer 5, 1–19.

Jönsson, L.J. and Martín, C., 2016. Pretreatment of lignocellulose: Formation of inhibitory by-products and strategies for minimizing their effects. *Bioresource Technology*, 199, 103–112.

Kainthola, J., Podder, A., Fechner, M., and Goel, R., 2021. Bioresource Technology An overview of fungal pretreatment processes for anaerobic digestion: Applications, bottlenecks and future needs. *Bioresource Technology*, 321 (September 2020), 124397.

Karimi, K., 2015. *Biofuel and Biorefinery Technologies (Lignocellulose-Based Bioproducts)*. Springer, Berlin.

Karimi, K. and Taherzadeh, M.J., 2016. A critical review of analytical methods in pretreatment of lignocelluloses: Composition, imaging, and crystallinity. *Bioresource Technology*, 200, 1008–1018.

Khare, V., Khare, C., Nema, S., and Baredar, P., 2019. Introduction to Energy Sources. In Murray Moo-Young (Ed.) *Tidal Energy Systems*, Academic Press.

Khatri, V., Meddeb-Mouelhi, F., Adjallé, K., Barnabé, S., and Beauregard, M., 2018. Determination of optimal biomass pretreatment strategies for biofuel production: Investigation of relationships between surface-exposed polysaccharides and their enzymatic conversion using carbohydrate-binding modules. *Biotechnology for Biofuels*, 11 (1), 1–16.

Kim, K.H., Dutta, T., Sun, J., Simmons, B., and Singh, S., 2018. Biomass pretreatment using deep eutectic solvents from lignin derived phenols. *Green Chemistry*, 20 (4), 809–815.

Knott, B.C., Crowley, M.F., Himmel, M.E., Ståhlberg, J., and Beckham, G.T., 2014. Carbohydrate-protein interactions that drive processive polysaccharide translocation in enzymes revealed from a computational study of cellobiohydrolase processivity. *Journal of the American Chemical Society*, 136 (24), 8810–8819.

Kögel-Knabner, I. and Amelung, W., 2013. Dynamics, chemistry, and preservation of organic matter in soils. In Turekian, K. and Holland, H. (Eds.), *Treatise on Geochemistry*, Second Edition, Elsevier, Oxford, 157–215.

Liu, K., Wang, Z., Shi, L., Jungsuttiwong, S., and Yuan, S., 2021. Ionic liquids for high performance lithium metal batteries. *Journal of Energy Chemistry*, 59, 320–333.

Liu, Y., Guo, B., Xia, Q., Meng, J., Chen, W., Liu, S., Wang, Q., Liu, Y., Li, J., and Yu, H., 2017. Efficient cleavage of strong hydrogen bonds in cotton by deep eutectic solvents and facile fabrication of cellulose nanocrystals in high yields. *ACS Sustainable Chemistry and Engineering*, 5 (9), 7623–7631.

Liu, Y., Nie, Y., Lu, X., Zhang, X., He, H., Pan, F., Zhou, L., Liu, X., Ji, X., and Zhang, S., 2019. Cascade utilization of lignocellulosic biomass to high-value products. *Green Chemistry*, 21 (13), 3499–3535.

Liu, Z., Si, B., Li, J., He, J., Zhang, C., Lu, Y., Zhang, Y., and Xing, X.H., 2018. Bioprocess engineering for biohythane production from low-grade waste biomass: Technical challenges towards scale up. *Current Opinion in Biotechnology*, 50, 25–31.

Ma, C., Liu, C., Lu, X., and Ji, X., 2018. Techno-economic analysis and performance comparison of aqueous deep eutectic solvent and other physical absorbents for biogas upgrading. *Applied Energy*, 225 (January), 437–447.

Ma, J., Yue, H., Li, H., Zhang, J., Zhang, Y., Wang, X., Gong, S., and Liu, G.Q., 2021. Selective delignification of poplar wood with a newly isolated white-rot basidiomycete Peniophora incarnata T-7 by submerged fermentation to enhance saccharification. *Biotechnology for Biofuels*, 14 (1), 1–16.

Mahmood, H., Moniruzzaman, M., Iqbal, T., and Khan, M.J., 2019. Recent advances in the pretreatment of lignocellulosic biomass for biofuels and value-added products. *Current Opinion in Green and Sustainable Chemistry*, 20, 18–24.

Malaeke, H., Housaindokht, M.R., Monhemi, H., and Izadyar, M., 2018. Deep eutectic solvent as an efficient molecular liquid for lignin solubilization and wood delignification. *Journal of Molecular Liquids*, 263, 193–199.

Martínez, C.M., Cantero, D.A., and Cocero, M.J., 2018. Production of saccharides from sugar beet pulp by ultrafast hydrolysis in supercritical water. *Journal of Cleaner Production*, 204, 888–895.

Millati, R., Wikandari, R., Ariyanto, T., Putri, R.U., and Taherzadeh, M.J., 2020. Pretreatment technologies for anaerobic digestion of lignocelluloses and toxic feedstocks. *Bioresource Technology*, 304, 122998. (October 2019).

Morais, A.R.C., Da Costa Lopes, A.M., and Bogel-Łukasik, R., 2015. Carbon dioxide in biomass processing: Contributions to the green biorefinery concept. *Chemical Reviews*, 115 (1), 3–27.

Moriarty, P. and Honnery, D., 2017. Sustainable energy resources: Prospects and policy. In Mohammad G. Rasul, Abul Kalam Azad, Subhash C. Sharma (Eds.), *Clean Energy for Sustainable Development: Comparisons and Contrasts of New Approaches*, pp. 3–27. London, UK.

Meng, X., Parikh, A., Seemala, B., Kumar, R., Pu, Y., Wyman, C.E., Cai, C.M., and Ragauskas, A.J., 2019. Characterization of fractional cuts of co-solvent enhanced lignocellulosic fractionation lignin isolated by sequential precipitation. *Bioresource Technology*, 272 (September 2018), 202–208.

Marta G. Plaza, Covadonga Pevida, 2019. Current status of CO_2 capture from coal facilities. In Isabel Suárez-Ruiz, Maria Antonia Diez, Fernando Rubiera (Eds.) *New Trends in Coal Conversion*, Woodhead Publishing, 31–58.

Narayanaswamy, N., Faik, A., Goetz, D.J., and Gu, T., 2011. Supercritical carbon dioxide pretreatment of corn stover and switchgrass for lignocellulosic ethanol production. *Bioresource Technology*, 102 (13), 6995–7000.

Owusu, P.A. and Asumadu-Sarkodie, S., 2016. A review of renewable energy sources, sustainability issues and climate change mitigation. *Cogent Engineering*, 3 (1), 1–14.

Pan, M., Zhao, G., Ding, C., Wu, B., Lian, Z., and Lian, H., 2017. Physicochemical transformation of rice straw after pretreatment with a deep eutectic solvent of choline chloride/urea. *Carbohydrate Polymers*, 176 (May), 307–314.

Paudel, S.R., Banjara, S.P., Choi, O.K., Park, K.Y., Kim, Y.M., and Lee, J.W., 2017. Pretreatment of agricultural biomass for anaerobic digestion: Current state and challenges. *Bioresource Technology*, 245 (September), 1194–1205.

de Paula, R.G., Antoniêto, A.C.C., Ribeiro, L.F.C., Srivastava, N., O'Donovan, A., Mishra, P.K., Gupta, V.K., and Silva, R.N., 2019. Engineered microbial host selection for value-added bioproducts from lignocellulose. *Biotechnology Advances*, 37 (6), 107347.

Rastogi, M. and Shrivastava, S., 2017. Recent advances in second generation bioethanol production: An insight to pretreatment, saccharification and fermentation processes. *Renewable and Sustainable Energy Reviews*, 80 (May), 330–340.

Reading, F., 2020. Further Reading. *Literature Compass*, 998–1007.

Rocha, M.V.P., Rodrigues, T.H.S., de Albuquerque, T.L., Gonçalves, L.R.B., and de Macedo, G.R., 2014. Evaluation of dilute acid pretreatment on cashew apple bagasse for ethanol and xylitol production. *Chemical Engineering Journal*, 243, 234–243.

Romano, N., 2018. Alternative and new sources of feedstuffs. In Nunes, C. S.; Kumar, V. (Eds.), *Enzymes in Human and Animal Nutrition: Principles and Perspectives*, pp. 381–401, London.

Roy, R., Rahman, M.S., and Raynie, D.E., 2020. Recent advances of greener pretreatment technologies of lignocellulose. *Current Research in Green and Sustainable Chemistry*, 3 (August), 100035.

Saini, J.K., Saini, R., and Tewari, L., 2015. Lignocellulosic agriculture wastes as biomass feedstocks for second-generation bioethanol production: Concepts and recent developments. *3 Biotech*, 5 (4), 337–353.

Shen, Z., Zhang, K., Si, M., Liu, M., Zhuo, S., Liu, D., Ren, L., Yan, X., and Shi, Y., 2018. Synergy of lignocelluloses pretreatment by sodium carbonate and bacterium to enhance enzymatic hydrolysis of rice straw. *Bioresource Technology*, 249 (August 2017), 154–160.

Smith, E.L., Abbott, A.P., and Ryder, K.S., 2014. Deep Eutectic Solvents (DESs) and Their Applications. *Chemical Reviews*, 114 (21), 11060–11082.

Soares, B., Tavares, D.J.P., Amaral, J.L., Silvestre, A.J.D., Freire, C.S.R., and Coutinho, J.A.P., 2017. Enhanced Solubility of Lignin Monomeric Model Compounds and Technical Lignins in Aqueous Solutions of Deep Eutectic Solvents. *ACS Sustainable Chemistry and Engineering*, 5 (5), 4056–4065.

Sun, Z., Fan, Q., Zhang, M., Liu, S., Tao, H., and Texter, J., 2019. Supercritical Fluid-Facilitated Exfoliation and Processing of 2D Materials. *Advanced Science*, 6 (18), 1901084.

Tan, Y.T., Chua, A.S.M., and Ngoh, G.C., 2020. Deep eutectic solvent for lignocellulosic biomass fractionation and the subsequent conversion to bio-based products – A review. *Bioresource Technology*, 297 (October 2019), 122522.

Tanpichai, S., Witayakran, S., and Boonmahitthisud, A., 2019. Study on structural and thermal properties of cellulose microfibers isolated from pineapple leaves using steam explosion. *Journal of Environmental Chemical Engineering*, 7 (1), 102836.

Verma, M., Sharma, S., and Prasad, R., 2009. International Biodeterioration & Biodegradation Biological alternatives for termite control : A review. *International Biodeterioration & Biodegradation*, 63 (8), 959–972.

Wei, L., Li, K., Ma, Y., and Hou, X., 2012. Dissolving lignocellulosic biomass in a 1-butyl-3-methylimidazolium chloride-water mixture. *Industrial Crops and Products*, 37 (1), 227–234.

Wu, D., Wei, Z., Mohamed, T.A., Zheng, G., Qu, F., Wang, F., Zhao, Y., and Song, C., 2022. Lignocellulose biomass bioconversion during composting: Mechanism of action of lignocellulase, pretreatment methods and future perspectives. *Chemosphere*, 286 (July 2021), 131635.

Xia, Q., Liu, Y., Meng, J., Cheng, W., Chen, W., Liu, S., Liu, Y., Li, J., and Yu, H., 2018. Multiple hydrogen bond coordination in three-constituent deep eutectic solvents enhances lignin fractionation from biomass. *Green Chemistry*, 20 (12), 2711–2721.

Xia, Z., Li, J., Zhang, J., Zhang, X., Zheng, X., and Zhang, J., 2020. Processing and valorization of cellulose, lignin and lignocellulose using ionic liquids. *Journal of Bioresources and Bioproducts*, 5 (2), 79–95.

Xu, H., Peng, J., Kong, Y., Liu, Y., Su, Z., Li, B., Song, X., Liu, S., and Tian, W., 2020. Key process parameters for deep eutectic solvents pretreatment of lignocellulosic biomass materials: A review. *Bioresource Technology*, 310 (March), 123416.

Yang, S., 2007. Chapter 1. Bioprocessing – from Biotechnology to Biorefinery, In Shang-Tian Yang (Ed.) *Bioprocessing for Value-Added Products from Renewable Resources*, Elsevier, 1–24.

Zhang, L., Zhao, D., Feng, M., He, B., Chen, X., Wei, L., Zhai, S.R., An, Q. Da, and Sun, J., 2019. Hydrogen Bond Promoted Lignin Solubilization and Electrospinning in Low Cost Protic Ionic Liquids. *ACS Sustainable Chemistry and Engineering*, 7 (22), 18593–18602.

Zhanhui S., Kejing Z., Mengying Si, Mingren Liu, Shengnan Zhuo, Dan Liu, Lili Ren, Xu Yan, Yan Shi, Synergy of lignocelluloses pretreatment by sodium carbonate and bacterium to enhance enzymatic hydrolysis of rice straw, *Bioresource Technology*, 249, 2018, 154-160. https://doi.org/10.1016/j.biortech.2017.10.008.

Zhao, L., Sun, Z.F., Zhang, C.C., Nan, J., Ren, N.Q., Lee, D.J., and Chen, C., 2022. Advances in pretreatment of lignocellulosic biomass for bioenergy production: Challenges and perspectives. *Bioresource Technology*, 343 (August 2021), 126123.

8 Ionic Liquids as Solvents for Separation of Biobutanol

Kalyani A. Motghare, Diwakar Shende, and Kailas L. Wasewar
Visvesvaraya National Institute of Technology (VNIT)

CONTENTS

DOI: 10.1201/9781003203452-8

8.1 INTRODUCTION

8.1.1 Biofuels and Biochemicals (Global Energy Scenarios)

As a result of reducing oil reserves and fossil fuel reserves, global energy demand is continuously rising. The use of petroleum and other liquid fuels has been projected to increase from 95 million barrels per day (b/d) in 2015 to 104 million b/d in 2030 and 113 million b/d by 2040 (US IEA 2017). Figure 8.1 shows the primary energy consumption worldwide.

Fossil fuel consumption will reach its peak in India by 2035 (Figure 8.2). By 2030, India will surpass China as the largest growing market for energy. The rapid depletion of natural resources, and the rising price of conventional fuels over the past few decades have to led to increased concern over replacing conventional fuels with alternative biofuels in the developing world. A biofuel is a renewable energy source produced from organic matter. Adding biofuels to gasoline can reduce the amount of carbon dioxide (CO_2) released into the atmosphere when sources are burned. Petroleum resources are becoming scarcer all over the world and the economy of the nation benefits from it. Therefore, there is a need to replace traditional fuel stock with alternative biofuels of the second generation. Global energy demands are increasing in the industrial age. According to reports, 80% of the primary energy supply in the world is supplied by renewable sources. Fossil fuels contribute to global warming. As of now, natural gas, petroleum, and coal make up more than 95%. Thirty-three percent of the world's energy demand, corresponding to 34%, 24% and 19%, respectively, is generated from oil, gas, and coal (Huber, Corma 2007).

Overdependence on petroleum-based fuels has a negative impact on the environment. Global warming is believed to result from the emission of greenhouse gases from these fuels (Trenberth et al. 2003). A study published recently found that Antarctica is losing 160 billion tons of ice per year to the sea due to global warming

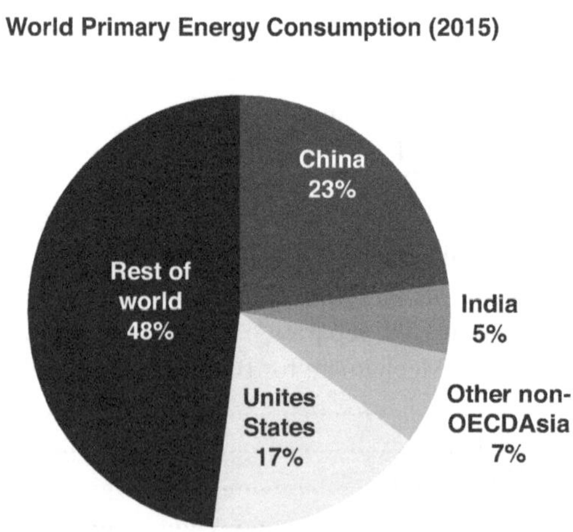

FIGURE 8.1 World primary energy consumption 2015 (US IEA 2017).

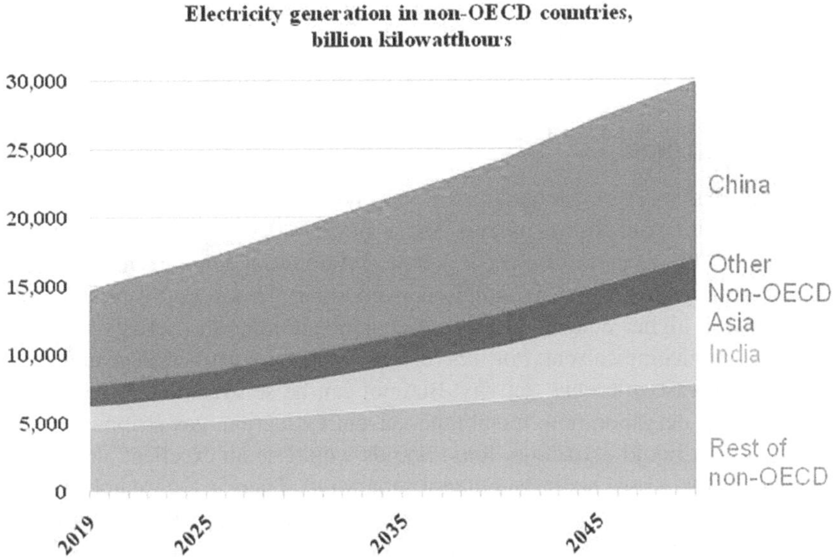

FIGURE 8.2 Electricity generation in non-OECD countries. (Energy Information Administration, International Energy Outlook 2019.)

and greenhouse gas emissions. McMillan and colleagues (2014) report that this amount is twice as much as during the last survey, between the years 2005 and 2010. Using fossil fuels to produce greenhouse gases, the study shows the negative effect of these gases on the world's environment. Furthermore, fossil fuel reservoirs deplete over time and eventually run out (Aftabuzzaman and Mazloumi 2011). Current circumstances demand that renewable resources be used to produce fuel and chemicals.

A biofuel is a fuel produced directly or indirectly from organic matter, such as biomass or animal waste. As long as they do not exceed the quantities utilised by photosynthesis, they are carbon neutral. Furthermore, biochemicals are chemicals that are directly or indirectly derived from biomass. Public and scientific interest in biofuels has grown significantly over the last few years due to uncertainties related to oil prices and greenhouse gas emissions (Demirbas 2009). Many biofuels, including biogas, biodiesel, bioethanol, have been researched or used in the energy sector. FAME-type (fatty acid methyl ester) biodiesel is one of the various biofuels used in the transportation sector. Biodiesel is considered a possible replacement for conventional diesel. The results of many studies have shown that biodiesel has similar properties to conventional diesel fuel and can be blended with it in any proportion.

Additionally, biodiesel and its blends produce almost as much power as conventional diesel, and the amount of carbon monoxide (CO), total hydrocarbons (THD), carbon dioxide (CO_2), and soot produced has reduced. The oxygen content in biodiesel has been implicated as the reason for the complete and efficient combustion of biodiesel (Fang et al. 2009; Qin et al. 2007b; Qin et al. 2007a). A well-known biofuel that is widely used is bio-alcohol. Renewable fuels, such as bioethanol and biobutanol, can be produced by fermentation of sugars derived from biomass, e.g. corn, sugar beets, sugar cane, and agricultural residue (Ezeji et al. 2007; Hansen et al. 2005).

8.1.2 Biobutanol

Biobutanol shares several characteristics with petroleum fuels. There are two ways to make butanol: a chemical process and a biochemical process. Biobutanol is a by-product of the fermentation process called Acetone-Butanol-Ethanol (ABE), in which anaerobic Clostridium *Sp.* are used as microbes. However, this process is limited because a higher concentration of biobutanol inhibits the activity of bacteria involved in producing solvent. For this reason, biobutanol extraction should be continuous to increase microbial activity. Butanol can be separated from fermentation broth by using pervaporation, membrane solvent extraction, gas stripping, adsorption, and liquid-liquid extraction. Ionic liquids would be an excellent alternative to conventional toxic solvents in biobutanol separation. Therefore, reșearch should be directed toward applications of room-temperature ionic liquids (RTILs) for biobutanol recovery. There are four isomers of butanol which include two stereoisomers of sec-butanol, n-butanol, tert-butanol and isobutanol. As per the information given on Alternative Fuels Records Center (AFDC), biobutanol has ASTM D7862 gas quality standard and can be used as a transportation gas and additionally as an additive in vehicles as much as 5% with traditional fuels. which according to smooth Air Act provisions reduces greenhouse gas emissions. Wide applications of butanol are shown in Figure 8.3.

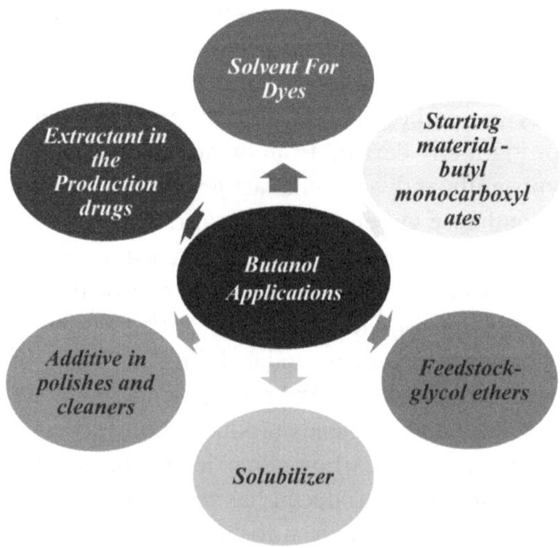

FIGURE 8.3 Applications of butanol.

TABLE 8.1
Characteristics of Butanol over Other Fuels (Freeman et al. 1988; Dean et al. 1992)

Fuel Type	Heating Value (mJ/dm^3)	Octane Number	Cetane Number	Evaporation Heat (mJ/kg)
Butanol	29.2	96	25	29.2
Gasoline	32	80–99	0–10	0.36
Methanol	16	111	3	16
Ethanol	19. 6	108	8	0.92

8.1.3 Comparison of Butanol over Other Fuels

In the global biofuel market, biobutanol has the potential to replace conventional fuels such as gasoline, ethanol, petrol and diesel. The most commonly used biofuel is ethanol. Biobutanol has superior properties over ethanol and gasoline, making it a potential biofuel for transportation use. Physicochemical characteristics are similar to gasoline and ethanol. Biobutanol has a high energy content (29.2 MJ/dm^3), non-hygroscopic, efficient blending properties, and less corrosive than conventional fuels. The solubility of butanol is about 7% by weight (Davis et al. 2008). The characteristics of butanol fuels (Freeman et al. 1988; Dean 1990) over other fuels are given in Table 8.1.

8.2 PRODUCTION APPROACHES FOR BUTANOL

Butanol (butyl alcohol) can be produced via two routes: biochemical and petrochemical. Oxo synthesis (Bohnet 2003) and aldol condensation are the two methods that come under petrochemical routes. It includes a chemical reaction between propylene and carbon monoxide, hydrogen. Rhodium or cobalt as a catalyst can be used. The mixture of n-butyraldehyde and iso-butyraldehyde is obtained is to be hydrogenated to the resultant n-butanol and isobutyl alcohols (Park et al. 1996). In 1913, Jones developed a bacterial fermentation process for producing butanol (Jones and Woods, 1986). Acetone, butanol and ethanol (ABE) are the key products (Qureshi and Maddox 1995) of ABE fermentation process using anaerobic *Clostridium* bacteria, which is a biochemical route for the production of biobutanol and the product is to be formed in the ratio of 3:6:1. 1-butanol (or simply butanol) is 4-carbon straight-chain alcohol with the chemical formula of C_4H_9OH. Butanol is an important industrial chemical, which can be used as a solvent for paints, dyes, varnishes, coating etc. It is also a precursor or intermediate for the chemical synthesis of many plastics and chemicals e.g. hydraulic fluids and safety glass (Jin et al. 2011). In addition, butanol is a very promising fuel with desirable fuel properties (Freeman et al. 1988).

8.2.1 Chemical Synthesis

Oxo synthesis, Reppe synthesis and crotonaldehyde are the most traditional process for preparing butanol (Hahn et al. 2012). Maximum butanol production is done by

Oxo synthesis. Following are the chemical reactions that result in the formation of butanol.

8.2.1.1 Oxo Synthesis

$$CH_3CH = CH_2 + H_2 + CO \rightarrow CH_3(CH_2)_2 CHO + (CH_3)_2 CHCHO \quad (8.1)$$

Equation 8.1 takes place in the presence of a catalyst, which produces a mixture of n- and i-butyraldehyde. Catalysts used in this process include cobalt, rhodium, and ruthenium.

$$CH_3(CH_2)_2 CHO + H_2 \rightarrow CH_3(CH_2)_3 OH \quad (8.2)$$

Hydrogenation of the butyraldehydes yields isobutanol (Equation 8.2) and n-butanol (Equation 8.3).

$$(CH_3)_2 CHCHO + H_2 \rightarrow (CH_3)_2 CHCH_2OH \quad (8.3)$$

8.2.1.2 Reppe Synthesis

In the Reppe synthesis (Eq. 8.4), water is used as a catalyst to react with propylene and carbon monoxide in the presence of tertiary ammonium salts of polynuclear iron carbonyl hydrides.

Carbon dioxide is produced besides *n*-butanol.

$$CH_3CH = CH_2 + 2H_2O + 3CO \rightarrow CH_3(CH_2)_3 OH + 2CO_2 \quad (8.4)$$

8.2.1.3 Crotonaldehyde Hydrogenation

Before the Oxo process was developed, chloronaldehyde hydrogenation was widely used.

$$CH_3CHO \rightarrow CH_3CH(OH)CH_2CHO \quad (8.5)$$

$$CH_3CH(OH)CH_2CHO \rightarrow CH_3CH = CHCHO + H_2O \quad (8.6)$$

$$CH_3CH = CHCHO + 2H_2 \rightarrow CH_3(CH_2)_3 OH \quad (8.7)$$

Aldolisation of acetaldehyde leads to the production of acetaldehyde, which produces crotonaldehyde, dehydrate. The crotonaldehyde was obtained from this process (Eq. 8.7). N-Butanol is formed by hydrogenation. Among the most common catalysts, chromium, copper, and nickel are the solid, gaseous and liquid phases, respectively.

8.2.2 Fermentation

The most common fermentation methods are batch, fed-batch, and continuous fermentations. Feeding the substrate according to the fermentation strategy, in batch fermentation, all substrates and nutrients are added into the fermentor at the beginning of the fermentation, while a fed-batch mode is started with a small volume of substrates and more sugars are added later. Substrates are added constantly to the fermentor in continuous fermentation, thereby keeping the fermentation broth volume constant (Qureshi & Blaschek 2006). The reaction rate is usually limited by substrate quantity. Hence, better fermentation yields can be obtained by the feed control, whereas the formation of products is more regulated and avoiding the production of undesirable intermediates and by-products is easier. Besides the substrate concentration, other important variables to be controlled during the fermentation include the oxygen level, pH, temperature, and concentration of nutrients and inhibitors. As both ethanol and butanol are inhibitors for the microorganisms producing them, combining fermentation and separation techniques as in situ product removal (ISPR) is favourable (Maddox et al. 2000; Qureshi & Blaschek 2006).

The reactor types include e.g. a batch reactor (BR), a continuous stirred tank reactor (CSTR), a packed bed reactor (PBR), and a fluidised bed reactor (FBR) (Qureshi et al. 2005) for the production of biobutanol. Further, hydrolysis and fermentation can be performed either as separate steps in different reactors, known as separate hydrolysis and fermentation (SHF), or as combined operation i.e. simultaneous saccharification and fermentation (SSF). In addition, a consolidated bioprocessing (CBP) method is under development, aiming to convert lignocelluloses to products in one single process step. Further, researchers have also developed more advanced fermentation strategies, including using concentrated sugar solutions, suspended cells, cell immobilisation with or without a cell recycling, and modification of cells by metabolic engineering (Qureshi & Ezeji 2008).

Apart from conventional distillation, several studies have focused on techniques for real-time ABE recovery during fermentation due to the butanol toxicity to the culture. These techniques include vacuum recovery, adsorption, gas stripping, steam stripping, freeze crystallisation, liquid-liquid extraction, perstraction (membrane extraction), reverse osmosis, membrane distillation, sweeping gas pervaporation, thermopervaporation, vacuum pervaporation and hybrid processes in which distillation-pervaporation and distillation-vapour permeation are used (Ezeji et al. 2003; Vane 2005; Vane 2008; Mariano et al. 2013; Hecke et al. 2013; Lin et al. 2013; Abdehagh et al. 2015; Kujawska et al. 2015).

These non-conventional methods for the recovery of ABE solvents are being developed to increase butanol concentration and productivity and reduce energy and butanol production costs. Some *situ* recovery methods, including membrane-based systems, such as pervaporation, perstraction, reverse osmosis and adsorption, might not be good recovery options for fibrous LBHs due to potential membrane fouling. However, gas stripping, liquid-liquid extraction, and vacuum-assisted stripping appear compatible with LBH. Nonetheless, membrane-based systems, molecular sieve processes included, may dehydrate ABE after recovery using vacuum, gas stripping or liquid-liquid extraction. The energy requirement to recover butanol (and ethanol) from fermentation broth is influenced by their initial concentration in the broth, expected concentration in the distillate, and the type of the recovery systems. Detail reviews of ABE fermentation and in situ recovery methods can be found elsewhere (Groot & Luyben 1986; Dürre 1998; Ezeji et al. 2004; Vane 2005; Vane 2008; Kujawska et al. 2015).

8.2.2.1 Acetone, Butanol and Ethanol (ABE) Fermentation

The ABE fermentation process using *Clostridium acetobutylicum* or *Clostridium-beijerinckii* strains is characterised by two phases. In the first phase, sugars, raw material for fermentation, are converted to acetic and butyric acid, decreasing the pH value of the culture. This phase is known as 'acidogenic phase'. In the second phase, known as the 'solventogenic phase', sugars and some of the acids are converted into solvents i.e. acetone, butanol and ethanol accompanied by an increase in pH. Sometimes, an excess production of acid takes place in the acidogenic phase without the microorganisms switching to the solventogenic phase. This phenomenon is known as 'acid crash'. Acid crash during ABE fermentation lower the production of solvents and instead butyric and acetic acids are produced in larger quantities (Maddox et al. 2000).

8.3 SEPARATION OF VALUABLE BIOCHEMICALS

The decision concerning the best method to separate the compounds of interest from the raw material is dependent on several aspects, such as the characteristics of the target extracts and raw material (physical-chemical properties), available technology, required purity, electivity, stability and, more importantly here, the greenness of the whole process. As can be seen in Figure 8.5, the most cited techniques in these research papers were based on solvent/maceration (25% of the total), microwave (19%), ultrasonication (14.7%) and supercritical fluid processing (13%), followed by methods using ionic liquids (7%), enzymatic and subcritical fluid treatment (6%), as well as the association of two or more techniques.

According to the literature, the most widespread approaches for separating natural products from a number of matrices are based on liquid-liquid or solid-liquid extraction (LLE and SLE). Several greener alternatives have been proposed by replacing toxic or non-renewable organic solvents and the extraction times. In some cases, solid-phase extractions (SPE) were also carried out and decreased both the amount of solvent and the number of extraction cycles, offering high enrichment factors. The mass transfer enhancement for SLE has mainly been studied and applied, contributing to technology innovation, process intensification and integration, and energy saving, especially important for microwave, ultrasound, and high-pressure processing, for instance. An overview of these techniques and related examples will be discussed in this section. Main green and sustainable techniques used to separate products from waste are given in Figure 8.4.

8.3.1 Biobutanol Separation

Distinctive partition methods can recuperate this green biofuel from ageing stock to improve biobutanol creation. Nonetheless, centralisation of biobutanol (ordinarily > 10 g/L) (Ha et al. 2010) acquired in a maturation stock will be liable for restraint of microbial development and influences its yield. For detachment of biobutanol, refining course isn't appropriate as it structures butanol-water azeotrope and requires

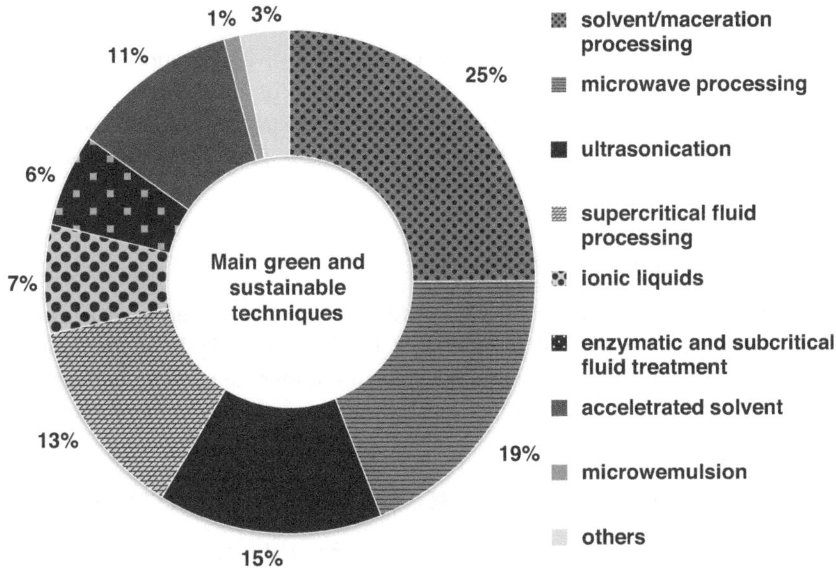

FIGURE 8.4 Main green and sustainable techniques used to separate products from waste. (ISIS Web of knowledge. https://link.springer.com/article/10.1007/s11192-017-2622-5)

more energy. There are distinctive detachment methods such as fluid extraction, adsorption, layer dissolvable, gas stripping, pervaporation which are reasonable monetary cycles contrasted with refining (Abdehagh et al. 2014; Kujawska 2015) for partition of butanol from ageing stock.

8.3.2 Adsorption

Activated carbon (AC) F-400 was the best adsorbent for butanol recovery after Abdehagh et al. (2015) compared six different types of adsorbents Using XD-41, H-511, and KA-I resin, Lin et al. (2012) performed selective separation of butanol and found the maximum recovery of butanol with KA-I resin. However, low adsorption capacity can make the process more economically efficient at industrial scales (Liu et al. 2013).

8.3.3 Gas Stripping

Using nitrogen gas stripping, it is possible to recover butanol from fermentation broth (Lodi & Pellegrini 2016). By implementing antifoam inside fermentation broth, inhibition rates to microbes can be reduced. This method observes maximum separation efficiency towards butanol compared to other products (Lu et al. 2012) and reduces operational cost.

8.3.4 Pervaporation

Separation by pervaporation is also known as pervaporative separation. Pervaporation is a method for butanol separation in which mixtures of components are recovered from fermentation broth through partial vaporization using a nonporous or porous membrane. Using a polydimethylsiloxane/ceramic composite membrane (Liu et al. 2013) separated butanol-aqueous solution.

8.3.5 Liquid-Liquid Extraction

In liquid-liquid extraction, hydrophobic solvents are to be used as extractants to separate biobutanol from the fermentation broth. Due to hydrophobic character, biobutanol is selectively extracted into a solvent but there are a few challenges with liquid-liquid extraction, that is sometimes extractants are toxic to the microbes.

8.4 GREEN METHODS FOR THE SEPARATION OF BUTANOL

A variety of techniques have been proposed to separate biobutanol from the fermentation broth, but the use of liquid extracting agent would be a cost-effective method. Ionic liquids can be used as a novel extractant and could be a prospective solution to replace conventional volatile, toxic solvents. Nowadays, more research interest is emphasised on these novel ionic liquids as the designer green solvents, nonvolatile and environment-friendly for the separation industry. Recovery of biobutanol by ionic liquids could be utilised in gasoline-driven combustion systems because of its excellent blending properties with fewer emissions. Therefore, ionic liquids, which recently gained some notable recognition in science and industry, may turn out to be highly appropriate for the recovery of biobutanol. Ammonium and Phosphonium-based ionic liquids (Cascon et al. 2011) and imidazolium-based ionic liquids (ILs) (Kubizek et al. 2013; Garcia-Chavez et al. 2012) have been reported in the literature and applied as an extractant for separation of butanol from the fermentation broth and this method was found to have maximum separation efficiency compared to other conventional processes.

8.4.1 Ionic Liquids: A Brief History

The term "ionic liquids" usually refers to salts of ionic ions with melting points below 100°C. There are many interesting systems whose melting point is near or below room temperature. Several key scientists have discussed the history of ionic liquids elsewhere, but it is worth recalling briefly here. Walden described in 1914 the physical properties of ethyl ammonium nitrate (mp: 12°C–14°C) produced by the reaction of ethylamine with concentrated nitric acid. Few structures of ionic liquid as cations and anions are given as follows (Figure 8.5).

The next half-century saw sporadic reports of ionic liquids as media for electrochemical studies and, less commonly, as solvents for organic reactions. This work involved eutectic mixtures of chloroaluminate-based salts such as $AlCl_3$–NaCl and pyridinium hydrochloride. In the 1980s, this period also saw the first use of ionic liquids as reaction media for organic synthesis, and, in 1990, for biphasic catalysis.

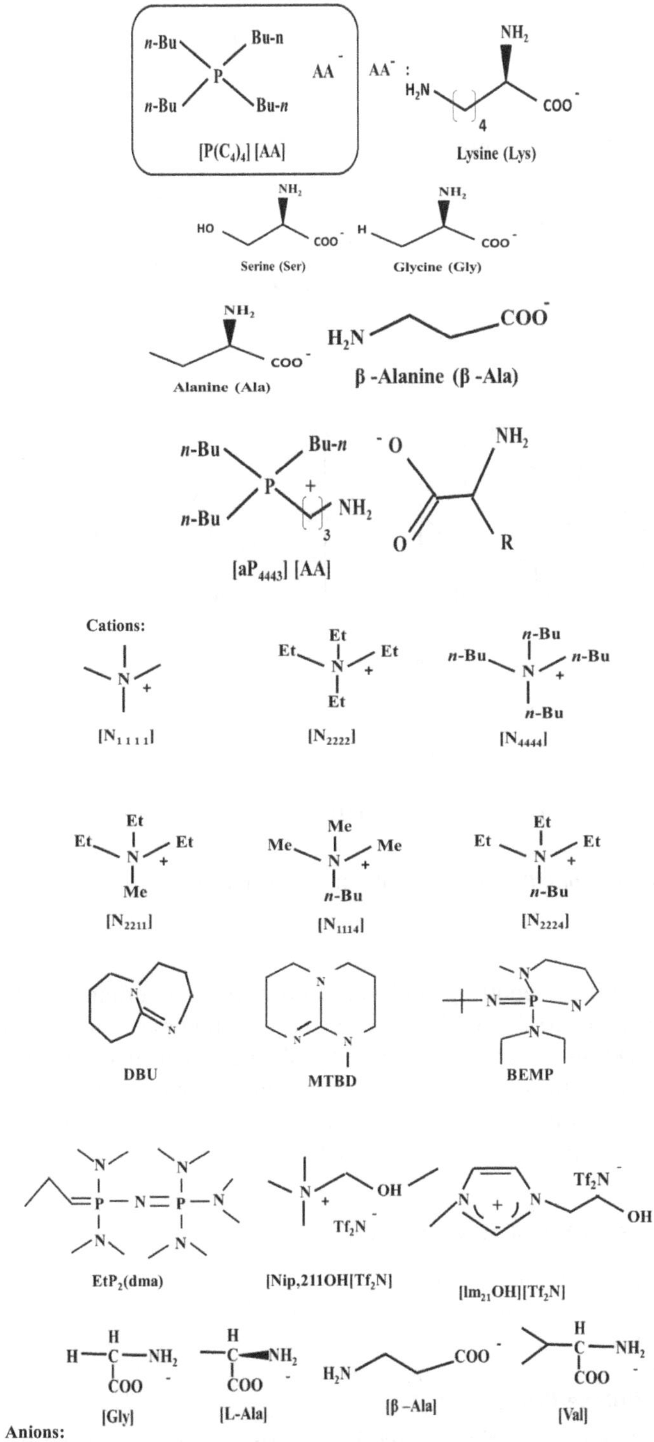

FIGURE 8.5 Structures of cations and anions.

(Continued)

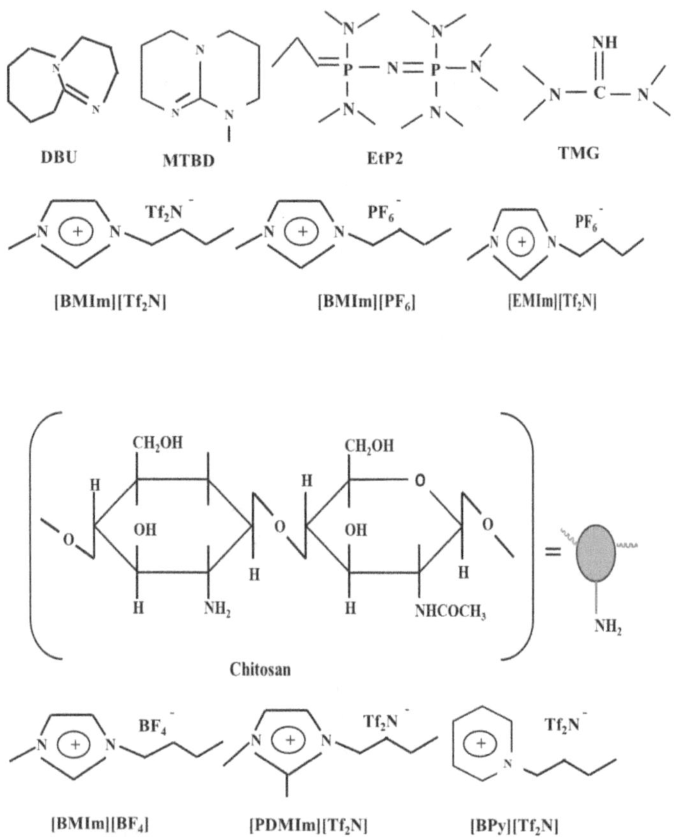

FIGURE 8.5 (*Continued*) Structures of cations and anions.

In the early 1990s, a report by Wilkes and co-workers describing the first air and moisture stable imidazolium salts, based on tetrafluoroborate, [BF4]-, and hexafluorophosphate, [PF6]-fuelled further interest in the field. This interest has seen explosive growth during the past decade. Reflecting this, the number of papers published on ionic liquids has increased from approximately 40 per year in the early 1990s to multiple hundreds per year today (Table 8.2).

8.4.2 Applications of Ionic Liquids

Because ionic liquids have unique properties, they are widely used in various fields of chemical research. Ionic liquids as reaction solvents have been applied in many types of reactions.

8.4.2.1 Cellulose Processing

Cellulose as a bio-renewable resource is the most widely distributed in the world. For the first time, ionic liquids in cellulose processing were applied by Graenacher in

TABLE 8.2
Comparison of Organic Solvents with Ionic Liquids (ILs), Updated from Work By Plechkova and Seddon (2008). Range of Values Were Analysed from the NIST IL Thermo Database at 25°C and Atmospheric Pressure

Property	Organic Solvents	Ionic Liquids
Number of solvents	>1,000	$>10^6$
Applicability in a given process	Single function	Multifunction
Cost	Generally cheap	2–100 times more expensive than organic solvents
Recyclability/toxicity	Green imperative-survey of toxicity of organic solvents is controlled by REACH	Economic imperative-toxicity and biodegradability are often not well known
Vapour pressure	Solvents have vapour pressure greater than limit used in the classification of volatile organic compounds (VOCs)	For aprotic ILs,:negligible vapour pressure under normal conditions
Flammability	Usually flammable	Usually non-flammable, but some ILs are used as propellants
Tuneability	Limited range of solvents available	Virtually unlimited range means "designer solvents"
Chirality	Rare	Common and tuneable
Catalytic ability	Rare	Common and tuneable
Viscosity/m $Pa^{-1} s^{-1}$	0.2–100	20–97,000
Density/g cm^{-3}	0.6–1.7	0.8–3.3
Refractive Index	1.3–1.6	1.3–2.2
Electrical conductivity/ms cm^{-1}	Usually insulator	Up to 120
Thermal conductivity/W m^{-1} K^{-1}	0.1–0.6	0.1–0.3

1934 when ionic liquids were made of 1-ethylpyridinium chloride and free nitrogen-containing bases. Now Robin Rogers and co-workers have found that real cellulose solutions can be produced at technically useful concentration using ionic liquids. Although the technology that ionic liquids dissolve cellulose has been presented as a new idea, this technology opens up great potential for cellulose processing (Swatloski et al. 2002). Cellulose processing is the content dissolve of cellulose and the subsequent disposal. For the dissolving of cellulose, the great volumes of various chemical auxiliaries can be reduced by using ionic liquids such as carbon disulphide (CS_2). For the subsequent disposal, the process can be significantly simplified by the use of ionic liquids. They, as a solvent, can be recycled entirely (Raoa et al. 2007).

8.4.2.2 Hydrogenation Reaction

Many ionic liquids have already been used in hydrogenation reactions and have two advantages. The rate of reaction rate by using ionic liquids is several times faster than using conventional solvents. The mixture solution of ionic liquids and catalysts that has already been used can be recycled. As research has shown, ionic liquids can play two roles: one of solvents and the other as catalysts in the hydrogenation reaction.

Ionic liquids can dissolve parts of the transition metals. Homogeneous reactions use ionic liquids which dissolve transition metals as a catalyst. The homogeneous reaction has been mostly applied in the hydrogenation reactions of ionic liquids fields. In addition, the ionic liquids are easily separated and sublimated when they are used in diesel fuel which mainly contained aromatics of the hydrogenation reaction. At the same time, ionic liquids will not pollute the environment.

8.4.2.3 Biobutanol Separation

In addition, ionic liquids have very interesting physical and chemical properties, making them suitable for separation and purification technologies. It is possible to extract butanol from the fermentation broth using ionic liquids. CO_2 supercritical dissolved in an ionic liquid can extract the nonvolatile organic compounds. According to Deng (2006, 232–235), Chinese chemists have studied the application of ionic liquids in solid-solid separation. Separation of taurine from sodium sulfate solids mixture was achieved by using [BMIM]PF6 as a leaching agent. The ionic liquid recovery rate is 97%, making it an excellent option for many applications. Schematic representation of biobutanol separation from ABE fermentation broth using ionic liquids is shown in Figure 8.6.

Ionic liquids are characterised by very low volatility and a wide temperature range (often exceeding 200°C) (Johnson 2007), which may allow relatively easy regeneration (via low-pressure distillation) and recirculation. Additionally, there are no toxic fumes. Based on their study of butanol and various ionic liquids, Wu et al. (2003) investigated binary liquid-liquid equilibrium. Additionally, Bendova and Wagner (2006) examined the binary equilibrium between [BMIM][PF6] and 1-butanol, and they were in good agreement with previous studies.

Fadeev and Meagher (2001) investigated the solubilities of 1-butyl-3-methylimidazolium ([BMIM][PF6]) and 1-octyl-3-methylimidazolium hexafluorophosphate ([OMIM][PF6]) with butanol and water at very low concentrations of butanol (<5% by weight).

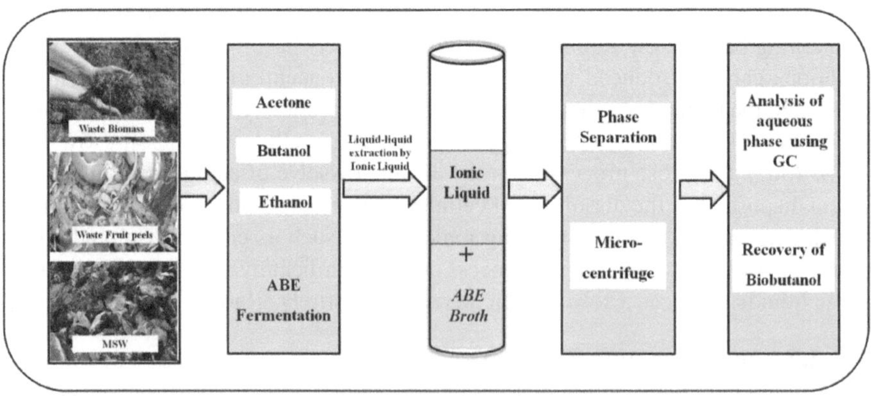

FIGURE 8.6 Schematic representation of biobutanol production and separation from ABE fermentation broth using ionic liquids.

Ionic liquids with a hydrophobic nature were evaluated as liquid extraction agents in this study. Recently, liquid-liquid extraction using ionic liquids at room temperature has become a very popular topic. The use of ionic liquids at room temperature for liquid-liquid extraction has become increasingly popular over the last decade. Liquid-liquid extraction using room-temperature ionic liquids has become increasingly popular during the past decade. The results presented in this study demonstrated the potential of hydrophobic ionic liquids as liquid extraction agents.

Among the advantages of RTILs the following are most often mentioned:

- Negligible vapour pressure (Earle et al. 2000; Ji et al. 2012; Sowmiah et al. 2009)
- Thermal stability (Kosmulski et al. 2004)
- Wide liquidus regions (Brennecke and Maginn 2001; Johnson 2007)
- Ability to dissolve both organic and inorganic compounds (Simoni et al. 2010), as well as hydrophobicity or hydrophilicity

Depending on their properties, ionic liquids differ in their separation performance towards butanol. As a function of anion and cation structures, the distribution ratios and selectivities can vary considerably (Domańska and Królikowski 2012; Garcia-Chavez et al. 2012; Ha et al. 2010; Huang et al. 2014; Marciniak et al. 2016; Rabari and Banerjee 2013; Simoni et al. 2010; Stoffers and Górak 2013).

Their high prices are currently one of the main obstacles preventing ionic liquids from being used in extraction processes. ABE fermentation involves regenerating the used-up solvent and separating the mixture of extracted substances, which can be done most efficiently through distillation. According to Ezeji et al. (2004) and Qureshi and Blaschek (2001), butanol constitutes up to 80% (Ezeji et al. 2004; Green 2011). ABE is primarily composed of acetone, while the fraction of ethanol in the total ABE is usually no more than 10%.

When designing a commercial extraction process, it is necessary to understand how the thermodynamic equilibrium of the mixture of the feed solution and the extracting agent is influenced by its composition at the beginning and by physical parameters such as temperature and pressure. This knowledge will allow integration of extraction and distillation in multi-stage extraction in the future. The equilibrium may be approximated using one of several mathematical models available. In the literature, the most popular equations to approximate liquid-liquid equilibrium are NRTL and UNIQUAC (based on the so-called "local composition" concept), which provide estimated activity coefficients of the individual components in each liquid phase. NRTL equations often yield more accurate results and has the ability to model both water and organic compounds such as ionic liquids (Cheruku and Banerjee 2012; Domańska and Lukoshko 2015; Haghnazarloo et al. 2013; Haghtalab and Paraj 2012; Królikowski 2016; Liu et al. 2016; Mohsen-Nia et al. 2008; Rabari and Banerjee 2013; Zhang et al. 2010). Phosphonium-based ionic liquids and imidazolium-based ionic liquids and their blends have also been explored for the separation of butanol from aqueous phase (Motghare et al. 2019a; Motghare et al. 2019b; Motghare et al. 2021). NRTL equations can only be applied if their binary parameters are defined in relation to the interactions between their constituents. The list of ionic liquids used in the separation of biobutanol is presented in Table 8.3.

TABLE 8.3
List of Ionic Liquids in the Separation of Biobutanol

Ionic Liquids	Reference	Type of Study
HMIM [PF_6] BMIM [Tf_2N]	Kubiczek and Kamiński (2013)	In the present paper, distribution coefficients, extraction efficiency and selectivity have been studied for the extraction of n-Butanol from aqueous solutions with the help of phase equilibrium of five-component systems. The highest efficiency was found for both the ionic liquids, 77% and 85%, at 50°C and at room temperature, efficiency was 50%–65%.
BMIM [Tf_2N] MIM [Tf_2N]	Davis et al. (2008)	In this study, the ionic liquids used were synthesised by two slightly different methods. [bmim][Tf2N] was synthesised first by preparing [bmim][Br] using an ultrasonic approach similar to that utilised by Namboodiri and Varma (2001). [hmim][Tf2N] was synthesised via a "slow-reaction" method. This method has the added advantage of reducing the initial degree of discoloration of the ionic liquid. The ternary liquid-liquid equilibrium data for two ionic liquid+ water+ butanol systems has been presented. The data shows that [bmim][Tf2N] and [hmim][Tf2N] exhibit high selectivity for butanol when low concentrations of butanol are present in the initial aqueous phase.
BMIM [PF6]– HMIM [PF6]– OMIM [PF6]– BMIM [Tf2N]– HMIM [Tf2N]– OMIM [Tf2N]– BMIM [TfO]– HMIM [TfO]– OMIM [TfO]– BMIM [BF4]– HMIM [BF4]–	Sung Ho Ha et al. (2010)	In the present paper, 11 kinds of imidazolium-based hydrophobic ILs (with four different types of anions) were tested for the extraction behaviour of butanol from aqueous solution. Among the tested ILs, [Hmim] [TfO] was the most and [Omim][Tf2N] was the least hygroscopic IL. However, butanol extraction efficiency and selectivity depend on the polarity of ILs. Considering extraction efficiency and selectivity, [Tf2N]-based ILs among the tested ILs showed to be the best extract solvent for the recovery of butanol from aqueous media. Among the studied ILs, [Omim][Tf2N] showed the highest butanol distribution coefficient (1.939), selectivity (132) and extraction efficiency (74%) at 323.15 K, respectively. Toxicity of ILs = Fluorinated anions such as [BF4]– and [PF6]– undergo hydrolysis in aqueous media and these hydrolytic products are toxic to microorganisms (Weissermel et al. 2008, Qureshi et al. 2006). Thus, ILs based on these anions might not be a suitable choice for in situ recovery of butanol from fermentation broth in practice. It means that [Tf2N]-based ILs seem to be more suitable for in situ butanol recovery by liquid-liquid extraction among these tested ILs, although toxicity tests of these ILs against microorganisms need to be carried out.

(*Continued*)

TABLE 8.3 (*Continued*)
List of Ionic Liquids in the Separation of Biobutanol

Ionic Liquids	Reference	Type of Study
BMIM [NTf2]– PnMIM [NTf2]– BOOMMIM [NTf2]– BPYMIM [NTf2]– PnMIM [NTf2]– P_{66614}[CL]– P_{66614}[SCN]– P_{66614}[DCA]– P_{66614}[salicylate(sal)]– P_{66614} [diisooctylphosphinate(DIOP)] P_{66614} [1-Octane-Sulfonate(C_8SO_3)]	Liu et al (2017)	**Extraction of 1,3-Propanediol and 2,3-Butanediol** In the present work, Bistriflimide-based ILs at 25°C to extract 1,3-propanediol and [P66614]+-based ILs at 25°C to extract 2,3-butanediol from aqueous solution were investigated. A range of hydrophobic ionic liquids were prepared and compared. The long chain phosphonium salts containing the trihexyltetradecyl phosphonium [P66614] cation exhibited promising extraction behaviour. Among them, [P66614] [octanesulfonate (C8SO3)] combined high hydrophobicity, good stability, and relatively high extraction efficiency with the distribution coefficient DBDO = 0.390 and extraction selectivity SBDO = 4.83 (2,3-butanediol) and DPDO = 0.219 and SPDO = 2.65 (1,3-propanediol) at 25°C. Additionally, this material exhibited good compatibility with the fermentation process, facilitating its use in bioprocesses. Along with extraction efficiency and selectivity, stability of ionic liquids using TGA analysis, biocompatibility of ionic liquids has been studied in the present paper.
DMIM [TCB] P6,6,6,14 [TCB] DMIM10,1 [FAP]	Heitmann et al. (2012) supported ionic liquid based pervaporation membranes (SILMs)	In this work, extraction of n-butanol using the pervaporation performance of SILMs with tetracyanoborate and tris (pentafluoroethyl) trifluorophosphate ionic liquids (ILs) was investigated. Pervaporation was carried out at 37°C using binary mixtures of n-butanol and water with n-butanol concentrations lower than 5 wt.%. Two concepts for immobilisation of ILs were tested using nylon or polypropylene as support material. ILs were immobilised by inclusion between silicone layers or by dissolution in poly(ether block amide).
[BMIM] [PF6]– [OMIM] [PF6]	Fadeev et al. (2001)	In the present work, pervaporative BuOH recovery from 1 wt.% aqueous solution and [OMIM][PF6] was investigated using commercial polydimethylsiloxane membrane MEM-100 (MemPro Co., www.MemPro.com).
[Ph3t] [NTF] [THA] [DHSS] [OMA] [NTF] [Ph3t] [DCN] OA	Cascon et al. (2011)	In 2011, Cascon et al. used [N8888][dihexylsulfosuccinate (DHSS)] and [P66614]- [dicyanoamide (DCA)] to extract 1-butanol and achieved a high DBuOH of 7.99 and 7.49, respectively.

(*Continued*)

TABLE 8.3 (*Continued*)
List of Ionic Liquids in the Separation of Biobutanol

Ionic Liquids	Reference	Type of Study
		Garcia-Chavez et al. (2012) adopted [N8888][2-methyl-1-naphthoate] and reported a DBuOH of up to 21.26. Nann et al. (2013) studied the anion of tetracyanoborate ([TCB]−) coupled with 3-decyl-1-methylimidazolium and applied it to 1-butanol extraction. The IL exhibited excellent extraction performance with 8.0 < DBuOH < 12.0.27.
[HMIM] [TCB] [DMIM][TCB] [P14,6,6,6][TCB]	Domańska et al. (2012)	Domańska et al. (2012) studied the extraction of *n*-butanol from aqueous solutions at 308.15K using three ionic liquids, and obtained very encouraging results. The authors measured the compositions of both liquid phases using gas chromatography. They correlated the experimental LLE data by means of the NRTL equation respectively.
[TDTHP] [Phosph]	Rabari and Banerjee (2013)	In this work, a phosphonium-based ionic liquid (IL) trihexyl(tetradecyl) Phosphonium bis(2,4,4-trimethylpentyl) phosphinate ([TDTHP] [Phosph]) has been used for the recovery of 1-propanol and 1-butanol at $T = 298.15$ K and $p = 1$ atm. High values of selectivity ranging from 53 to 252 and 80 to 305 were observed for 1-propanol and 1-butanol, respectively. The distribution coefficient is greater than unity which indicates easier diffusion of solute from aqueous phase to extract phase. NRTL and UNIQUAC models show deviation (RMSD) in the range of 0.1%–0.5% for both systems.
[COC2mPIP] [NTf2] 1-(2-methoxyethyl)-1-methylpiperidinium bis(trifluoromethyl sulfonyl)imide [COC2mPYR] [NTf2] 1-(2-methoxyethyl)-1-methylpyrrolidinium bis(trifluoromethyl sulfonyl)imide [bmPIP] [NTf2] 1-butyl-1-methyl piperidinium bis(trifluoromethyl sulfonyl)imide	Marciniak et al. (2016)	Marciniak et al. (2016) studied phase equilibria in three ternary systems containing *n*-butanol, water.

8.5 POSSIBLE HYPOTHETICAL MECHANISM FOR BIOBUTANOL SEPARATION USING IONIC LIQUIDS

Polarity is another factor that may affect butanol recovery by liquid-liquid extraction since it affects the solvation behaviours. Solvents are polar or nonpolar based on their polarity. Solvents with a higher dielectric constant are generally more polar. Polarity is one of the common parameters. Further, as the alkyl chain of a cation increases, the dielectric constant decreases. In addition, butanol extraction efficiency is strongly influenced by the polarity of ILs. As the polarity of ILs increased, the butanol extraction efficiency decreased linearly as dielectric constant increased. Figure 8.7 illustrates the possible hypothetical mechanisms in the separation of biobutanol using ionic liquids. Tetradecyl molecule attached with phosphonium cation paired with hydroxyl group from butanol forms a covalent bond with an anion of ionic liquid.

8.6 COMMERCIAL ASPECT

Charles Weizmann developed the biochemical process for producing butanol in the 1910s (Gabriel 1928). Fermentation was used industrially until the 1960s, when the feedstock price increased and chemical production overtook the market. Both academic and industrial researchers collaborated on research and development projects. Ni and Sun (2009) reported that 11 ABE plants are now operating in China. Furthermore, several companies such as Butamax™ Advanced Biofuels (founded by BP and DuPont), Cobalt Technologies, Gevo, and Metabolic Explorer are intensively

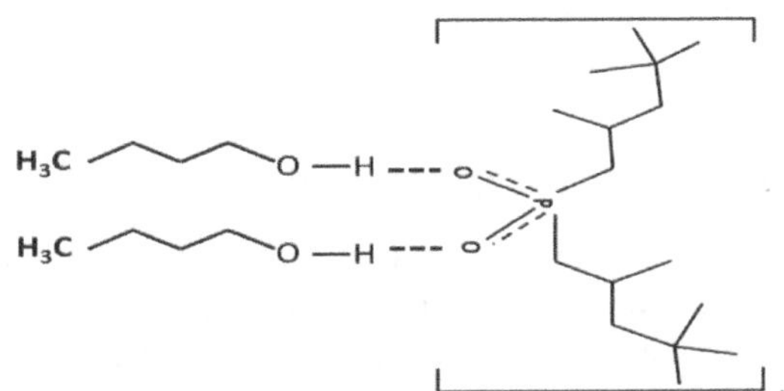

1-butanol

bis 2,4,4 –trimethylpentyl-phosphinate anion

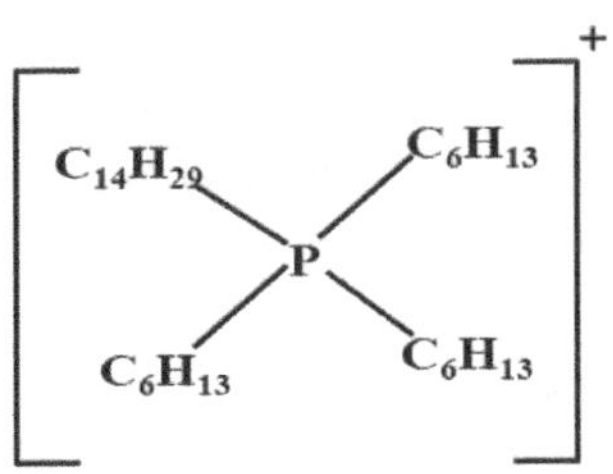

Tetradecyl(trihexyl) phosphonium cation

FIGURE 8.7 Possible hypothetical mechanism in the separation of biobutanol using ionic liquids.

TABLE 8.4
List of Companies with Location and Application of Butanol

Sr. No.	Name of Company	Location	Application
1	Sovert	UK-based company n-butanol production via fermentation	Feedstock is spoiled and waste food from domestic and commercial sources.
2	Saudi Butanol Company (SaBuCo)	TASNEE Petrochemical Complex in Jubail Industrial City	Produces 330,000t per annum of n-butanol and 11,000t per annum of iso-butanol.
3	ButamaxGevo	Joint venture: BP and DuPont	Commercialise butanol by 2012-2013. Process yields isobutanol.
4	Cathay Industrial Biotech	Jilin Province of Northeast China	Produces and sells n-butanol.
5	Cobalt Technologies	California, USA	A very detailed business approach is available which will promote product efficiently for funding.

developing biobutanol technologies. The Gevo (2012) study also showed that an old ethanol plant could be retrofitted for butanol production. In addition to being a biofuel, biobutanol is also a renewable energy source. The list of companies with location and application of butanol is given in Table 8.4.

8.7 DISCUSSION AND CONCLUSION

Biobutanol is the next-generation alternative biofuel that can replace conventional petroleum fuels, as its properties are similar. Anaerobic fermentative approach for biobutanol production has a few challenges, as butanol concentration is toxic to microbes and affects the fermentation process. Various techniques have been discovered for separation. A number of methods are available for separation, such as gas strippHowever, ionic liquids can be used as a novel extractant and could be a prospective solution to replace conventional volatile, toxic solvents. Recovery of biobutanol by ionic liquids could be utilised in gasoline-driven combustion systems due to its excellent blending properties with fewer emissions. Therefore, ionic liquids, which recently gained some notable recognition in science and industry, may be highly appropriate for the recovery of biobutanol.

ACKNOWLEDGEMENTS

Authors acknowledge the Department of Biotechnology, Government of India, India, for financial support under the project titled "Design and synthesis of ionic liquids for separation of biobutanol from fermentation broth to enhance its production" (BT/PR16803/PBD/26/511/2016) (Project Investigator: Dr. Kailas L. Wasewar).

REFERENCES

Abdehagh, N., Gurnani, P., Tezel, F.H. and Thibault, J., 2015. Adsorptive separation and recovery of biobutanol from ABE model solutions. *Adsorption*, 21(3), pp. 185–194.

Abdehagh, N., Tezel, F.H. and Thibault, J., 2014. Separation techniques in butanol production: challenges and developments. *Biomass and Bioenergy*, 60, pp. 222–246.

Aftabuzzaman, M. and Mazloumi, E., 2011. Achieving sustainable urban transport mobility in post peak oil era. *Transport Policy*, 18(5), pp. 695–702.

Bendova, M. and Wagner, Z., 2006. Liquid–liquid equilibrium in binary system [bmim][PF6]+ 1-butanol. *Journal of Chemical & Engineering Data*, 51(6), pp. 2126–2131.

Bohnet, M., 2003. *Ullmann's Encyclopedia of Industrial Chemistry*, 6th, completely rev. Ed.; Wiley-VCH: Weinheim.

Brennecke, J.F. and Maginn, E.J., 2001. Ionic liquids: innovative fluids for chemical processing. *American Institute of Chemical Engineers. AIChE Journal*, 47(11), p. 2384.

Cascon, H.R., Choudhari, S.K., Nisola, G.M., Vivas, E.L., Lee, D.J. and Chung, W.J., 2011. Partitioning of butanol and other fermentation broth components in phosphonium and ammonium-based ionic liquids and their toxicity to solventogenic clostridia. *Separation and Purification Technology*, 78(2), pp. 164–174.

Cheruku, S.K. and Banerjee, T., 2012. Liquid–liquid equilibrium data for 1-ethyl-3-methylimidazolium acetate–thiophene–diesel compound: experiments and correlations. *Journal of Solution Chemistry*, 41(5), pp. 898–913.

Davis, S.E. and Morton Iii, S.A., 2008. Investigation of ionic liquids for the separation of butanol and water. *Separation Science and Technology*, 43(9–10), pp. 2460–2472.

Dean, J.A., 1990. Lange's handbook of chemistry. *Material and Manufacturing Process*, 5(4), pp. 687–688.

Demirbas, A., 2009. Political, economic and environmental impacts of biofuels: a review. *Applied Energy*, 86, pp. S108–S117.

Domańska, U. and Królikowski, M., 2012. Extraction of butan-1-ol from water with ionic liquids at T = 308.15 K. *The Journal of Chemical Thermodynamics*, 53, pp. 108–113.

Domańska, U. and Lukoshko, E.V., 2015. Separation of pyridine from heptane with tricyanomethanide-based ionic liquids. *Fluid Phase Equilibria*, 395, pp. 9–14.

Deng, Y. 2006. Green chemistry and chemical engineering: Ionic liquid properties, preparation and application. *Petrochemical Press of China*, 24, pp. 232–235.

Dürre, P., 1998. New insights and novel developments in clostridial acetone/butanol/isopropanolfermentation. *Applied Microbiology and Biotechnology*, 49(6), pp. 639–648.

Earle, M.J., Seddon, K.R. and Sheldon, R., 2000. Catalytic reactions in ionic liquids. *Pure and Applied Chemistry*, 72, pp. 1391–1398.

Ezeji, T.C., Qureshi, N. and Blaschek, H.P., 2003. Production of acetone, butanol and ethanol by Clostridium beijerinckii BA101 and in situ recovery by gas stripping. *World Journal of Microbiology and Biotechnology*, 19(6), pp. 595–603.

Ezeji, T.C., Qureshi, N. and Blaschek, H.P., 2004a. Acetone butanol ethanol (ABE) production from concentrated substrate: reduction in substrate inhibition by fed-batch technique and product inhibition by gas stripping. *Applied Microbiology and Biotechnology*, 63(6), pp. 653–658.

Ezeji, T.C., Qureshi, N. and Blaschek, H.P., 2004b. Butanol fermentation research: upstream and downstream manipulations. *The Chemical Record*, 4(5), pp. 305–314.

Ezeji, T.C., Qureshi, N. and Blaschek, H.P., 2007. Bioproduction of butanol from biomass: from genes to bioreactors. *Current Opinion in Biotechnology*, 18(3), pp. 220–227.

Freeman, J., Williams, J., Minner, S., Baxter, C., DeJovine, J., Gibbs, L., Lauck, J., Muller, H. and Saunders, H., 1988. Alcohols and ethers: a technical assessment of their application as fuels and fuel components. API publication, 4261.

Gabriel, C.L., 1928. Butanol fermentation process1. *Industrial & Engineering Chemistry*, 20(10), pp. 1063–1067.

Garcia-Chavez, L.Y., Garsia, C.M., Schuur, B. and de Haan, A.B., 2012. Biobutanol recovery using nonfluorinated task-specific ionic liquids. *Industrial & Engineering Chemistry Research*, 51(24), pp. 8293–8301.

Christopher R., Dave M. and Gary Bevers, Gevo® White Paper Transportation Fuels Renewable Solution ISOBUTANOL—A RENEWABLE SOLUTION for THE TRANSPORTATION FUELS VALUE CHAIN, May 2011. https://www.etipbioenergy.eu/images/wp-isob-gevo.pdf. Green, E.M., 2011. Fermentative production of butanol—the industrial perspective. *Current Opinion in Biotechnology*, 22(3), pp. 337–343.

Groot, W.J. and Luyben, K.C.A., 1986. In situ product recovery by adsorption in the butanol/isopropanol batch fermentation. *Applied Microbiology and Biotechnology*, 25(1), pp. 29–31.

Ha, S.H., Mai, N.L. and Koo, Y.M., 2010. Butanol recovery from aqueous solution into ionic liquids by liquid–liquid extraction. *Process Biochemistry*, 45(12), pp. 1899–1903.

Haghnazarloo, H., Lotfollahi, M.N., Mahmoudi, J. and Asl, A.H., 2013. Liquid–liquid equilibria for ternary systems of (ethylene glycol+ toluene+ heptane) at temperatures (303.15, 308.15, and 313.15) K and atmospheric pressure: experimental results and correlation with UNIQUAC and NRTL models. *The Journal of Chemical Thermodynamics*, 60, pp. 126–131.

Haghtalab, A. and Paraj, A., 2012. Computation of liquid–liquid equilibrium of organic-ionic liquid systems using NRTL, UNIQUAC and NRTL-NRF models. *Journal of Molecular Liquids*, 171, pp. 43–49.

Hahn, H.D., Dämbkes, G., Rupprich, N. and Bahl, H. 2012. Butanols. In *Ullmann's Encyclopedia of Industrial Chemistry*, vol. 6, Wiley-VCH Verlag GmbH & Co, Weinheim, pp. 417–430.

Hansen, A.C., Zhang, Q. and Lyne, P.W., 2005. Ethanol–diesel fuel blends—a review. *Bioresource Technology*, 96(3), pp. 277–285.

Huber, G.W. and Corma, A., 2007. Synergies between bio-and oil refineries for the production of fuels from biomass. *Angewandte Chemie International Edition*, 46(38), pp. 7184–7201.

Huang, H.J., Ramaswamy, S. and Liu, Y., 2014. Separation and purification of biobutanol during bioconversion of biomass. *Separation and Purification Technology*, 132, pp. 513–540.

Ji, X., Held, C. and Sadowski, G., 2012. Modeling imidazolium-based ionic liquids with ePC-SAFT. *Fluid Phase Equilibria*, 335, pp. 64–73.

Jones, D.T. and Woods, D.R., 1986. Acetone-butanol fermentation revisited. *Microbiological Reviews*, 50(4), pp. 484–524.

Kujawska, A., Kujawski, J., Bryjak, M. and Kujawski, W., 2015. ABE fermentation products recovery methods—a review. *Renewable and Sustainable Energy Reviews*, 48, pp. 648–661.

Kosmulski, M., Gustafsson, J. and Rosenholm, J.B., 2004. Thermal stability of low temperature ionic liquids revisited. *Thermochimica Acta*, 412(1–2), pp. 47–53.

Królikowski, M., 2016. Liquid–liquid extraction of p-xylene from their mixtures with alkanes using 1-butyl-1-methylmorpholinium tricyanomethanide and 1-butyl-3-methylimidazolium tricyanomethanide ionic liquids. *Fluid Phase Equilibria*, 412, pp. 107–114.

Kubiczek, A. and Kamiński, W., 2013. Ionic liquids for the extraction of n-butanol from aqueous solutions. *Ecological Chemistry and Engineering A*, 20(1), pp. 77–87.

Kujawska, A., Kujawski, J., Bryjak, M. and Kujawski, W., 2015. ABE fermentation products recovery methods—a review. *Renewable and Sustainable Energy Reviews*, 48, pp. 648–661.

Liu, W., Zhang, Z., Ri, Y., Xu, X. and Wang, Y., 2016. Liquid−liquid equilibria for ternary mixtures of water + 2- propanol + 1-alkyl-3-methylimidazolium bis(trifluoromethylsulfonyl) imide ionic liquids at 298.15K. *Fluid Phase Equilibria*, 412, pp. 205–210.

Lin, X., Li, R., Wen, Q., Wu, J., Fan, J., Jin, X., Qian, W., Liu, D., Chen, X., Chen, Y. and Xie, J., 2013. Experimental and modeling studies on the sorption breakthrough behaviors of butanol from aqueous solution in a fixed-bed of KA-I resin. *Biotechnology and Bioprocess Engineering*, 18(2), pp. 223–233.

Lin, X., Wu, J., Jin, X., Fan, J., Li, R., Wen, Q., Qian, W., Liu, D., Chen, X., Chen, Y. and Xie, J., 2012. Selective separation of biobutanol from acetone–butanol–ethanol fermentation broth by means of sorption methodology based on a novel macroporous resin. *Biotechnology Progress*, 28(4), pp. 962–972.

Liu, H., Wang, G. and Zhang, J., 2013. The promising fuel-biobutanol. *Liquid, Gaseous and Solid Biofuels-Conversion Techniques*, pp. 175–198.

Liu, W., Zhang, Z., Ri, Y., Xu, X., Wang, Y., 2016. Liquid–liquid equilibria for ternary mixtures of water + 2- propanol + 1-alkyl-3-methylimidazolium bis(trifluoromethylsulfonyl) imide ionic liquids at 298.15K. *Fluid Phase Equilibria*, 412, pp. 205–210. DOI: 10.1016/j.fluid.2015.12.051.

Lodi, G. and Pellegrini, L.A., 2016. Recovery of butanol from ABE fermentation broth by gas stripping. *Chemical Engineering Transactions*, 49, pp. 13–18.

Lu, C., Zhao, J., Yang, S.T. and Wei, D., 2012. Fed-batch fermentation for n-butanol production from cassava bagasse hydrolysate in a fibrous bed bioreactor with continuous gas stripping. *Bioresource Technology*, 104, pp. 380–387.

Maddox, I.S., Steiner, E., Hirsch, S., Wessner, S., Gutierrez, N.A., Gapes, J.R. and Schuster, K.C., 2000. The cause of "Acid Crash" and "Acidogenic Fermentations" during the batch Acetone-Butanol-Ethanol(ABE-) fermentation process. *Journal of Molecular Microbiology and Biotechnology*, 2(1), pp. 95–100.

Mariano, A.P., Dias, M.O.S. and Junqueira, T.L., 2013. Utilization of pentoses from sugarcane biomass: techno-economics of biogas vs. butanol production. *Bioresource Technology*, 142, pp. 390–399.

Marciniak, A., Wlazło, M. and Gawkowska, J., 2016. Ternary (liquid+ liquid) equilibria of {bis (trifluoromethylsulfonyl)-amide based ionic liquids+ butan-1-ol+ water}. *The Journal of Chemical Thermodynamics*, 94, pp. 96–100.

McMillan, M., Rodrik, D. and Verduzco-Gallo, Í., 2014. Globalization, structural change, and productivity growth, with an update on Africa. *World Development*, 63, pp. 11–32.

Mohsen-Nia, M., Nekoei, E. and Mohammad Doulabi, F.S., 2008. Ternary (liquid + liquid) equilibria for mixtures of(methanol + aniline + n-octane or n-dodecane) at T=298.15 K. *The Journal of Chemical Thermodynamics*, 40, 330–333. DOI: 10.1016/j.jct.2007.05.018.

Motghare, K.A., Rajkumar, N. and Wasewar, K.L., 2019a. Separation of butanol using natural non-toxic solvents and conventional chemical solvents. *Chemical Data Collections*, 21, p. 100225.

Motghare, K.A., Shende, D. and Wasewar, K.L., 2021. Butanol recovery Ionic liquids as green solvents. *Journal of Chemical Technology & Biotechnology*, 97, pp. 873–884.

Motghare, K.A., Wasewar, K.L. and Shende, D.Z., 2019b. Separation of Butanol Using Tetradecyl (trihexyl) phosphonium Bis (2, 4, 4-trimethylpentyl) phosphinate, Oleyl Alcohol, and Castor Oil. *Journal of Chemical & Engineering Data*, 64(12), pp. 5079–5088.

Nann, A., Held, C. and Sadowski, G. 2013. *Industrial & Engineering Chemistry Research* 52(51), 18472–18481. DOI: 10.1021/ie403246e

Ni, Y. and Sun, Z., 2009. Recent progress on industrial fermentative production of acetone–butanol–ethanol by Clostridium acetobutylicum in China. *Applied Microbiology and Biotechnology*, 83(3), pp. 415–423.

Park, C.H., 1996. Pervaporative butanol fermentation using a new bacterial strain. *Biotechnology and Bioprocess Engineering*, 1, 1–8.

Plechkova, N.V. and Seddon, K.R., 2008. Applications of ionic liquids in the chemical industry. *Chemical Society Reviews*, 37(1), pp. 123–150.

Qin, J., Liu, H.F., Yao, M.F. and Chen, H, 2007a. Influence of biology oxygenated fuel component on diesel engine characteristics. *Transactions of CSICE*, 3, p. 014.

Qin, J., Liu, H.F., Yao, M.F. and Chen, H., 2007b. Experiment study on diesel engine fueled with biodiesel and diesel fuel. *Journal of Combustion Science and Technology*, 4, p. 009.

Qureshi, N., Annous, B.A., Ezeji, T.C., Karcher, P. and Maddox, I.S., 2005. Biofilm reactors for industrial bioconversion processes: employing potential of enhanced reaction rates. *Microbial Cell Factories*, 4(1), pp. 1–21.

Qureshi, N. and Blaschek, H.P., 2001. Recent advances in ABE fermentation: hyper-butanol producing Clostridium beijerinckii BA101. *Journal of Industrial Microbiology and Biotechnology*, 27(5), pp. 287–291.

Qureshi, N. and Blaschek, H.P., 2006. Butanol production from agricultural biomass. In Shetty K, Pometto A, Paliyath G & Levin RE (eds) *Food Biotechnology*. CRC Press, Boca Raton, FL, pp. 525–551.

Qureshi, N. and Ezeji, T.C., 2008. Butanol, 'a superior biofuel'production from agricultural residues (renewable biomass): recent progress in technology. *Biofuels, Bioproducts and Biorefining: Innovation for a Sustainable Economy*, 2(4), pp. 319–330.

Qureshi, N. and Maddox, I.S., 1995. Continuous production of acetone-butanol-ethanol using immobilized cells of Clostridium acetobutylicum and integration with product removal by liquid-liquid extraction. *Journal of Fermentation and Bioengineering*, 80(2), pp. 185–189.

Rabari, D. and Banerjee, T., 2013. Biobutanol and n-propanol recovery using a low density phosphonium based ionic liquid at T= 298.15 K and p = 1 atm. *Fluid Phase Equilibria*, 355, pp. 26–33.

Simoni, L.D., Chapeaux, A., Brennecke J.F., Stadtherr M.A., 2010. Extraction of biofuels and biofeedstocks from aqueous solutions using ionic liquids. *Computers and Chemical Engineering*, 34, 1406–1412. DOI: 10.1016/j.compchemeng.2010.02.020.

Sowmiah, S., Srinivasadesikan, V., Tseng, M.C. and Chu, Y.H., 2009. On the chemical stabilities of ionic liquids. *Molecules*, 14(9), pp. 3780–3813.

Stoffers, M. and Górak, A., 2013. Continuous multi-stage extraction of n-butanol from aqueous solutions with 1-hexyl-3-methylimidazolium tetracyanoborate. *Separation and Purification Technology*, 120, pp. 415–422.

Swatloski, R.P., Spear, S.K., Holbrey, J.D. and Rogers, R.D., 2002. Dissolution of cellose with ionic liquids. *Journal of the American Chemical Society*, 124(18), pp. 4974–4975.

Trenberth, K.E., Dai, A., Rasmussen, R.M. and Parsons, D.B., 2003. The changing character of precipitation. *Bulletin of the American Meteorological Society*, 84(9), pp. 1205–1218 (Available at www.cgd.ucar.edu/cas/adai/papers/rainChBamsR.pdf).

US Department of Energy, 2019. Alternative Data Fuel Centre. https://afdc.energy.gov/fuels/emerging_biobutanol.html.

US Energy Information Administration (US IEA), 2017. International Energy Outlook.

Varma, R.S. and Namboodiri, V.V., 2001. An expeditious solvent-free route to ionic liquids using microwaves. *Chemical Communications*, 7, pp. 643–644.

Vane, L.M., 2005. A review of pervaporation for product recovery from biomass fermentation processes. *Journal of Chemical Technology & Biotechnology: International Research in Process, Environmental & Clean Technology*, 80(6), pp. 603–629.

Vane, L.M., 2008. Separation technologies for the recovery and dehydration of alcohols from fermentation broths. *Biofuels, Bioproducts and Biorefining*, 2(6), pp. 553–588.

Weissermel, K. and Arpe, H.J., 2008. *Industrial Organic Chemistry*. Federal Republic of Germany: John Wiley &Sons.

Wu, C.-T., Marsh, K.N., Deev, A.V. and Boxall, J.A., 2003. Liquid-liquid equilibria of room-temperature ionic liquids and butan-1-ol. *Journal of Chemical & Engineering Data*, 48, pp. 486–491.

Zhang, W., Hou, K., Mi, G., Chen, N., 2010. Liquid-liquid equilibria of the ternary system thiophene + octane + dimethyl sulfoxide at several temperatures. *Applied Biochemistry and Biotechnology*, 160, pp. 516–522. DOI: 10.1007/s12010-008-8382-1.

9 Intensification in Bioethanol Production and Separation

Kailas L. Wasewar
Visvesvaraya National Institute of Technology (VNIT)

CONTENTS

DOI: 10.1201/9781003203452-9

9.1 INTRODUCTION

It can be observed that global warming has been increasing at a rate of around 0.2°C per year, and it has been reported that the Earth and its ecosystem can withstand a temperature rise of up to 4.5°C. The worldwide present growth of human population is linked to urbanization and industrial development, raising serious concerns about future generations' energy security. Energy from fossil sources meets approximately 86% of the world's primary energy demand (Sharma and Saini, 2020). Currently, fossil fuels are the major source for the production of fuels, and it has been shifted toward more and more sustainable alternative options for fossil-based fuel. Alternative energy sources can be used instead of conventional or fossil-based energy. These energy sources are nuclear, solar, electric, hydroelectric, biofuels, geothermal, and wind. Biofuels based on renewable biomass transformation are the sustainable approach to mitigating the issues arising from fossil-based fuels (Kumakiri et al., 2021).

Ethanol obtained from biomass as bioethanol is one of the renewable sources of energy. Bioethanol has been more promising as it can be easily mixed with petroleum and used as an equivalent fuel without changing the combustion engine. Based on the survey of available biomass for energy generation, it is estimated that biomass can contribute up to 90% of the requirement of world's energy (Sharma and Saini, 2020). Fermentation of biomass (energy crops, agriculture residues, etc.) with sugars is the popular route for the production of bioethanol. Bioethanol has been considered as one of the most sustainable and green transport fuels, as it can mix with fossil-based fuels and hence be a probable substitute for fossil-based fuels. Bioethanol obtained through fermentation has a low concentration in the aqueous phase, resulting in downstream processing with high energy requirements. Furthermore, the formation of azeotrope of water–ethanol makes more complex and challenging downstream processing (Errico, 2017).

Ethanol fermentation is the bioprocess for the production of bioethanol, which has been considered as a potential biofuel. The main raw materials for the production of biobutanol are sugar, starch, and lignocellulose-based materials. End-product toxicity, which inhibits growth, is the major challenge in commercializing the production of bioethanol. The approach of process intensification (PI) can result in improvements in existing equipment/technologies/processes or create new equipment/technologies/process alternatives, which are required to produce specific products and provide the service with more innovation and sustainable methods and practices. PI is broadly classified into equipment and methods. Intensifying methods and equipment may be considered as a promising approach for the separation of biobutanol. In the present chapter, basic aspects of bioethanol, PI, and its application in production processes have been systematically discussed.

9.2 BIOETHANOL

Ethanol contains almost 35% oxygen, which provides cleaner combustion with reduced emissions. The octane number assesses the quality of gasoline or the performance in engines, and it is 108 for ethanol, so it can be replaced completely or partially by mixing with transport fuel. Bioethanol is a renewable and environmentally friendly source of energy and high-value derivatives. Also, bioethanol is one of the alternative renewable solvents and transport fuels, and its demand has been growing continuously.

Bioethanol can be produced from sugar-rich feedstock of the first generation, such as corn grains, sugar beets, potatoes, wheat, cassava, sugar cane, and other crops, and from second generation as lignocellulose-based feedstock. The first-generation biofuels such as bioethanol are produced from crops that are competitive with the production of food for human beings and cause damage to the environment. Hence, the second- and third-generation feedstock have been mostly considered nowadays for bioethanol production. Around 10% ethanol can be blended with transportation fuel, especially petrol, as per the Bureau of Indian Standards of 2004, which can be an excellent anti-freezing and cleaning agent (Sakthivel et al., 2018; Kaminski et al., 2008).

The fermentation of sugars mainly produces bioethanol. The typical steps involved in ethanol production from biomass are pretreatment of the biomass, saccharification to sugar monomers, fermentation, and finally, separation and purification (Kumakiri et al., 2021). The ethanol concentration in fermentation broth is generally not more than 10% due to product and substrate inhibition. For the fuel-grade ethanol, more than 99.5 mol% (99.8 wt%) purity is required, achieved mainly by distillation, where more than one-third energy of the total product is utilized for distillation.

The top eight courtiers for bioethanol production in 2019 based on available literature data is presented by Kumakiri et al. (2021). The top bioethanol-producing countries are the United States, Brazil, the European Union, China, India, Canada, Thailand, and Argentina (Aizarani, 2023).

9.3 CLASSIFICATION OF BIOETHANOL

The feedstocks for bioethanol production have been classified as first-generation, second-generation, and third-generation bioethanol.

9.3.1 First-Generation Bioethanol

First-generation bioethanol is produced from the feedstock of food crops. The major crops used in the production of first-generation bioethanol are sugar crops (sugar beet, sugarcane, and sweet sorghum) and starch crops (maize, cassava, wheat, corn, potato, and barley) (Balat, 2009; Sharma and Saini, 2020).

9.3.2 Second-Generation Bioethanol

The first-generation bioethanol has the problem of sustainability due to the utilization of food crops for ethanol production. For second-generation bioethanol production, lignocellulosic feedstock is used. This lignocellulosic feedstock is non-edible renewable material, including residual parts of food crops, other non-edible crops, and lignocellulosic wastes from municipal and industrial sources (Antizar-Ladislao and Turrion-Gomez 2008; Sharma and Saini, 2020). It benefits the use of renewable feed, reduces the dependency on fossil sources, and reduces emissions and footprints. Thermochemical and biochemical approaches are used in the production of second-generation bioethanol.

9.3.3 Third-Generation Bioethanol

The aquatic biomass feedstock (such as algae and cyanobacteria) is used for the production of third-generation bioethanol. The limitations of earlier generation feedstock are overcome in third-generation feedstock. Algae have cell walls with holocellulose and starch as storage materials used for bioethanol production (Sharma and Saini, 2020).

9.4 PROCESS INTENSIFICATION

PI provides the technologies and alternatives for the process industry, including the bioprocess industry, to reduce energy consumption and production costs for a wide range of products with a variety of feedstocks by the synergic effect of multifunctional approaches at the spatial and timescale to improve heat, mass, and momentum transports. Ramshaw (1983) defined PI for the first time, and he has been considered as a pioneer of PI. Various definitions and approaches for PI have been presented in the literature and also for a variety of applications, including process industry, wastewater treatment, energy, and so on (Tian et al., 2018; Wasewar et al., 2020; Wasewar, 2021a, b, c). PI has been successfully implemented in the process industry for reaction, separation, hybrid processes, and unconventional energy sources.

9.4.1 Principles of Process Intensification

Van Gerven and Stankiewicz (2009) have presented four principles for PI that can be summarized as: (1) maximization of the efficacy of the interacting or non-interacting actions, (2) allowing each and every molecule in a specific operation for the same processing experience to achieve uniform properties of the products, (3) optimization of the various driving forces involved in operation at each and every scale and also the specific contact surface area for more effectiveness, and (4) maximization of synergetic effects obtained from multitasking operations.

9.4.2 Process Intensification for Bioethanol Production

Fermentation is the main pathway for the production of bioethanol using various biomasses. Lignocellulosic biomass is available in huge quantities as forest waste, agriculture waste, and so forth. Many inhibitors are present in lignocellulosic biomass, which must be removed after hemicellulose extraction. Innovative and sustainable detoxification steps using PI approaches have been successfully developed, such as the membrane pervaporation bioreactor, extractive fermentation, and vacuum membrane distillation bioreactor (QTR, 2015). Due to the low ethanol concentration (5–12 wt%) in fermentation broth and the formation of water–ethanol azeotrope at 96.5 wt% ethanol, the separation and concentration of ethanol from water to fuel-grade (99.99%) ethanol is a highly energy-intensive process that requires high energy. Novel and sustainable low-energy demanding technologies are considered, such as the use of ionic liquids in extractive distillation or the use of hyperbranched polymers for separation (Huang et al., 2008). Also, other less energy-demanding and more efficient equipment and technologies (molecular sieve adsorption, pervaporation, etc.) are considered but have the limitation of low capacity (Frolkova and Raeva, 2010). Furthermore, the divided wall column can significantly reduce the energy requirement for separation. Around 10%–20% energy can be saved by using an optimized two-column extractive distillation with a single divided wall column, and 20% energy can be saved in the case of an optimized two-column azeotropic distillation with a single divided wall column (Kiss & Suszwalak, 2012).

The steps involved in the production of bioethanol can be broken down into five main steps, such as pretreatment of biomass, hydrolysis of cellulose, concentration and detoxification of substrate, ethanol fermentation, and separation and purification of ethanol (Errico, 2017). PI techniques are being developed and implemented to obtain safer processes with greater equipment efficiency, reduce their size and operating costs, incorporate retrofitting, consume a minimum of energy, generate the least possible amount of waste, and obtain as many products with the least possible amount of raw material, which can be implemented for the production and separation of bioethanol (Vaghari et al., 2015; Gonzalez-Contreras et al., 2020).

9.5 BIOMASS PRETREATMENT ALTERNATIVES

Different technologies can be employed to treat biomass, which depends on the nature and type of feedstock. The pretreatment technologies (physical, chemical, physicochemical, and biological processes) include conventional and advanced technologies (dilute acid, steam explosion, etc.) (AchinasandEuverink, 2016; Gonzalez-Contreras et al., 2020).

9.5.1 Physical Treatment Methods

Physical treatment methods increase the surface area and decrease the degree of polymerization and crystallinity of lignocellulosic-based feedstock for ethanol production. This treatment method is relatively eco-friendly compared to other methods due to the low or almost zero release of toxins. Still, more energy is required, increasing the ethanol production cost. The various physical treatment methods are mechanical methods, extrusion, grinding, milling, comminution, chipping, alternative energy sources (ultrasound, microwaves, and electron beams), and so forth (Baruah et al., 2018; Sharma and Saini, 2020). This also involves a few thermal methods such as hydrothermal, high-pressure steaming pyrolysis, and so on.

9.5.1.1 Mechanical Comminution

In mechanical comminution, the size of biomass feedstock is reduced by milling, chipping, shredding, and grinding. Grinding and milling reduce the size of biomass up to 0.2 mm and chipping up to 10–30 mm. The small biomass particle size resulted in high energy consumption (Sharma and Saini, 2020). The milling can be performed using various processes, which include ball milling (dry, wet, and vibratory), two-roll milling, hammer milling, fluid energy milling, disk milling, and colloid milling (Kumari and Singh, 2018). Compared to other methods, energy consumption is higher in mechanical comminution, which further increases the overall cost of producing bioethanol from biomass.

9.5.1.2 Extrusion

Extrusion is the thermophysical method for lignocellulosic biomass pretreatment. In such methods, biomass undergoes compression, heating, shearing, and mixing effects initially used for metal processing. As compared to mechanical comminution methods for pretreatment of biomass, extrusion is more effective because of the fiber shortening and defibrillation effects and requires less energy. Also, the extrusion method for pretreatment of biomass has additional advantages such as better process monitoring and control, adaptability and scalability, short residence times, and production of lesser inhibitors (Duque et al., 2017; Kumar and Sharma 2017; Sharma and Saini, 2020).

9.5.1.3 Microwave Irradiation (Dielectric Heating)

Microwaves are electromagnetic waves as non-ionizing radiations with a frequency of 300 MHz–300 GHz and a wavelength of 1 mm–1 m (Hassan et al., 2018). It has been considered for the pretreatment of lignocellulosic biomass as an intensified approach (Sharma and Saini, 2020). The biological and non-biological transformation processes are accelerated effectively by providing non-thermal and thermal effects using microwaves. Heating biomass using microwaves reduces the requirement of chemicals for pretreatment and also enhances the rate of reaction. The microwave heating of biomass does not involve any material for uniform heat transfer as compared to conventional heating. The microwave irradiation heating leads to changes in the ultrastructure of cellulose, complete or partial removal of hemicellulose and lignin, or degradation of these materials, and an increase in the porosity of lignocellulose biomass. This method has been getting more attention due to its many promising advantages, including ease of operation, shorter heating time, faster heat transfer, volumetric heating, selectivity, low-energy requirements, instant control and accurate regulation, and minimum inhibitor release (Kumari and Singh, 2018; Sharma and Saini, 2020).

9.5.1.4 Ultrasonication

The suspended biomass in the aqueous phase is exposed to ultrahigh frequency in the range of 100 kHz–1 MHz to provide sound waves in the ultrasonication method of pretreatment. Due to the exposure of a sound wave, a cavitation process occurs through the formation and collapse of bubbles at high intensities, where the complex mechanism of chemical and mechanical processes is responsible for the degradation of biomass pretreatment. The local hot spots are formed due to the collapsing of bubbles, and this wields the thermal effects that contribute significantly to the treatment of biomass. In this process, water vapors present in cavities are dissociated into H, O, and OH radicals (Saini et al., 2015; Kumari and Singh, 2018; Sharma and Saini, 2020). The disruptive shear forces are generated due to bubble implosion taking place from the surface to the interior of the biomass via free radicals (Saini et al., 2015). The oxidation and solubilization of organic molecules presents in biomass is achieved by oxidative radicals (Bundhoo and Mohee, 2018). In this process, the particle size is reduced, increasing the surface area of the biomass. The major steps observed in ultrasonication treatment of lignocellulose biomass involve deconstruction of lignocellulose, rupture of hemicelluloses, and cellulose crystallinity reduction. These changes improve the cellulolytic enzymes' accessibility for the formation of sugars through the conversion of carbohydrates by the hydrolysis process (Sharma and Saini, 2020).

9.5.1.5 Electron Beam Irradiation

In the electron beam irradiation process, lignocellulosic biomass is exposed to a highly charged and accelerated electron stream from an electron beam accelerator. The moving electrons carry high energy, and a beam of electrons is bombarded at the biomass surface, where changes in biomass occur due to various effects. These include depolymerization of cellulose because of breaking of chain, oxidative chemical changes in biomass, hydrogen bond splitting in cellulose and hence reduction in crystallinity, and also changes in structure due to disruption using free radicals. These effects are responsible for the enhancement of enzymatic hydrolysis of lignocellulose (Saini et al., 2015; Sharma and Saini, 2020).

9.5.2 Physicochemical Pretreatment

Chemical and physicochemical pretreatment methods for biomass include pretreatment with acids (hydrochloric acid, sulfuric acid, phosphoric acid, etc.), pretreatment with alkalis (sodium hydroxide and other hydroxides, ammonia, ammonium sulfite, etc.), explosion methods (ammonia fiber explosion, steam explosion, SO_2 explosion, CO_2 explosion, etc.), using oxidizing agents (wet oxidation, hydrogen peroxide, ozone, etc.), pretreatment with organosolv (ethanol, butanol, benzene, ethylene glycol, etc.), and pretreatment with ionic liquids (Sharma and Saini, 2020).

9.5.2.1 Alkali Pretreatment

In the application of lignocelluloses for the production of bioethanol, the alkaline treatment method as a pretreatment has been intensively considered, which alters the lignocellulose chemically and structurally. Alkalis dissolve lignin present in biomass via saponification (de-esterification), and further, this disintegrates the ester linkages (lignin–carbohydrate linkages) between various monolignols and also between hemicellulose and lignin. This increases lignin removal and biomass digestibility (Kucharska et al., 2018; Sharma and Saini, 2020).

9.5.2.2 Alkaline Peroxide Pretreatment

Biomass digestibility is increased with the treatment of hydrogen peroxide by biomass delignification. Hydrogen peroxide (superoxide and hydroxyl radical) is a strong oxidizing agent and provides better pretreatment than the alkali treatment method. This technique has been extensively and successfully applied for the pretreatment of different types of biomass, such as husk (rice), straws (wheat, rice, barley, etc.), bagasse (sweet sorghum, sugarcane, etc.), bamboo, corn stover, softwood, and other parts of plants (Dutra et al., 2018; Kumari and Singh, 2018; Sharma and Saini, 2020).

9.5.2.3 Acid Pretreatment

Acid hydrolysis of biomass is another most popular process for pretreating biomass to reduce recalcitrance in the process of bioethanol production. Various organic and inorganic acids that have been used for the treatment of biomass feedstock for ethanol production include acetic acid, maleic acid, oxalic acid, propionic acid, formic acid, sulfuric acid (H_2SO_4), hydrochloric acid (HCl), phosphoric acid (H_3PO_4), and nitric acid (HNO_3) (Kumari and Singh, 2018; Baruah et al., 2018; Sharma and Saini, 2020). In typical pretreatment of biomass, diluted sulfuric acid is used at 190°C for reaction with the biomass, and then separation of the hydrolysate mixture is performed by filtration. After neutralization and reacidification, liquids and solids are sent to the saccharification section (Gonzalez-Contreras et al., 2020).

9.5.2.4 Organosolv Pretreatment

Organic solvents are used for the extraction of lignin from biomass, and this method is like organosolv pulping but with a lower level of delignification. The typical solvents applied for organosolv treatment of biomass feedstock for ethanol production are ethanol, acetone, 4-hydrogenation furfuryl alcohol, methanol, and ethylene glycol. Also, catalysts can be used in this process to enhance the process and reduce the operating temperature. A few inorganic and organic acids and bases and their salts used in this process are HCl, H_2SO_4, oxalic acid, salicylic acid, and formic acid, $MgCl_2$, NaOH, $Fe_2(SO_4)_3$ (Chen et al., 2017; Nitsos et al., 2017; Kumar and Sharma, 2017; Sharma and Saini, 2020). It has many advantages, including the easy recovery of solvent by distillation. In the alkali pretreatment method, high-purity lignin is recovered in the form of a solid precipitate and the hemicellulosic as a liquid fraction, which is considered and employed as potential and promising substrates for the production of various value-added products, including bioethanol in biorefineries (Zhao et al., 2009).

9.5.2.5 Steam Explosion Pretreatment

The steam explosion method for pretreatment of biomass has been studied most extensively among various physicochemical pretreatment approaches for deconstruction of biomass for enhancement of the digestibility of biomass (Sharma and Saini, 2020). It is a kind of physicochemical process for biomass pretreatment in which high-pressure steam at 200°C is used. Subsequently, the material is expanded adiabatically to atmospheric pressure in a flash unit, and then the hydrolyzed biomass is sent to the saccharification section (Gonzalez-Contreras et al., 2020).

9.5.2.6 Wet Oxidation

In the wet oxidation process for pretreatment of biomass feedstock, simple thermal oxidation of the biomass is performed in the presence of pure oxygen or air using water, which acts as an oxidizing agent. Biomass is soaked in an aqueous solution before oxidation and then oxidized at higher temperatures (125°C–370°C) and pressures of 0.5–2 MPa for around 30 min. The wet oxidation method for pretreatment of biomass is an efficient approach to unwrap or release the cellulose structure. The hemicellulose present in biomass is hydrolyzed to form various organic acids and sugar monomers, and other processes include partial degradation of cellulose and cleaving and oxidation of lignin. The efficiency of the wet oxidation approach for pretreatment of biomass depends on various process parameters, such as operating temperature, retention or residence time, and oxygen pressure (Chen et al., 2017; Kumar and Sharma, 2017; Den et al., 2018; Kumari and Singh, 2018; Sharma and Saini, 2020).

9.5.2.7 Ammonia Fiber Explosion Method

The ammonia fiber explosion method combines alkali treatment and steam explosion using anhydrous ammonia with a 1:1 ratio in biomass, which is carried out at 60°C–100°C and 1–5.2 MPa for around 5–60 min with the rapid release of pressure. Almost 97% of the ammonia is recovered and recycled in this process. During this process, the combined effects of temperature and explosion trigger the following changes: (1) biomass swelling, (2) disruption of structure of lignocellulose at ultra- and macro-levels, (3) breaking of ester linkages into complexes of lignin–carbohydrate, (4) lignin removal (partial), (5) hydrolysis of hemicellulose, (6) deacetylation of acetyl group compounds, and (7) decrystallization of cellulose, as a result it enhances the biomass digestibility (Kumar and Sharma, 2017; Chen et al., 2017; Baruah et al., 2018; Sharma and Saini, 2020).

9.5.2.8 CO_2 Explosion (Supercritical CO_2)

Carbon dioxide has been considered as a green solvent, and it can be potentially used under critical conditions for various applications, including the pretreatment of biomass feedstock for ethanol production. Carbon dioxide has a critical temperature of 31.1°C and a critical pressure of 1,071 psi, as compared to 374.2°C and 3,208 psi for water. The carbon dioxide explosion or supercritical carbon dioxide approach is a kind of multifunctional process where steam explosion and supercritical CO_2 explosion are performed together to obtain synergic effect for the pretreatment of biomass. Supercritical carbon dioxide has the mass transfer properties of gases and the salvation properties of liquids, which help to speed up the carbon dioxide diffusion in the interior of biomass and then the dissolution of various components present in biomass. A higher yield of sugar is obtained due to an increased rate of penetration, which is achieved because of the easy penetration of carbon dioxide at high pressure. The deconstruction of the matrix of lignocellulosic biomass takes place due to the release of pressure, which further results in an increase in surface area and, subsequently, the rate of enzymatic hydrolysis. The formation of carbonic acid by combining the CO_2 with water catalyzes the hemicellulose hydrolysis. The carbon dioxide explosion approach has advantages of low cost of solvent, non-toxicity, no degradation of sugars, non-flammability, effectiveness for high biomass loadings, low pretreatment temperatures, environmental acceptability, and easy recovery after extraction (Chen et al., 2017; Baruah et al., 2018; Sharma and Saini, 2020).

9.5.2.9 SO_2 Explosion

It is similar to a CO_2 explosion where biomass is treated using steam in the presence of SO_2 (1%–4%). The presence of SO_2 mediated lower temperatures for treatment and reaction times and decreased inhibitor formation (Bura et al., 2002; Chen et al., 2017; Sharma and Saini, 2020). This method was mostly employed for the pretreatment of softwoods, which resulted in cellulose-rich biomass.

9.5.2.10 Ionic Liquids

Ionic liquids have been considered as a green solvent and are widely used in many applications such as separation, catalysts, and solvents. It has negligible vapor pressure, thereby being mostly non-toxic and safe. Ionic liquids, which mainly consist of cations and anions, are employed for the pretreatment of rice straw, poplar, and bamboo (Weerachanchai and Lee, 2014; Kumar and Sharma, 2017; Wang et al., 2017; Chen et al., 2017; Kumari and Singh, 2018). However, the use of ionic liquids for the pretreatment of biomass is under research, and it can be extended to an industrial scale. There have been many challenges that must be overcome to adapt on a commercial scale. These major challenges are ionic liquid costs, viscosity of ionic liquids, which restricts mass transfer, ionic liquid toxicity toward enzymes and microorganisms, and energy requirement for regeneration and recycling of ionic liquids (Kumari and Singh, 2018; Sharma and Saini, 2020).

9.5.3 Biological Pretreatment

Biological pretreatment and lignocellulolytic microorganisms (fungi [brown rot, white rot, and soft rot] and bacteria) are employed to deconstruct biomass to produce bioethanol. A wide variety of natural microorganisms is available, mostly for breaking lignin. Depolymerization of lignin occurs through the hydrolytic action of extracellularly secreted ligninolytic enzymes, which results in the easy accessibility of cell wall carbohydrate polymers (Sharma and Saini, 2020). Various biomass feedstock, such as rice straw, *Parthenium hysterophorus* weed, wheat straw, sugarcane bagasse, rapeseed straw, corn stover, and sorghum husk, have been treated with different microorganisms such as *Cerrena unicolor*, *Phanerochaete chrysosporium*, *Chaetomium brasiliense*, *Phlebia brevispora*, *Inonotus tropicalis*, *Irpex lacteus*, *Trametes hirsuta*, *Gloeophyllum trabeum*, *Chaetomium globosum*, and *Phanerochaete chrysosporium* (Sharma and Saini, 2020).

9.6 BIOMASS HYDROLYSIS OR SACCHARIFICATION

Biomass hydrolysis or saccharification is the next step after pretreatment of biomass used for converting the carbohydrates polymers (hemicelluloses, cellulose, etc.) into fermentable sugars by an enzymatic or chemical process (Achinas and Euverink, 2016; Sharma and Saini, 2020).

9.6.1 Acid Hydrolysis

In some cases, it has been considered as a pretreatment of carbohydrate polymers. The hydrolysis process produces two streams as liquid and solid fractions. Liquid fractions have products (sugars) from hydrolysis and mainly cellulolignin (cellulose and lignin) as a solid fraction. The acids can be easily penetrated by lignin without much pre-processing of biomass. Sulfuric acid is the most commonly used acid for biomass hydrolysis than other acids such as nitric acid, hydrochloric acid, phosphoric acid, and formic acid. The acids are more effective at breaking down the hemicelluloses and cellulose polymers through hydrolyzing the hydrogen bonds, covalent bonds, van der Waals forces, and various other intermolecular bonds between constituent sugars, which are catalyzed by H^+ ions of the acid (Ghaffar et al., 2017; Sharma and Saini, 2020).

9.6.2 Enzymatic Hydrolysis

The enzymatic hydrolysis of biomass for ethanol production has typically been performed at a pH of 4.5–5 and a temperature of 40°C–50°C without adding any acids. This method has the advantages of higher selectivity, higher yield, lower energy consumption, no corrosion, no degradation products, etc. Comparatively, the enzymatic hydrolysis process is slower than the chemical hydrolysis process, but it is more eco-friendly than the acid hydrolysis process (Kennes et al., 2016; Sharma and Saini, 2020). Various cellulase-producing microorganisms such as fungi (*P. janthinellum, Humicola insolens, P. occitanis, H. grisea, Trichoderma reesei, Melanocarpus albomyces*, etc.), bacteria (*B. cereus, Cellvibriogilvus, Cellulomonas biazotea, Eubacterium cellulosolvens, Paenibacillus curdlanolyticus, Geobacillus* sp., etc.), and actinomycetes (*C. flavigena, Streptomyces* sp., *C. cellulans, S. nitrosporeus, S. lividans, S. flavogrisus, S. drozdowiczii, S. nitrosporus*, etc.) are used for cellulose hydrolysis (Sharma and Saini, 2020).

9.7 FERMENTATION

In ethanol fermentation, lignocellulosic sugars are transformed into ethanol and other bioproducts by a biological process using suitable microorganisms. Pentoses and hexoses are the significant sugars produced by the hydrolysis of lignocellulosic biomass (hemicelluloses and cellulose, respectively) by hydrolysis (Achinas and Euverink, 2016; Sharma and Saini, 2020). Typical reactions for the production of ethanol from these sugars by fermentation are presented below (Kennes et al., 2016):

$$C_6H_{12}O_6 \rightarrow 2C_2H_5OH + 2CO_2$$

$$C_5H_{10}O_5 \rightarrow 1.67C_2H_5OH + 1.67CO_2$$

Theoretically, around 511 g of ethanol is produced per kg of glucose or xylose. In actuality, this may be lower because of many reasons, such as metabolic processes. The sugars as substrate can be transformed to bioethanol by fermentation process using different types of microorganisms from the fungal (eukarya domain) and bacterial (prokarya domain) groups. A few of the microorganisms employed for ethanol fermentation are *Saccharomyces cerevisiae, Kluyveromyces marxianus, Zymomonas mobilis, Escherichia coli, Klebsiella oxytoca*, and so on. The ethanol fermentation can be operated in batch, fed-batch, or continuous mode with numerous variations such as recycling of cells, batch culture repetition, and so forth (Kang et al., 2014; Sharma and Saini, 2020). The *Saccharomyces cerevisiae* microorganism is used worldwide for the production of bioethanol by fermentation. Conventionally, fermentation–distillation–dehydration steps are involved, and this plant is mostly not profitable for small capacity plants, say less than 15,000 kL/year (Kumakiri et al., 2021).

9.8 INTEGRATION OF HYDROLYSIS AND FERMENTATION

The integration of hydrolysis and fermentation benefits from enhanced energy efficiency and a reduced number of units, which directly reduce capital and operating costs, process time, and energy consumption in ethanol production. The integration of various operations and processes helps to improve the overall performance of the pretreatment, fermentation, and downstream processes. Different approaches to ethanol production include separate hydrolysis and co-fermentation, separate hydrolysis and fermentation, simultaneous saccharification and co-fermentation, simultaneous saccharification and fermentation, and consolidated bioprocessing (Kang et al., 2014; Devarapalli and Atiyeh, 2015; Sharma and Saini, 2020).

9.8.1 Separate Hydrolysis and Fermentation

In the case of this approach to ethanol fermentation, hydrolysis (saccharification) at around 40°C–50°C and fermentation at around 30°C are performed in separate bioreactors. From pretreated steps, solid and liquid fractions are obtained. The liquid fraction is the hydrolysate, which contains pentoses that further undergo hydrolysis to form sugars and then undergo fermentation. This process has the advantage of maintaining optimum conditions during hydrolysis and fermentation and can be used continuously with the recycling of cells. Apart from these advantages, it has certain disadvantages such as product and substrate inhibition, contamination risk, and being expensive (Kang et al., 2014; Devarapalliand Atiyeh, 2015; Kennes et al., 2016; Sharma and Saini, 2020).

9.8.2 Separate Hydrolysis and Co-Fermentation

In this approach, hydrolysis and fermentation are performed as integrated steps in the same vessel using C_6 and C_5 sugars. Saccharification of cellulose and hemicelluloses produces C_6 and C_5 sugars separately, and then co-fermentation is performed in the same, single vessel to produce ethanol. This process has the advantages of reducing additional vessel costs due to single reactor, operating at optimum conditions, reducing or eliminating viscosity issues, requiring fewer enzymes, and producing higher ethanol concentrations and rates. This process also has certain disadvantages such as inhibition by products and substrate, and limitations of co-fermenting microbes (Kang et al., 2014; Devarapalliand Atiyeh, 2015; Sebayang et al., 2016; Kennes et al., 2016; Sharma and Saini, 2020).

9.8.3 Simultaneous Saccharification and Fermentation

In this approach, only one bioreactor is used for simultaneous enzymatic hydrolysis and fermentation, which improves the yield of ethanol as cellulases do not encounter end-product inhibition (Gauss et al., 1976). Pretreated biomass and enzymes (cellulases) allow for hydrolysis to glucose monomers (mostly hexoses), which are further converted into ethanol by fermentation simultaneously without accumulating glucose and cellobiose. It has a low cost as compared to the separated hydrolysis and co-fermentation approach. Also, it has other advantages such as a lower contamination risk because of ethanol's presence, an enhanced rate, and a higher concentration of ethanol. In simultaneous hydrolysis and fermentation, the optimum conditions of hydrolysis and fermentation are compromised and also has the issues of product inhibition at high concentrations, low mixing, high viscosity, and low heat transfer; cell recycling cannot be performed due to biomass mixing (Kang et al., 2014; Devarapalliand Atiyeh, 2015; Sharma and Saini, 2020).

9.8.4 Simultaneous Saccharification and Co-Fermentation

The integration of biomass hydrolysis and fermentation of C_5 and C_6 sugars is considered as simultaneous saccharification and co-fermentation. This approach has many advantages such as a faster and enhanced rate, a high concentration of ethanol due to co-fermentation, the avoidance of end-product inhibition by sugars as substrate, reduced production costs, and the ability to operate with high solid loadings. Apart from its numerous advantages, it has a few drawbacks such as the difficulty of maintaining optimum operating conditions for hydrolysis and fermentation, as well as co-fermenting microbe limitations (Kang et al., 2014; Devarapalliand Atiyeh, 2015; Sharma and Saini, 2020).

9.8.5 Consolidated Bioprocessing

In the approach of consolidated bioprocessing, cellulase production, biomass hydrolysis, and fermentation or co-fermentation are integrated. A single bioreactor is used in this process for the direct conversion of biomass after pretreatment to ethanol, which further leads to the maximum utilization of the bioreactor for the bioconversion of biomass to ethanol. This type of fermentation configuration for the production of bioethanol from lignocellulosic biomass is also known as direct microbial conversion (Sebayang et al., 2016). This intensifying approach has many advantages, including the requirement of a minimum number of bioreactors, a simplified approach that is economical and energy-efficient, a reduced quantity of exogenous enzyme and substrate, and hence a reduced cost. This process has a significant disadvantage in terms of a lack of suitable consolidated bioprocessing-facilitated microbes with essential features and characteristics, as well as lower yield of ethanol due to the formation of various fermentation co-products (lactic acid, acetic acid, etc.) (Kang et al., 2014; Devarapalliand Atiyeh, 2015; Sharma and Saini, 2020).

9.9 INTEGRATION OF PRODUCTION AND SEPARATION

9.9.1 Conventional Distillation

To purify the ethanol to fuel grade, distillation is the only separation method that has been employed frequently. In distillation, the separation of two or more compounds is obtained based on the relative volatilities of these compounds. Distillation is the oldest and conventional approach for the separation of bioethanol, having certain advantages and disadvantages. In the traditional distillation method for separating ethanol, the fermentation products containing water and ethanol are heated in a column to an elevated temperature to separate them. Due to its lower boiling point as compared to water, ethanol evaporates faster than water and is collected in containers after condensation (Sharma and Saini, 2020). By distillation, ethanol may be concentrated up to 95% using a series of distillation columns, and later, it will form an azeotrope with water. Ethanol–water azeotrope is formed at 95.6 mass percent ethanol; hence, conventional distillation is not useful for getting the fuel grade of 99.6% pure ethanol (Clark, 2005). Hence, it is necessary to look for other methods or require certain modifications in the distillation process to break the azeotrope or bypass it.

9.9.2 High-Temperature Fermentation with Vacuum Distillation

For small capacity plants, conventional approaches are not suitable, and innovative approaches are needed. High-temperature (313 K) fermentation technology has simultaneous saccharification and fermentation for ethanol production with the benefits of reducing microbial contamination, cooling costs, and hydrolytic enzymes (Matsushita et al., 2016; Kosaka et al., 2019; Kumakiri et al., 2021). Furthermore, coupling with vacuum distillation improves the production and separation of bioethanol in high-temperature fermentation, lowering equipment costs and production time (Murata et al., 2015; Kumakiri et al., 2021). This process has primary and secondary recovery units, along with a vacuum pump and drain system. This approach has the additional benefit of continuous operation with increased yields and avoids the exposure of microbes to high concentrations of fermentation products.

9.9.3 Azeotropic Distillation

Azeotropic distillation can be employed to achieve more than 96 wt% (89 mol%) purity of ethanol. In azeotropic distillation, another component is added to form a new low boiling azeotrope with two immiscible liquid phases. The components of these two phases are then separated further. Due to the addition of distillation columns for further separation, azeotropic distillation is energy intensive for fuel-grade ethanol production. Ethylene diamine is the most commonly employed separating agent (Timofeev et al., 2003; Sharma and Saini, 2020).

9.9.4 Extractive Distillation

At industrial scale, fuel-grade ethanol is obtained by dehydration of ethanol using an extractive distillation approach. In this approach, a solvent is required as an extractant, which is added to an ethanol–water mixture to change the relative volatility and restrict the formation of azeotropes (Seader and Henley, 1998). The added extractant generally has a higher boiling point than either of the compounds to be separated. More volatile compounds are obtained in a more purified form, and less volatile or high boiling components with extractant as the bottom product. Extractive distillation is a fractional vaporization technique (Sharma and Saini, 2020). In the extractive distillation of ethanol, a solvent is used as an extractant and a separating agent that does not form an azeotrope with ethanol and water. Glycerol is the most successful solvent for ethanol dehydration by extractive distillation (Lara-Montano et al., 2019; Gonzalez-Contreras et al., 2020). In typical extractive distillation, two columns are used. In the first column, 99.98% pure ethanol is obtained as the top product and an extractant–waste mixture as the bottom product, which further regenerates extractant and recycles it to the first column. Ethylene glycol is the famous and most successful solvent used for the separation of ethanol by extractive distillation.

9.9.5 Pressure-Swing Distillation

The azeotropic composition of any mixture changes with respect to operating pressure, and this advantage has been considered in pressure-swing distillation (Seader and Henley, 1998). A lower or higher pressure than atmospheric pressure is used, which shifts the azeotrope. Pressure-swing distillation approach is the most commonly used, modest, and less expensive method for ethanol purification. This method is a better option than extractive distillation and azeotropic distillation for ethanol dehydration to obtain fuel-grade ethanol (Arifeen et al., 2007; Sharma and Saini, 2020). The major drawbacks are listed as follows: only a few percentage improvements in concentration, still expensive, requiring large capacity towers with vacuum or high-pressure conditions, and a large number of stages. Hence, it has not been the good choice for commercial-scale ethanol dehydration.

9.9.6 Reactive Distillation

In this approach of separation, a reaction takes place after each addition of a compound to the mixture of compounds to be separated. The added compound reversibly reacts with one of the compounds to be separated from the mixture, resulting in products with different relative volatility without forming azeotrope, which is separated by using conventional distillation (Seader & Henley, 1998). A suitable separating reactant is required in this process, and a minimum of three distillation columns are needed for complete separation. Because of the requirement for inexpensive material for reactive distillation and the complex nature of the process, it is not yet used for the separation of ethanol–water, but it can be explored for the same.

9.9.7 Adsorption

Adsorption is mostly used to get anhydrous ethanol from 95% ethanol obtained by distillation or other separation methods. Natural adsorbents are considered as the most effective adsorbents. Many adsorbents prepared from agricultural wastes (different seeds, beans, types of cereals, and other feedstocks) have been investigated. The ethanol is dehydrated by using adsorptive molecular sieves, with some drawbacks, such as a large liquid volume requirement (Sharma and Saini, 2020).

9.9.8 Adsorption–Distillation

Adsorption and distillation can be combined for effective separation of ethanol from water. The adsorption columns are operated at an elevated pressure than the desorption column. After the first cycle, the compression of the columns is altered because the adsorber converts the desorber, and so on. The heat required for distillation is utilized from the heat generated by adsorption–desorption process. The distillate obtained from this column is near the azeotropic conformation and is ultimately transferred to the vaporizer. The accumulated product at the bottom comprises only hints of ethanol (Sharma and Saini, 2020).

9.9.9 Molecular Sieve Adsorption Distillation

The microporous beads are used as molecular sieves to pack the column in the molecular sieve adsorption–distillation approach. In this approach, small molecules (water) from the water–ethanol mixture are trapped inside the sieve, while larger molecules (ethanol) pass through the sieves unimpeded. This approach gives almost 99.6 wt% pure ethanol. For regeneration of sieve, the bed is heated to take off the water (Kwiatkowski et al., 2006). Molecular sieves of zeolites are used in beds for adsorptive purification of ethanol. In this process, ethanol (4.4 Å) is retained in the pores of the molecular sieves, while water (2.8 Å) passes through sieves with pores of 3 Å. In this process, almost 99.98% wt/wt pure ethanol is obtained. It has two columns with dehydration and regeneration cycles. Hot air is passed through the loaded and saturated molecular sieve for ethanol recovery and regeneration of molecular sieves (Gonzalez-Contreras et al., 2020).

9.9.10 Reverse Osmosis

In this membrane separation process, osmotic pressure is applied on the feed side (upward side) of the reverse osmosis membrane to achieve ethanol separation from fermentation or a product mixture. Reverse osmosis in the ethanol separation process is a necessary step before distillation (Wenten, 2016). Various studies on the separation of ethanol from the aqueous phase are presented in the literature (Sharma and Saini, 2020). Two units of different membranes for reverse osmosis have been investigated to obtain 96% ethanol from 10% ethanol with more than 95% recovery (Nakao, 1994). This process is more energy-efficient than pervaporation and requires only 1/1000th the energy of distillation (Nakao, 1994; Kumakiri et al., 2021).

9.9.11 Pervaporation

Pervaporation can be applied to separate binary or multicomponent mixtures by fractional vaporization across a porous or non-porous membrane. The ethanol–water mixture is separated by using a hydrophilic or hydrophobic membrane with the driving force of sweeping gas, a vacuum pump, or a temperature change followed by condensation and separation. Pervaporation is suitable to separate an azeotropic mixture of many aqueous and organic mixtures. Pervaporation has been considered as an eco-friendly alternative to conventional methods. Various operating conditions, such as feed quality and quantity, temperature and pressure of feed and permeate, product flow rates, especially permeate, membrane cutoff, nature of the membrane, and so on, directly affect the pervaporation system's performance. Pervaporation using various mostly organic membranes has been widely employed to separate ethanol from fermentation broth and dehydrate ethanol to obtain 99.9% pure ethanol (Khalid et al., 2019).

9.9.12 Fermentation–Pervaporation

The integration or hybrid method can be employed for bioethanol production. Fermentation coupled with pervaporation can be considered an intensifying integration approach with reduced inhibition of microorganisms and increased conversion (Errico, 2017). In such integration, a filtration unit is placed between fermentation and pervaporation units for the removal of suspended solids from the broth. The retentate is recycled to the fermentor, and the permeate is subjected to pervaporation for further concentration of ethanol.

9.9.13 Distillation–Pervaporation

The hybrid process of integration of distillation and pervaporation may result in less energy being required to produce ethanol with more than 99.5 wt% purity (Errico, 2017). Various integration alternatives are possible for ethanol separation and purification to achieve fuel-grade ethanol (Errico, 2017). Distillation with two pervaporation units is used to obtain 99.8 wt% ethanol from 8.8 wt% ethanol in fermentation broth (Tusel and Ballweg, 1983). The major benefit of distillation with two pervaporation processes is reducing the steam requirement for the 1.6–5 kg/L ethanol produced (Tusel and Ballweg, 1983). The different intensifying configurations have been investigated to enhance separation efficiency and reduce overall cost. These are the distillation with heat integration of two distillation and pervaporation systems, the pervaporation unit of plate type with the condenser in a single compact unit (Lurgipervaporator), the pervaporation system between two distillation units, and so forth (Sander and Soukup, 1988; Luyben, 2009; Nagy et al., 2015). The integration of the pervaporation system between two distillation units shows a 28% saving in the capital cost and a 40% reduction in the operating cost compared with separation by distillation using an entrainer (Brüschke and Tusel, 1986).

9.9.14 Membrane Liquid Extraction

In the membrane liquid extraction approach, the solvent phase is on the permeate side, which enhances the transport of ethanol to permeate through a microporous membrane based on the solubility difference. In such systems, a nearly three-fold increase in substrate utilization is observed for various batch and fed-batch cycles in ethanol production and has also been investigated by many researchers (Núñez-Gómez et al., 2014; Snochowska et al., 2015; Aslam et al., 2017).

9.9.15 Vapor Permeation

Vapor permeation can be used in situ for avoiding product inhibition in ethanol fermentation. Vapor permeation means transporting molecules from a vapor feed mix solution to a vapor permeate via a non-porous membrane. Vapor permeation is carefully connected to gas permeation contrary only in that a vapor mixture comprises substances that are condensable at typical conditions. In contrast, a gas combination comprises only permanent gases. It can be integrated with other methods such as pervaporation and distillation to enhance the productivity and separation of ethanol (Sharma and Saini, 2020).

9.9.16 Distillation–Membrane Separation

Distillation coupled with membrane separation is one of the energy-saving alternatives to azeotropic distillation. Membrane separations such as membrane distillation, pervaporation, nanofiltration, and reverse osmosis have been employed to separate ethanol.

9.9.17 Mechanical Vapor–Recompression Distillation with Membrane Vapor Permeation

In mechanical vapor compression, the overhead vapor is compressed and used as the auxiliary fluid in the column reboiler, where it exchanges its latent heat of condensation. Part of the condensed vapor is used as liquid reflux, and the other part is the final product. This technology alone is not able to reach fuel-grade purity, so other separation units are normally required. In this case, coupling distillation with membrane vapor permeation has the double benefit of increasing the achievable purity while reducing the energy consumption. The main issue is related to the resistance of the membrane at a temperature of 130°C, which is required to keep the water–ethanol mixture above the dew point. The energy savings in the integrated system were quantified as meeting half the requirement when only distillation was considered for ethanol separation (Huang et al., 2010, Errico, 2017).

9.9.18 Liquid–Liquid Extraction

In the case of liquid–liquid extraction for the removal of ethanol from fermentation broth, a solvent is added. Based on solubility and affinity toward solvents, ethanol separation is achieved. In situ or ex situ modes of operation may be performed, which depend on the biocompatibility of the solvent. Paraffin oil is used for ethanol separation from fermentation broth (Offeman et al., 2008).

9.9.19 Supercritical Approach

Carbon dioxide at a supercritical state can be considered a promising approach for separating ethanol from a diluted stream, including fermentation broth. Supercritical carbon dioxide has been used for the separation of ethanol from dilute or low-quantity fermentation ethanol (0.1–1.7 wt%) (Schacht et al., 2008).

9.9.20 Salt Separations

Ethanol can be dehydrated using salts, where water is dried using salts. In this approach, an azeotropic solution is reached by distillation, and then a salt separation process is employed. Almost pure ethanol can be obtained by this method. Since salt will absorb water but not ethanol, nearly pure ethanol can be collected. The wet salt can then be heated to evaporate the water and dry the salt, which can then be recycled (Mathewson, 1980).

9.10 CONCLUSION

Bioethanol has the great potential to contribute significantly in the industrial and transportation sectors while also reducing the environmental impact in terms of global warming and footprints. A lot of literature is available on the intensification of production and separation of bioethanol to obtain fuel-grade ethanol. PI techniques are being developed and implemented to obtain safer processes with greater equipment efficiency, reduce their size and operating costs, incorporate retrofitting, consume a minimum of energy, generate the least possible amount of waste, and obtain as many products with the least possible amount of raw material, which can be implemented for the production and separation of bioethanol (Vaghari et al., 2015). However, to achieve sustainable designs, it is necessary to analyze aspects of energy integration, waste management, and their environmental impact. Kumakiri et al. (2021) investigated the energy required to obtain 80% ethanol from 10% ethanol. Various alternatives of membrane separation with air sweep have been compared with distillation, and the energy requirements are calculated as 1,199–1,450 and 1,803 W/kg, respectively. Still, there has been a wide scope for improving the pretreatment, fermentation, and separation processes for bioethanol production using process-intensifying equipment and methods.

REFERENCES

Achinas, S., & Euverink, G. J. W. (2016). Consolidated briefing of biochemical ethanol production from lignocellulosic biomass. *Electronic Journal of Biotechnology*, 23, 44–53.

Aizarani, J. (2023). Ethanol fuel production in top countries 2022. *Statista*. https://www.statista.com/statistics/281606/ethanol-production-in-selected-countries/

Antizar-Ladislao, B., & Turrion-Gomez, J. L. (2008). Second-generation biofuels and local bioenergy systems. *Biofuels, Bioproducts and Biorefining*, 2(5), 455–469.

Arifeen, N., Wang, R., Kookos, I. K., Webb, C., & Koutinas, A. A. (2007). Process design and optimization of novel wheat-based continuous bioethanol production system. *Biotechnology Progress*, 23(6), 1394–1403.

Aslam, M., Charfi, A., Lesage, G., Heran, M., & Kim, J. (2017). Membrane bioreactors for wastewater treatment: A review of mechanical cleaning by scouring agents to control membrane fouling. *Chemical Engineering Journal*, 307, 897–913.

Balat, M. (2009). Bioethanol as a vehicular fuel: A critical review. *Energy Sources, Part A: Recovery, Utilization, and Environmental Effects*, 31(14), 1242–1255.

Baruah, J., Nath, B. K., Sharma, R., Kumar, S., Deka, R. C., Baruah, D. C., & Kalita, E. (2018). Recent trends in the pretreatment of lignocellulosic biomass for value-added products. *Frontiers in Energy Research*, 6, 141.

Brüschke, H. E. A. & Tusel, G. F. (1986). Economics of industrial pervaporation processes.. In: Drioli, E. (ed). *Membranes and Membrane Process* (pp. 581–586). New York: Springer Science + Business Media.

Bundhoo, Z. M., & Mohee, R. (2018). Ultrasound-assisted biological conversion of biomass and waste materials to biofuels: A review. *Ultrasonics Sonochemistry*, 40, 298–313.

Bura, R., Mansfield, S. D., Saddler, J. N., & Bothast, R. J. (2002). SO_2-catalyzed steam explosion of corn fiber for ethanol production. *Applied Biochemistry and Biotechnology*, 98, 59–72.

Chen, H., Liu, J., Chang, X., Chen, D., Xue, Y., Liu, P., . . . , Han, S. (2017). A review on the pretreatment of lignocellulose for high-value chemicals. *Fuel Processing Technology*, 160, 196–206.

Clark, J. (2005). Non-ideal mixtures of liquids. Retrieved on 8th September 2021 from http://www.chemguide.co.uk/physical/phaseeqia/nonideal.html.

Den, W., Sharma, V. K., Lee, M., Nadadur, G., & Varma, R. S. (2018). Lignocellulosic biomass transformations via greener oxidative pretreatment processes: Access to energy and value-added chemicals. *Frontiers in Chemistry*, 6, 141.

Devarapalli, M., & Atiyeh, H. K.. (2015). A review of conversion processes for bioethanol production with a focus on syngas fermentation, *Biofuel Research Journal*, 7, 268–280

Duque, A., Manzanares, P., & Ballesteros, M. (2017). Extrusion as a pretreatment for lignocellulosic biomass: Fundamentals and applications. *Renewable En*ergy, 114, 1427–1441.

Dutra, E. D., Santos, F. A., Alencar, B. R. A., Reis, A. L. S., de Souza, R. D. F. R., da Silva Aquino, K. A., & Menezes, R. S. C. (2018). Alkaline hydrogen peroxide pretreatment of lignocellulosic biomass: Status and perspectives. *Biomass Conversion and Biorefinery*, 8(1), 225–234.

Errico, M. (2017). Process synthesis and intensification of hybrid separations. In Rong, B-G. (ed). *Process Synthesis and Process Intensification: Methodological Approaches* (pp. 182–212). De Gruyter. De Gruyter Textbook. https://doi.org/10.1515/9783110465068-005.

Frolkova, A. K., & Raeva, V. M. (2010). Bioethanol dehydration: State of the art. *Theoretical Foundations of Chemical Engineering*, 44(4), 545–556.

Gauss, W. F., Suzuki, S., & Takagi, M. (1976). Manufacture of alcohol from cellulosic materials using plural ferments (Vol. 3990944). Alexandria, VA: Office USPT, Bio Research Center Company Limited. Or United States Patent US3990944.

Ghaffar, A., Yameen, M., Aslam, N., Jalal, F., Noreen, R., Munir, B., . . . , Tahir, I. M. (2017). Acidic and enzymatic saccharification of waste agricultural biomass for biotechnological production of xylitol. *Chemistry Central Journal*, 11(1), 97.

Gonzalez-Contreras, M., Lugo-Mendez, H., Sales-Cruz, M., & Lopez-Arenas, T. (2020). Intensification of the 2G bioethanol production process. *Chemical Engineering Transactions*, 79, 121–126. DOI:10.3303/CET2079021.

Hassan, S. S., Williams, G. A., & Jaiswal, A. K. (2018). Emerging technologies for the pretreatment of lignocellulosic biomass. *Bioresource Technology*, 262, 310–318.

Huang, Y., Baker, R. W., & Vane, L. M. (2010). Low-energy distillation-membrane separation process. *Industrial & Engineering Chemistry Research*, 49, 3760–3768.

Huang, H. J., Ramaswamy, S., Tschirner, U. W., & Ramarao, B. V. (2008). A review of separation technologies in current and future biorefineries. *Separation and Purification Technology*, 62(1), 1–21.

Kaminski, W., Marszalek, J., & Ciolkowska, A. (2008). Renewable energy source-dehydrated ethanol. *Chemical Engineering Journal*, 135(1–2), 95–102.

Kang, Q., Appels, L., Tan, T., & Dewil, R. (2014). Bioethanol from lignocellulosic biomass: Current findings determine research priorities. *Scientific World Journal*, 2014, 13. DOI:10.1155/2014/298153.

Kennes, D., Abubackar, H. N., Diaz, M., Veiga, M. C., & Kennes, C. (2016). Bioethanol production from biomass: Carbohydrate vs syngas fermentation. *Journal of Chemical Technology & Biotechnology*, 91(2), 304–317.

Khalid, A., Aslam, M., Qyyum, M. A., Faisal, A., Khan, A. L., Ahmed, F., . . . , Bazmi, A. A. (2019). Membrane separation processes for dehydration of bioethanol from fermentation broths: Recent developments, challenges, and prospects. *Renewable and Sustainable Energy Reviews*, 105, 427–443.

Kiss, A. A. & Suszwalak, D. J. (2012). Enhanced bioethanol dehydration by extractive and azeotropic distillation in dividing-wall columns. *Separation and Purification Technology*, 86, 70–78.

Kosaka, T., Lertwattanasakul, N., Rodrussamee, N., Nurcholis, M., Dung, N. T., Keo-Oudone, C., Murata, M., Götz, P., Theodoropoulos, C., Maligan, J. M., et al. (2019). Potential of thermotolerant ethanologenic yeasts isolated from ASEAN countries and their application in high-temperature fermentation. Fuel Ethanol Prod. Sugarcane; BoD—Books on Demand: Norderstedt, Germany, pp. 121–154.

Kucharska, K., Rybarczyk, P., Hołowacz, I., Łukajtis, R., Glinka, M., & Kamiński, M. (2018). Pretreatment of lignocellulosic materials as substrates for fermentation processes. *Molecules*, 23(11), 2937.

Kumakiri, I., Yokota, M., Tanaka, R., Shimada, Y., Kiatkittipong, W., Lim, J. W., Murata, M., & Yamada, M. (2021). Process intensification in bio-ethanol production–Recent developments in membrane separation. *Processes*, 9, 1028. DOI:10.3390/pr9061028.

Kumar, A. K. & Sharma, S. (2017). Recent updates on different methods of pretreatment of lignocellulosic feedstocks: A review. *Bioresources and Bioprocessing*, 4(1), 7.

Kumari, D. & Singh, R. (2018). Pretreatment of lignocellulosic wastes for biofuel production: A critical review. *Renewable and Sustainable Energy Reviews*, 90, 877–891.

Kwiatkowski, J. R., McAloon, A. J., Taylor, F., & Johnston, D. B. (2006). Modeling the process and costs of fuel ethanol production by the corn dry-grind process. *Industrial Crops and Products*, 23, 288–296.

Lara-Montano, O. D., Melendez-Hernandez, P. A., Bautista-Ortega, R. Y., Hernandez, S., Amaya-Delgado, L., & Hernandez-Escoto, H. (2019). Experimental study on the extractive distillation based purification of second generation bioethanol. *Chemical Engineering Transactions*, 74, 67–72.

Luyben, W. L. (2009). Control of a column/pervaporation process for separating the ethanol/water azeotrope. *Industrial & Engineering Chemistry Research*, 48, 3484–3495.

Mathewson, S. W. (1980). *The Manual for the Home and Farm Production of Alcohol Fuel*. Ten Speed Press, Random House, J.A. Diaz Publications, Serrano.

Matsushita, K., Azuma, Y., Kosaka, T., Yakushi, T., Hoshida, H., Akada, R., & Yamada, M. (2016). Genomic analyses of thermotolerant microorganisms used for high-temperature fermentations. *Bioscience, Biotechnology, and Biochemistry*, 80, 655–668.

Murata, M., Nitiyon, S., Lertwattanasakul, N., Sootsuwan, K., Kosaka, T., Thanonkeo, P., Limtong, S., & Yamada, M. (2015). High temperature fermentation technology for low-cost bioethanol. *Journal of the Japan Institute of Energy*, 94, 1154–1162.

Nagy, E., Mizsey, P., Hancsok, J., Boldyryev, S., & Varbanov, P. (2015). Analysis of energy saving by combination of distillation and pervaporation for biofuel production. *Chemical Engineering and Processing: Process Intensification*, 98, 86–94.

Nakao, S. (1994). Optimization of membrane process for concentrating alcohol solution. *Membrane*, 19, 344.

Nitsos, C., Rova, U., & Christakopoulos, P. (2017). Organosolv fractionation of softwood biomass for biofuel and biorefinery applications. *Energies*, 11(1), 50.

Núñez-Gómez, K. S., López-Mendoza, L. C., López-Giraldo, L. J., & Muvdi-Nova, C. J. (2014). Study of acetone, butanol and ethanol liquid extraction from prepared aqueous solutions using membrane contactor technique. *CT&F-Ciencia, Tecnología y Futuro*, 5(4), 97–112.

Offeman, R. D., Stephenson, S. K., Franqui, D., Cline, J. L., Robertson, G. H., & Orts, W. J. (2008). Extraction of ethanol with higher alcohol solvents and their toxicity to yeast. *Separation and Purification Technology*, 63(2), 444–451.

QTR, Quadrennial Technology Review (2015). Chapter 6: Innovating clean energy technologies. In *Advanced Manufacturing, Technology Assessments*. US Department of Energy. https://www.energy.gov/sites/default/files/2017/03/f34/qtr-2015-chapter6.pdf.

Ramshaw, C. (1983). Higee' distillation-an example of process intensification. *Chemical Engineer*, 389, 13–14.
Saini, A., Aggarwal, N. K., Sharma, A., & Yadav, A. (2015). Prospects for irradiation in cellulosic ethanol production. *Biotechnology Research International*, 2015, 13.
Sakthivel, P., Subramanian, K. A., & Mathai, R. (2018). Indian scenario of ethanol fuel and its utilization in automotive transportation sector. *Resources, Conservation and Recycling*, 132, 102–120.
Sander, U & Soukup, P. (1988). Design and operation of a pervaporation plant for ethanol dehydration. *Journal of Membrane Science*, 36, 463–475.
Schacht, C., Zetzl, C., & Brunner, G. (2008). From plant materials to ethanol by means of supercritical fluid technology. *The Journal of Supercritical Fluids*, 46(3), 299–321.
Seader, J. D., & Henley, E. J. (1998). *Separation Process Principles*. 2nd Edition, Hoboken, NJ: Wiley.
Sebayang, A. H., Masjuki, H. H., Ong, H. C., Dharma, S., Silitonga, A. S., Mahlia, T. M. I., & Aditiya, H. B. (2016). A perspective on bioethanol production from biomass as alternative fuel for spark ignition engine. *RSC Advances*, 6(18), 14964–14992.
Sharma, D. & Saini, A. (2020). *Lignocellulosic Ethanol Production from a Biorefinery Perspective*. Springer Nature Singapore Pte Ltd. DOI:10.1007/978-981-15-4573-3.
Snochowska, K., Tylman, M., & Kamiński, W. (2015). Ethanol recovery from low-concentration aqueous solutions using membrane contactors with ionic liquids. *Ecological Chemistry and Engineering S*, 22(4), 565–575.
Tian, Y., Demirel, S. E., Hasan, M. M. F., & Pistikopoulos, E. N. (2018). An overview of process systems engineering approaches for process intensification: State of the art, *Chemical Engineering and Processing - Process Intensification*, 133, 160–210. DOI:10.1016/j.cep.2018.07.014.
Timofeev, V. S., Serafimov, L. A., & Timoshenko, A. V. (2003). Printsipy tekhnologiiosnovnogoorganicheskogoineftekhimicheskogosinteza. Uchebnoeposobiedlyavuzov.
Tusel, G. & Ballweg, A. Method and apparatus for dehydrating mixtures of organic liquids and water. US Patent 4,405,409 Sep. 20, 1983.
Vaghari, H., Eskandari, M., Sobhani, V., Berenjian, A., Song, Y., & Jafarizadeh-Malmiri, H. (2015). Process intensification for production and recovery of biological products. *American Journal of Biochemistry and Biotechnology*, 11, 37–43.
Van Gerven, T., & Stankiewicz, A. (2009). Structure, energy, synergy, time – The fundamentals of process intensification. *Industrial & Engineering Chemistry Research*, 48(5), 2465–2474.
Wang, F. L., Li, S., Sun, Y. X., Han, H. Y., Zhang, B. X., Hu, B. Z., . . . , Hu, X. M. (2017). Ionic liquids as efficient pretreatment solvents for lignocellulosic biomass. *RSC Advances*, 7(76), 47990–47998.
Wasewar, K. L., Singh, S., & Kansal, S. K. (2020). Process intensification of treatment of inorganic water pollutants. In *Inorganic Pollutants in Water* (pp. 245–271). Elsevier Inc.
Wasewar, K. L. (2021a). Intensifying approaches for removal of selenium. In Devi, P., Singh, P., Malakar, A., Snow, D. (eds). *Selenium Contamination in Water* (pp. 319–355). John Wiley & Sons Ltd. DOI:10.1002/9781119693567.ch16.
Wasewar, K. L. (2021b). Process intensification (PI) in wastewater treatments. In *Contamination of Water*. Elsevier, In-Press.
Wasewar, K. L. (2021c). Process Intensification (PI) in wastewater treatments : 1. Basics of PI and inorganic pollutants. In *Contamination of Water*. Elsevier, In-Press.
Weerachanchai, P. & Lee, J. M. (2014). Recyclability of an ionic liquid for biomass pretreatment. *Bioresource Technology*, 169, 336–343.
Wenten, I. G. (2016). Reverse osmosis applications: Prospect and challenges. *Desalination*, 391, 112–125.
Zhao, X., Cheng, K., & Liu, D. (2009). Organosolv pretreatment of lignocellulosic biomass for enzymatic hydrolysis. *Applied Microbiology and Biotechnology*, 82(5), 815.

10 Pervaporation as a Promising Approach for Recovery of Bioethanol

Kailas L. Wasewar
Visvesvaraya National Institute of Technology (VNIT)

CONTENTS

DOI: 10.1201/9781003203452-10

10.1 INTRODUCTION

In order to meet the soaring global energy demand, fossil energy and fuel sources have been exploited at an extraordinary rate. This has brought about general environmental, sociopolitical, and economic concerns regarding the impacts of using fossil fuels on climate change, energy dependency (security), and economic stability. In order to alleviate the issues confronted using non-renewable fossil sources, global attention has veered toward the production of alternative environmentally friendly renewable energy and fuel sources (Soufiani, 2019).

Among various resources for energy, fossil resources are still a major source for energy and chemicals. From the total fossil resources used, around 75% have been utilized for the production of energy and heat, around 20% have been utilized for the production of fuel, and only a few percent have been used for the production of a variety of materials and chemicals (Bušić et al., 2018). One-third of global energy and half of global oil have been utilized for transportation, which accounts for one-fifth of global greenhouse gas emissions (IRENA, 2016). Different alternatives can be found to replace oil in the transport sector, including electricity, natural gas, and biofuels (Pejó, 2020). Biofuels are considered as an interesting option to replace conventional fossil fuels for transport in the medium term due to their biodegradable, non-toxic, and carbon neutral nature (Toor et al., 2020). The production of biofuels (ethanol, butanol, etc.) has been increasing sharply since the last few years, even though the overall contribution of biofuels to total energy for the transport sector is around 4% (IEA, 2019a). Among various biofuels, bioethanol has been well established in terms of its production and utilization, with Brazil and the United States being the leading manufacturers of bioethanol (EPURE, 2019; Toor et al., 2020).

Bioethanol is the promising and potential substitute for fossil-based fuels, such as gasoline, and it has been used in proportions or in pure form in many countries. Also, bioethanol can be utilized for power generation, fuel cells, cogeneration systems, and as a building block for the production of other chemicals, polymers, and green solvents (Choi et al., 2015). A few physical properties of ethanol are presented in Table 10.1.

In bioethanol (95.63% by mass) production, two major separation steps are involved, which require high energy. The first step involves the concentration of ethanol near the azeotrope, which may further require two steps depending on the type of separation process. Generally, the first step is a conventional distillation to obtain ethanol with 92.4–94 wt% purity. The second step in obtaining anhydrous ethanol is dehydration of ethanol. Various methods for separation of ethanol include azeotropic distillation, extractive distillation (with dissolved salt, liquid solvent, hyperbranched polymers, ionic liquids, their mixture, etc.), pressure-swing distillation, membrane separation, and combination methods. Among the available membrane techniques, pervaporation is quite attractive due to its simplicity, low energy demands, and absence of extra chemicals; besides, the vacuum part of the process consumes the majority of energy (Bušić et al., 2018).

10.2 BIOETHANOL

Bioethanol has long been one of the main biofuels of interest that can be produced sustainably from a variety of feedstocks. However, the choice of raw material for bioethanol production has been a matter of controversy for decades. Although the commercial production of first-generation bioethanol benefits from a rather simple and well-matured process, its future application has been doubted as the utilized sugar and starch-based feedstocks (e.g., sugarcane and corn) compete with human food and animal feed, and its expansion increases the risk of indirect land-use change (Soufiani, 2019). Second-generation bioethanol is produced from lignocellulosic materials, such as agricultural and forest residues and municipal and industrial wastes. The complex lignocellulosic bioethanol production process is composed of pretreatment, hydrolysis, fermentation, and product recovery stages. Although lignocellulosic materials are considered relatively abundant, cheap, and not in competition with food and feed, their recalcitrant nature and process complications hinder the technological advancement necessary for worldwide commercialization (Soufiani, 2019).

10.2.1 Characteristics of Bioethanol (Pejó, 2020)

- Higher octane number of bioethanol (100–135) than gasoline (87–94).
- Better fuel efficiency and combustion.
- Reduced emission (hydrocarbons, NO_X, and particulate matter) due to higher oxygen content.
- Very low toxicity and explosion risk in case of spillage of ethanol.
- Requirement of higher temperature (78°C) for evaporation than gasoline.
- Less energy density (33% lower) and higher water miscibility than gasoline.
- Corrosion and degradation of the engines at high concentrations due to presence of water and oxygen.

TABLE 10.1
Few Physical Properties of Ethanol

Properties	Units	Values
Molecular formula	--	C_2H_5OH
Molecular weight	g/mol	46.07
Density (15°C)	g/mL	0.8
Density (20°C)	g/mL	0.9
Viscosity (20°C)	mPa·s	1.2
Viscosity (25°C)	mPa·s	1.074
Boiling point	°C	78
Auto ignition temperature	°C	161
Solubility in water (25°C)	g/L	Miscible
Octane number	--	100
Energy density	MJ/L	21.3
Cetane number	--	5–8
Laminar flame speed	Cm/s	39
C/H atom ratio	--	0.33
Oxygen content	wt%	34.78
Lower heating value	MJ/kg	26.8
Stoichiometric AFR	--	9.02
Latent heat (25°C)	kJ/kg	904
Saturation pressure (38°C)	kPa	13.8

10.2.2 Advantages of Bioethanol (Hilmioglu, 2009)

- Reduction in greenhouse gases emission by 37.1% by blending of ethanol with gasoline and other fuels (E85: 85% ethanol + 15% gasoline) and by 3.9% using E10.
- Decrease in overall ozone formation as ethanol combustion products are less reactive, hence less damage to ozone layer.
- Renewable energy as it is effectively obtained using energy from the sun into usable energy (photosynthesis, biomass feedstock, and processing).
- Benefits of energy security.
- Clean fuel (complete combustion).
- Reduction of high-octane additives.
- Biodegradable.
- Diluted to non-toxic concentrations.

10.2.3 Disadvantages of Bioethanol (Hilmioglu, 2009)

- Requires significant amount of energy and large land for production.
- For more than 10% ethanol fuels, not compatible for non-E85 fuel systems and may also be resulted into corrosion of ferrous components.

- Undesirable spark generation and increased internal wear in electric fuel pumps.
- Not compatible for level gauging indicators with capacitance fuel level system.

10.3 PRODUCTION OF BIOETHANOL

Direct or indirect chemical synthesis is the major production process for ethanol, which uses ethylene as a raw material. Another major biological approach is the fermentation of biomass containing sugars to obtain bioethanol. Apart from these approaches, others are also being developed, such as methanol carbonylation, synthesis gas conversion, homologation of methanol, and so on. These other processes for the production of ethanol are mostly not successful on a commercial scale, which may be due to certain drawbacks. Ethanol can be produced by thermochemical and bioprocess routes. However, fermentation as a bio-route is increasingly being considered due to its high conversion and selectivity, as well as low consumption of energy (Robak and Balcerek, 2018). First-generation biomass (sugarcane, cane juice or molasses, corn, wheat, cassava, or barley) and second-generation biomass (waste biomass that is non-completive for food) are used for the fermentative production of bioethanol. However, due to the food competitive nature of first-generation biomass, mostly waste biomass is preferred for advanced bioethanol production. The biomass types considered for bioethanol are lignocellulosic biomass, municipal solid wastes, industrial byproducts, or crude glycerol. The basic steps mainly involved in the production of bioethanol from various biomasses are pretreatment, hydrolysis, fermentation, concentration, and dehydration (Bušić et al., 2018; Dey et al., 2020). In fermentation, it can be performed in a batch, fed batch, or continuous manner. Separate hydrolysis and fermentation (SHF), simultaneous saccharification and fermentation (SSF), and consolidated bioprocessing (CBP) (also known as direct microbial conversion) are considered for the production of bioethanol from lignocellulosic-based biomass.

10.3.1 Pathways and Microorganisms for Bioethanol

The most popularly utilized microorganism for bioethanol production is the yeast *Saccharomyces cerevisiae*, which results in a high yield from glucose via the EMP (*Refactoring the Embden-Meyerhof-Parnas*) pathway due to its high tolerance to inhibitors and ethanol and is considered safe (GRAS: Generally Regarded As Safe). Other GRAS microorganisms are *kluveromyces marxianus* (thermotolerant) (Tomas-Pejo, 2011). *Zymomonas mobilis* from the bacterial group is also used in the Entner-Doudoroff (ED) pathway for bioethanol production with high ethanol and inhibitor tolerance, and it is also GRAS but more sensitive to inorganic ions (Tomas-Pejo, 2011). These are only able to ferment C6 sugars. Other strains from yeast, such as *Scheffersomyces stipitis*, *Candida shehatae*, and *Pachysolen tannophilus*, are able to convert C5 sugars into ethanol. *Escherchia coli* strains are able to produce ethanol from C6 and C5 sugars (Pejó, 2020; Tomas-Pejo, 2011).

10.3.2 Mode for Fermentation

Apart from substrate concentration, other variables are also important, such as oxygen level, temperature, pH, and concentration of inhibitors and nutrients. The level of control of these parameters is varied for different types of bioreactors considered for ethanol fermentation. The different types of bioreactors used for ethanol fermentation are batch reactors, fed batch reactors, semi-batch reactors, continuous stirred tank reactors, fluidized bed reactors, packed bed reactors, and so on (Niemistö, 2014). The two major steps, hydrolysis and fermentation, can be performed in separate or combined equipment. These include SHF, SSF, and CBP (Niemistö, 2014). The most basic method of ethanol fermentation is batch mode, in which all ingredients and additives are added prior to and during inoculation, with the exception alkali or acid for the adjustment to maintain the pH. In batch mode of ethanol fermentation, yeast cells can function comfortably at high initial substrate concentrations and high final concentrations of ethanol. In case of fed-batch ethanol fermentation mode, the yeast cells are worked at a low concentration of substrate with an increasing concentration of ethanol during the process. The fed-batch mode of operation has many advantages as compared to the batch fermentation mode of operation; these advantages include the ability to prolong culture life, reach the maximum viable cell concentration, and allow a higher concentration of accumulative fermentation product. For continuous ethanol fermentation, various bioreactors or a combination of reactors may be considered, such as a single stirred tank reactor or a reactor series. In the continuous process for ethanol fermentation, feed containing substrate along with culture medium, nutrients, and other additives is pumped continuously into an agitated vessel. As compared to batch fermentation for ethanol production, continuous ethanol fermentation shows higher productivity (Fan et al., 2019).

10.3.3 Typical Production Process

The typical bioethanol is produced by using mostly waste or non-competitive food crop biomass. The basic steps in the bioethanol production process are biomass handling and size reduction, pretreatment of biomass, dewatering of biomass, detoxification, saccharification, enzyme production, fermentation, and separation and purification of bioethanol (Vane, 2005). The lignocellulosic (lignin, cellulose, and hemicellulose) biomass is transformed into sugars of C5 and C6 (xylose and glucose), which are further fermented into ethanol, carbon dioxide, and other byproducts. Almost half a unit mass of bioethanol is produced per unit mass of sugar. In bioethanol production via fermentation, substrate is converted into ethanol and its concentration continuously increases and accumulates in the fermentation broth, which leads to inhibition of cell growth, resulting in low ethanol yield and productivity, low ethanol concentration in the broth, and a load on wastewater treatment due to the large amount of wastewater. As a result, a lot of energy is required for the recovery of ethanol from the broth. Accumulation of bioethanol in fermentation broth is the most challenging and concerning issue and, hence, ethanol must be removed in situ. In the case of typical batch fermentation for ethanol production, at 60–80 g/L, product inhibition for yeast cells starts, and the final concentration may reach a maximum of 140 g/L (Fan et al., 2019).

10.4 DOWNSTREAM PROCESSING FOR BIOETHANOL PRODUCTION

In bioethanol production, the main downstream processing comprises concentration and dehydration, which conventionally performed by distillation and subsequent dehydration by azeotropic distillation or extractive distillation. The primary step is to separate bioethanol from non-volatiles and solids present in fermentation broth by means of a filtration operation, and then concentrate it to an azeotropic point by distillation. Further concentration as a result of dehydration may be performed by azeotropic distillation, pervaporation, molecular sieve adsorption, and other advanced separation processes, such as membrane distillation, which have a lower energy requirement, and are more efficient and easier to operate. Almost 40% of total energy requirements are utilized by distillation in bioethanol production (Pejó, 2020; Muhammad and Rosentrater, 2020). Pretreatment of the biomass, fermentation or biological transformation, separation, and purification are the major steps performed for the production of bioethanol. The main constituent of the fermentation broth is water, then ethanol, which is required to separate and purify the bioethanol to a usable form and may be fuel grade (Zentou et al., 2019). In typical processes for the purification of ethanol from fermentation broth, distillation and molecular sieve adsorption are used, but they have a high capital cost and a high energy requirement (Liu et al., 2018). The improvement in production and separation efficiency and reduction in energy requirement for bioethanol production are the major concerning issues that can be possibly resolved by using suitable serration and purification technologies. The vast majority of them are either distillation systems or non-distillation systems. Few downstream processes for bioethanol recovery are summarized in the following section (Figure 10.1).

10.4.1 Membrane Filtration

Microfiltration (MF), nanofiltration (NF), and ultrafiltration (UF) are the major and primary membrane separation processes, which are basically pressure-driven separation approaches (Niemistö, 2014). The range of pore sizes of membranes is 0.05–10 μm, 1–100 nm, and less than 2 nm for MF separation, UF separation, and NF separation, respectively. The typical applications of these membrane separations for ethanol separation are discussed (Niemistö, 2014). MF membrane separation is used basically for the separation of bacterial cells or microbial yeast cells, solid particles (particle size > 0.02–0.1 μm), and soluble macromolecules. More than 500 Da of molecular size material, such as enzymes, cells, starch, and proteins, are separated using a MF membrane separation process. NF is effectively employed to retain materials having a 100–400 Da molecular weight cut-off size, such as lactose, glucose, and other sugars, from solutions containing ethanol. Also, these membrane separation processes are used along with solid–liquid separation earlier to the pervaporation

FIGURE 10.1 Few downstream processes for bioethanol recovery.

process to remove temperature-sensitive enzymes or cells and to reduce membrane fouling in pervaporation (Niemistö, 2014).

10.4.2 Distillation

Distillation is one of the oldest conventional separation methods useful to obtain ethanol concentrations up to the formation of azeotrope. Two distillation columns are needed for the separation and purification of bioethanol. It contributes almost 40% of the total energy required for the production of bioethanol and, hence, is equivalent in overall production cost.

10.4.3 Azeotropic Distillation

By the addition of a third volatile component or entrainer into an azeotropic mixture, a ternary azeotrope is formed and their relative volatilities are changed, thereby allowing the separation of these components is achieved. In many azeotropic separations, benzene and cyclohexane are the popular entrainers, but due to certain drawbacks, such as being carcinogenic (benzene) and very flammable (cyclohexane), they are not considered for bioethanol separations (Niemistö, 2014).

10.4.4 Extractive Distillation

Similar to azeotropic distillation, in this method a third component is added as a nonvolatile extractant, and it should not form an azeotrope with any of the compounds. A variety of solvents can be considered for the extractive distillation of ethanol with good solubility and a high distribution coefficient and efficiency. These extractants may be salts, organic solvents, ionic liquids, polymers, etc. (Niemistö, 2014).

10.4.5 Membrane Distillation

This hybrid approach is a very good alternative to conventional distillation or membrane separation. Generally, membrane distillation is carried out at temperatures lower than the normal boiling point using highly hydrophobic membranes. The issue of membrane pore wetting is avoided by using high hydrophobicity (Niemistö, 2014).

10.4.6 Pervaporation

Pervaporation is the best competitive method for conventional distillation to achieve 99.9% pure bioethanol. It can be applied with a hydrophobic membrane after fermentation for concentrating the bioethanol to a certain level, which may be more than 80%, and then a hydrophilic membrane for dehydrating the ethanol to 99.9%. It works on the concept of permeation and evaporation. Pervaporation requires less energy as compared to conventional distillation and reduces the overall production cost of bioethanol. More details are discussed in further sections.

10.4.7 Gas Stripping

In gas stripping, a gas is used as a stripping agent to strip out bioethanol from fermentation broth through the dissolution of the selective component into a gas. In gas stripping, anaerobic or inert gases (nitrogen, carbon dioxide, etc.) are circulated through the fermentor, followed by the release of the bioethanol by evaporation followed by condensation (Niemistö, 2014; Zentou et al., 2019). Gas stripping does not require any kind of complex process equipment setup and also does not require much modification to the plant. Aside from the benefits mentioned above, gas stripping is a simple process that requires low investment costs and is not harmful to cell culture. Fermentation integrated with gas stripping increases cell concentration, increases substrate consumption, increases ethanol productivity, reduces product inhibitory effect, and so forth (Niemistö, 2014; Zentou et al., 2019).

10.4.8 Vacuum Fermentation

In this process, continuous removal of bioethanol is performed by providing vacuum to the fermentor or bioreactor. This is a quick process, and complete fermentation is possible in this process using a high sugar concentration medium (Zentou et al., 2019). Vacuum fermentation was first developed to avoid product and substrate inhibition for bioethanol production by Cysewski and Wilke (1977). Bioethanol is evaporated at the fermentation temperature, and consequently, it is condensed through condensation using a chilling water system or cooling system. In the process of vacuum fermentation for ethanol production system, the concentration of ethanol is controlled at low levels to reduce the product inhibition effect on bioprocess effectively on cells or enzymes (Huang et al., 2015; Zentou et al., 2019).

10.4.9 Adsorption

Porous adsorbents with pore sizes similar to the molecular size of ethanol can be considered for the separation and recovery of bioethanol from fermentation broth. In adsorption, the fermentation broth is passed through an adsorbent bed, and the exit stream is recycled back to the fermentor. The adsorption processes typically engage two steps: adsorption and desorption for the recovery of a more concentrated ethanol, and then regeneration and reuse of the adsorbent (Fujita et al., 2011; Zentou et al., 2019). A variety of adsorbents are available for adsorption, but activated carbon is the most successful adsorbent for the recovery of bioethanol from fermentation broth and the aqueous phase.

10.4.10 Liquid–Liquid Extraction

In liquid–liquid extraction, solvent is used as an extractant to remove ethanol from fermentation broth. Mostly, it is carried out in a continuous plug flow manner in a countercurrent column. In the liquid–liquid extraction process for the separation of ethanol, the fermentation broth containing ethanol is passed through an extraction unit to have efficient contact for mass transfer with the extractant. The contact between the extractant and fermentation broth can be direct or indirect by using a packed bed or a non-wetted porous membrane, respectively. The effluent fermentation broth is recycled back to the fermentor. The concentrated ethanol from the loaded solvent is further separated, and the regenerated solvent is again used for extraction. The recovery of ethanol from fermentation broth in in situ mode provides many advantages, such as an enhancement in the rate of fermentation, the use of a high concentration of substrate, less consumption of water, and so on (Kollerup and Daugulis, 1986; Niemistö, 2014; Zentou et al., 2019).

10.4.11 Reverse Osmosis

Reverse osmosis is also the membrane process where the liquid feed is pressurized to separate water on the permeate side through a semipermeable membrane. The removal of water increases the concentration of ethanol, which reduces the energy requirement for distillation. Reverse osmosis is generally placed before distillation. For reverse osmosis, membranes of cellulose acetate (CA), polyamide, and non-cellulose acetate have been considered for the separation of ethanol (Niemistö, 2014).

10.4.12 Vapor Permeation

Vapor permeation is also one of the membrane processes having the same principle as pervaporation but differing in the feed mixture in the vapor phase. Hence, in vapor permeation, a change in phase does not occur. In vapor permeation, compressor or blower is used to pass a vapor feed mixture through the membrane module. In this case, driving force is not affected by feed side temperature, but flux is affected by vapor pressure on the feed side (Niemistö, 2014).

10.4.13 Comparison of Separation Processes

The various microorganisms employed for the production of bioethanol can typically survive up to 10 wt% of ethanol concentration in fermentation broth. Hence, it is essential to continuously remove ethanol from fermentation broth to overcome this situation. Also, to make the overall process more economical, ethanol must be concentrated, possibly at a high concentration, before considering the next refining or purification steps (Wei et al., 2014; Dey et al., 2020).

Distillation is the most popular and widely used conventional method for the separation of bioethanol from fermentation broth. However, distillation has the major drawback of high energy requirement and is difficult to integrate directly with fermentation due to its high temperature. Advanced separation processes, such as membrane distillation, membrane extraction, membrane pervaporation, and so on, have the potential to overcome the various issues that arise in the production of bioethanol. In membrane distillation, ethanol vapors are separated based on the partial pressure difference between ethanol and water. In the case of pervaporation and membrane extraction, the main driving force for separation is the concentration difference between feed and permeate through mass transfer. The resistance of the membrane is one of the limitations in the case of membrane extraction for ethanol.

Pervaporation appears to be more promising than other membrane-based processes for the separation of bioethanol because it can use feed with very low ethanol concentrations. It minimizes the energy requirement of other thermal operations used, such as distillation. The membrane distillation has the advantage of an operating temperature that is generally lower than conventional distillation, and the driving force is not affected by the thermal difference. Pervaporation may have significant limitations, such as temperature sensitivity to microorganisms and the requirement of heat source and recovery system (Dey et al., 2020). A few advantages and disadvantages of various downstream processes for the separation of ethanol from fermentation broth or aqueous streams are listed in Figure 10.2 (Niemistö, 2014).

The selection of a suitable recovery process is based on its techno-economic feasibility on an industrial scale. Furthermore, the production cost of bioethanol is not only dependent on separation costs but also on feedstock costs, enzyme costs, operating costs, pretreatment costs, and fermentation costs (Mendoza-Pedroza and Segovia-Hernandez, 2018). Hence, all parameters, including operating and capital costs, must be considered for a reliable comparison. Based on the studies available in the literature for pilot-scale production of bioethanol, vacuum fermentation has the lowest cost of around 0.6 USD/liter of bioethanol as compared to pervaporation (0.8 USD/liter) and distillation (1.3 USD/liter) (Kaewkannetra et al., 2012; Zhao et al., 2015; Kongkaew et al., 2018). The operating costs for purifying ethanol from 94 to 99.8 wt% have been investigated using various alternative approaches (Kaminski et al., 2008). For a 30-tonne-per-day ethanol production capacity, the total costs of

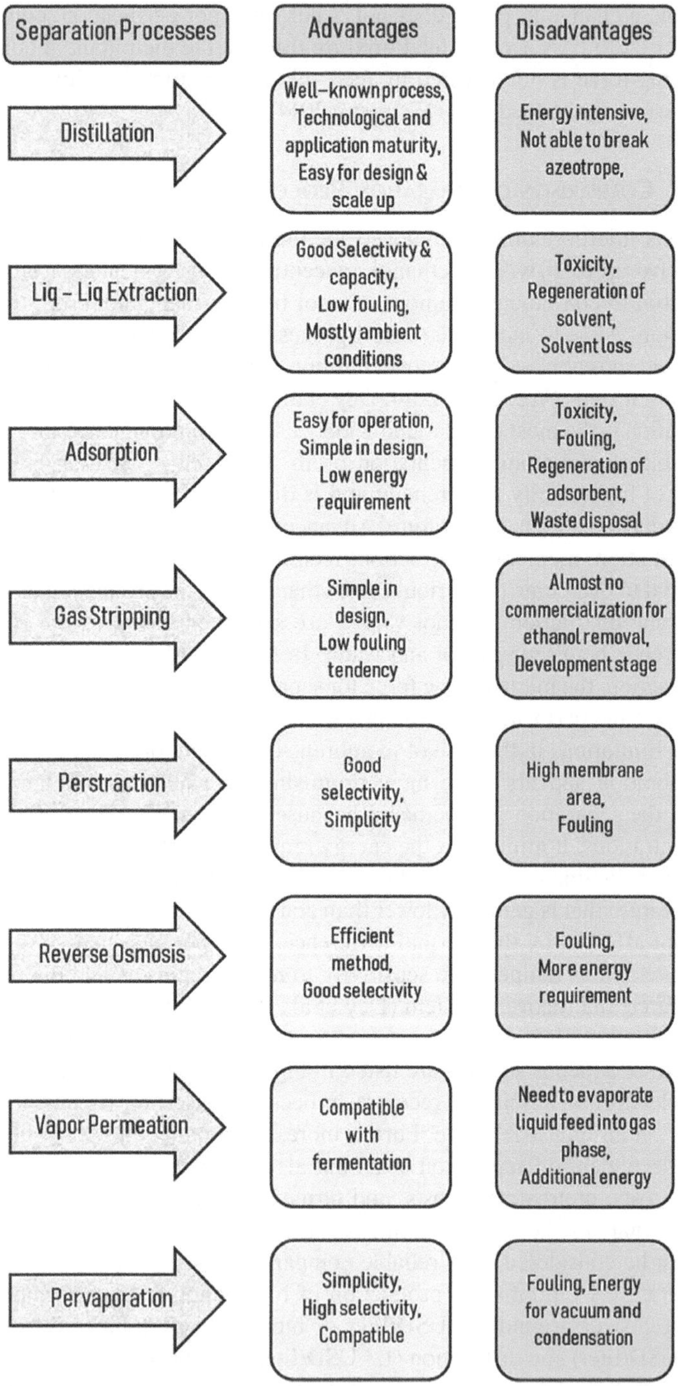

FIGURE 10.2 The advantages and disadvantages of few downstream processes for ethanol recovery from fermentation broth.

USD 31.95–45.65, 36.3, 12.6–16.6, and 15.75 per tonne of dehydrated ethanol have been found using azeotropic distillation, molecular sieve adsorption, pervaporation, and vapor permeation, respectively.

10.5 PERVAPORATION FOR BIOETHANOL PRODUCTION

10.5.1 Basics of Pervaporation

The first time the term pervaporation was coined was by Kober, and it comprises 'permeation' and 'evaporation', which were investigated for the selective removal of water by permeation using cellulose nitrate film from an aqueous solution of albumin and toluene (Kober, 1995). Pervaporation is one of the membrane separation methods where separation is achieved by partial vaporization through a porous or non-porous membrane. It has two steps: the permeation of one of the components through the membrane and its permeation into the vapor phase. The component separation is obtained due to individual component transport rate differences through the pervaporation membrane. Pervaporation membrane process, usually the upstream side (feed side) of the pervaporation membrane, is at ambient pressure and the downstream side (permeate side) is kept at vacuum to evaporate the selective component receiving on permeate side of the pervaporation membrane (Figure 10.3) (Bermudez et al., 2014).

The main driving force for the separation in the pervaporation process is the difference between feed and permeate side partial pressures. The relative volatility difference is not the driving force in this case as compared to distillation. Partial pressure is directly related to the chemical potential, and hence the transport of various feed constituents is achieved by a chemical potential difference between the liquid feed and vapor permeate across the membrane. The separation of mixtures not possible by typical conventional separation processes (extraction, absorption, distillation, adsorption, etc.) can be achieved by pervaporation. In pervaporation, exchange between two phases is achieved without direct contact between the phases. Pervaporation is one of the most energy-efficient membrane processes among various membrane processes and has great potential for the separation and purification of ethanol. Pervaporation has been widely employed mainly for the separation of organic–organic, organic–water, azeotropic solutions, and other organics dissolved in the aqueous phase.

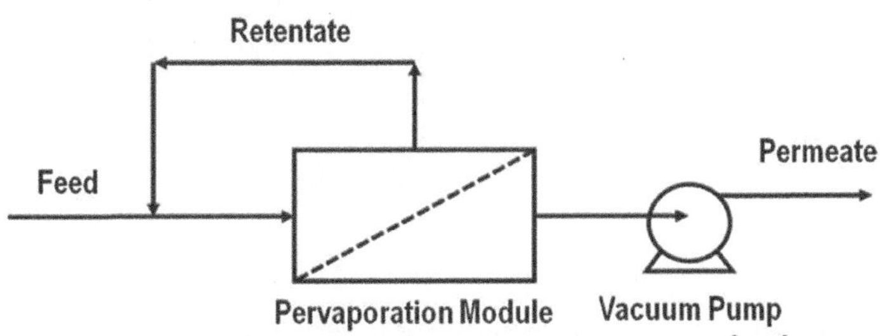

FIGURE 10.3 Typical schematic for pervaporation.

10.5.2 Application

As such, there are many applications of pervaporation in separation and purification, including ethanol separation. A few applications are listed below:

- Dehydration of solvents.
- Azeotropic separation.
- Continuous removal of ethanol from fermentation broth to avoid product inhibition.
- Water removal from esterification reactions to enhance productivity.
- Wastewater treatment by removing organic solvents.
- Pervaporative mass spectrometry.
- Flavors and perfumes concentration.
- Aromatics content reduction from refinery streams.
- Purification of extraction media.
- Purification of product stream after extraction.
- Purification of organic solvents.

10.5.3 Ethanol Mass Transfer in Pervaporation

The transfer of ethanol through the pervaporation membrane includes mainly four steps, which are represented in Figure 10.4 (Fan et al., 2019). These steps are as follows: convective ethanol mass transfer from bulk liquid through liquid boundary layer, where ethanol concentration decreases at interface of liquid boundary layer and

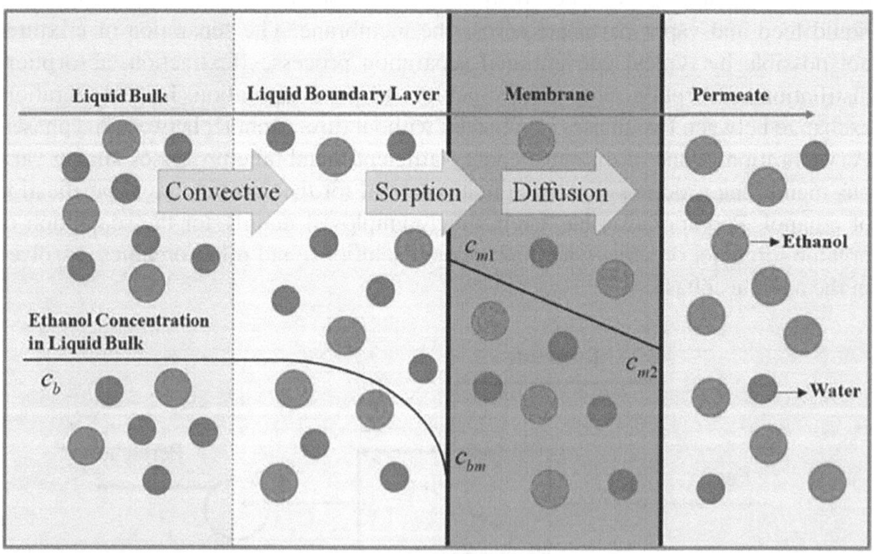

FIGURE 10.4 Ethanol mass transfer mechanism in pervaporation. (Adapted from Fan et al., 2019, Permission to be taken from **Elsevier.** Contact: https://www.elsevier.com/authors/permission-request-form.)

membrane; sorption of ethanol on the upstream side of the membrane; increase in ethanol concentration in the membrane, which is in equilibrium with ethanol concentration in the boundary layer at interface of boundary layer and membrane; diffusive transfer of ethanol (mainly) and other compounds such as water through the pervaporation membrane, where concentration of ethanol decreases due to membrane resistance; and the last step is the desorption of permeated ethanol and other compounds to the vapor phase on the downstream side of the pervaporation membrane, where permeate is further condensed into liquid. In most of the cases, the desorption step is not accounted for because it is very fast (Fan et al., 2019). The second and third steps, convection and diffusion through the boundary layer and membrane, respectively, are the major controlling steps for ethanol transfer in pervaporation.

10.5.4 Ethanol Mass Transport Model for Pervaporation

The pervaporation mechanism can be explained by both a kinetic and a thermodynamic approach. Solubility and diffusion play important roles in the thermodynamic and kinetic perspectives, respectively. Coupled transport mechanisms provide a better understanding of permeation in pervaporation. The mathematical models and approaches have been developed and used to describe the mass transfer from upstream to downstream of the membrane in pervaporation for ethanol separation. These models are pore flow model, solution–diffusion model, and pseudo-phase change model (Fan et al., 2019).

The typical classification of transport models applicable for pervaporation is presented in Figure 10.5. Mainly, pore flow models and solution–diffusion models have been considered for mass transport in pervaporation. The solution–diffusion model is the most well-proven and widely accepted model to describe the flow in the pervaporation process (Fan et al., 2019; Halakoo, 2019).

10.5.4.1 Solution–Diffusion Model

The solution–diffusion model is the mostly utilized model for representing the mass transfer mechanism in pervaporation, which was proposed by Thomas Graham (Lonsdale, 1982). Accordingly, the pervaporation processes can be described as the separation of one of the selective constituents from liquid feed to vapor permeate because of a concentration gradient across the membrane, with partial pressure acting as the driving force. The pervaporation membrane is selective and semipermeable, and vacuum is applied on the permeate side (Halakoo, 2019; Nagy, 2019).

The selective separation through pervaporation membrane basically depends on differences in penetrants of specific constituents through dissolution and diffusion at higher rates, and these are considered as the basis of separation in this model. In the solution–diffusion model, the transport across the pervaporation membrane has three consecutive steps (Figure 10.6) (Halakoo, 2019; Nagy, 2019):

- Step 1: Sorption of the component/s from liquid feed side into the membrane.
- Step 2: Diffusion of the component/s in the pervaporation membrane.
- Step-3: Desorption of the component/s to the vapor phase on the permeate side of the membrane.

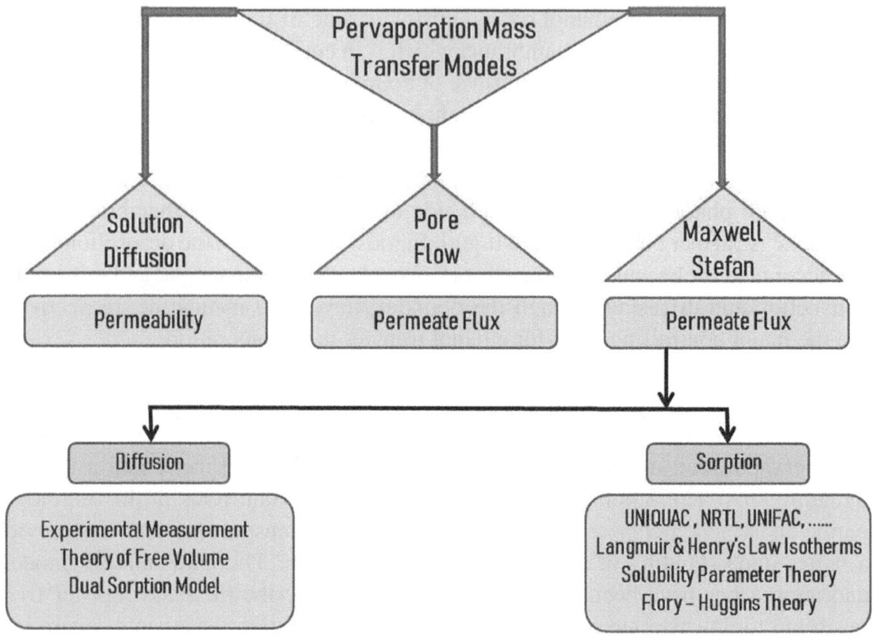

FIGURE 10.5 Classification of transport models applicable for pervaporation.

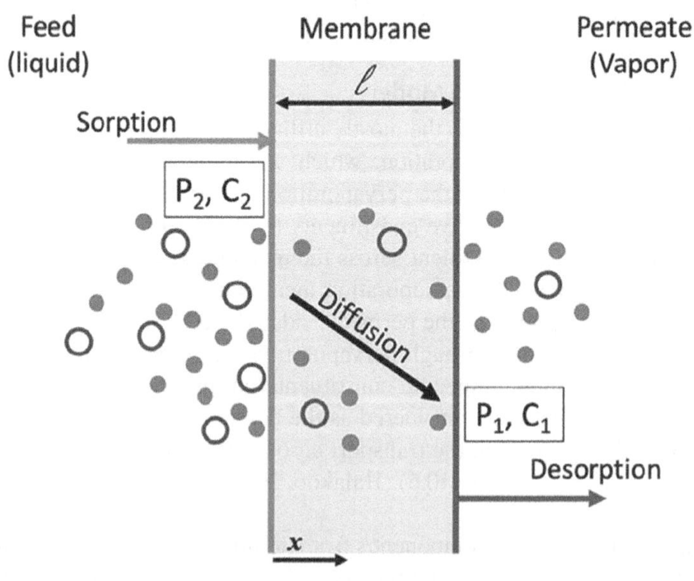

FIGURE 10.6 Schematic to represent solution–diffusion model for ethanol pervaporation.

10.5.4.2 Pore Flow Model

Pore flow model as an alternative transport model for the pervaporation process has been proposed (Okada and Matsuura, 1992). This model assumes straight cylindrical pores on the membrane surface that are perpendicular to the membrane surface. Also, it is assumed that the liquid is filled to a certain level in the pores of the membrane and remains as vapor (Figure 10.7). The liquid–vapor boundary is somewhere in the pore of the membrane. The mass transport according to the pore flow model involves three steps (Halakoo, 2019):

- Step 1: Transport of liquid feed to a liquid–vapor phase boundary from mouth of pore.
- Step 2: Evaporation of selective constituents at the phase boundary.
- Step 3: Transport of vapor from liquid–vapor boundary to the pore outlet at permeate side.

10.5.5 Pervaporation Configuration

In general, the pervaporation process system consists of a feed tank, a feed pump, a heater, a membrane module, a vacuum pump, and a condenser cold trap. The membrane material is selected based on the selective component to be separated. For organic compound selective separation, hydrophobic membranes are preferred, and for the separation of water, hydrophilic membranes are preferred (Zentou et al., 2019).

10.5.6 Pervaporation Membrane for Bioethanol Separation

The most popular organic membranes, such as CA, polyvinylacetate, polydimethylsiloxane (PDMS), silicalite, and polytetrafluro ethylene (PTFE), and polyvinylidene fluoride (PVDF) have been considered for concentrating bioethanol from fermentation broth with 3–15 wt% bioethanol produced by fermentation from biomass, such

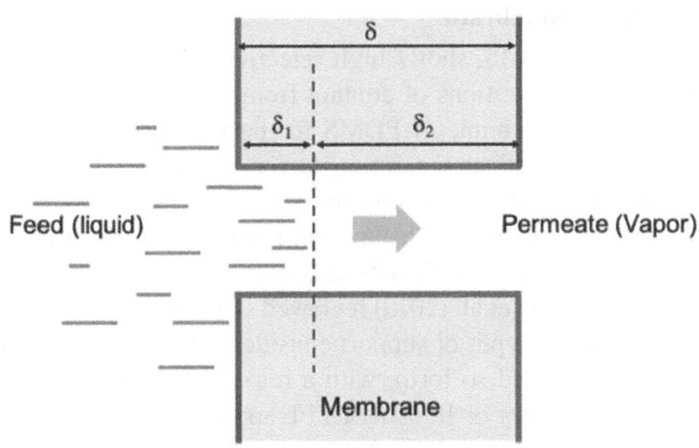

FIGURE 10.7 Schematic to represent pore flow model for ethanol pervaporation.

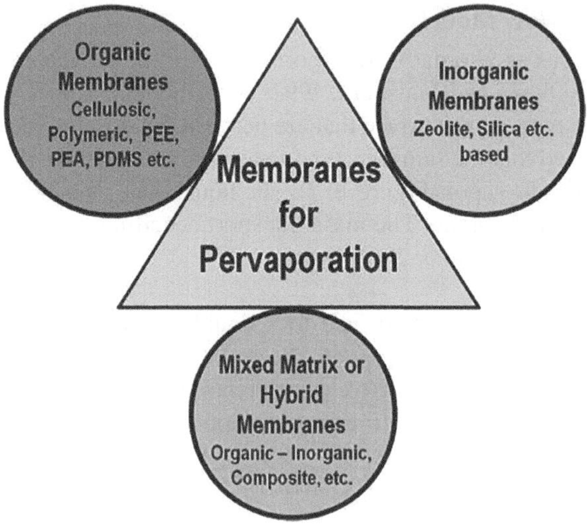

FIGURE 10.8 Different types of pervaporation membranes.

as sorghum juice and stem, willow wood chips, barley straw, banana waste, glucose, lignocellulosic, and newspaper waste (Zentou et al., 2019).

PDMS is the most commonly used polymeric membrane for the removal of bioethanol from fermentation broth due to its high selectivity and good stability. Other polymeric membranes are polyether block amide, poly-1-(trimethylsilyl)-1-propyne, porous polypropylene, polytetrafluoroethylene, and other modified polymeric membranes (Bušić et al., 2018). Zeolite-based inorganic membranes, polyamide–carbon nanotube composite members, and other composite membranes are also used for bioethanol separation. Different types of pervaporation membranes are depicted in Figure 10.8.

10.5.6.1 Polymeric Membrane

Silicon rubber, such as PDMS, shows high selectivity for ethanol and is most suitable to remove low concentrations of ethanol from fermentation broth (Peng et al., 2010). An unmodified membrane of PDMS for pervaporation provides flux in the range of 1–1,000 g/m^2h with a separation factor less than 10 for the removal of ethanol (Beaumelle et al., 1993). The commercial fermentation–pervaporation process is investigated with a 150 g/m^2h flux and a separation factor of 10.3 (O'Brien et al., 2000). The performance of PDMS can be enhanced by certain modifications by providing supports. Peng et al. (2010) reviewed several supported PDMS for ethanol removal. The different types of support considered for PDMS are PTFE CA PA, polystyrene (PS), PVDF, and so forth, with a maximum flux of 8000 g/m^2h using PVDF and a separation factor of 10 using PTFE support.

Poly(1-trimethylsilyl-1-propyne) (PTMSP) is a glassy polymer with a large free volume that has the potential to be another promising polymeric membrane for the removal of ethanol with a separation factor ranging from 9 to 26. This membrane has the advantage of a three-fold higher flux and a two-fold separation factor than PDMS membrane (Schmidt et al., 1997). PTMSP has not found much application in pervaporation as membrane material as it has the limitations of aging (physical and chemical), which deteriorates the properties of the membrane with time (Peng et al., 2010). The applicability of this membrane also can be improved by modifying pervaporation membrane using niobium pentachloride (NbC15) and tantalium pentachloride/triisobutylaluminum ($TaCl_5$/Al(i-Bu)$_3$) and grafting with PDMS (Peng et al., 2010).

Polydimethylsiloxane-imide (PSI) copolymers are synthesized using α-,ω-(bisaminopropyl) dimethylsiloxane oligomers (ODMS), aromatic dianhydrides (PMDA), and 1,3-bis(3-aminopropyl) tetramethyldisolxane (MDMS), with a separation factor of 10.6 and a flux of 560 g/m^{2}h (Krea et al., 2004). Also other membranes, such as polyoctylmethyl siloxane (POMS), poly(ether block amide) (PEBA 2533), perfluoropropane (PFP) thin films on porous polysulfone (PSf) support, polyphenylmethylsiloxane (PPMS)/CA, and dianhydrides:pyromellitic dianhydride (PMDA), are also used (Peng et al., 2010).

The nanocomposite dense membranes for pervaporation having a polystyrene-block-polybutadiene-blockpolystyrene matrix and high aspect ratio cellulose nanofibers are prepared and used for ethanol recovery (Kamtsikakis et al., 2021). Pervaporation experiments were performed with 10 wt% ethanol in water at 40°C, and almost three times more mass fluxes were observed.

10.5.6.2 Inorganic Membrane

Zeolite-based membranes are the main category of inorganic membranes for pervaporation. Zeolites have molecular-level pore sizes with regular crystalline structures that are constituted as hydrated aluminosilicates. The diffusion transport rate of components is different because of their different sizes and strengths of adsorption in zeolite crystals. Porous supports are used to deposit the polycrystalline layers on them to prepare zeolite membranes. The differences between adsorption and diffusion are used to separate components from mixtures. Zeolite membranes have a higher separation factor and flux as compared to polymeric membranes. Zeolites membranes have many advantages, such as high chemical resistance, low swelling, and high thermal stability. Zeolites membranes are one of the potential materials for membrane reactors and molecular sieve membranes (Peng et al., 2010). Zeolite membranes show a higher separation factor, higher efficiency, and higher flux to remove ethanol as compared to a polymeric membrane, PDMS (Vane, 2005). More details of different zeolite-based membranes for pervaporation to separate ethanol are mentioned in the literature (Bowen et al., 2004).

10.5.6.3 Mixed Matrix Membranes

The disadvantages of organic and inorganic membranes can be overcome by using mixed matrix membranes, which contain inorganic fillers such as silica dispersed in polymer. Various fillers, such as silicalite-1, zeolite, silicate, polyphosphazene nanotubes, carbon black, and fumed silica, are used as fillers in PDMS membrane, which may provide a separation factor up to 59 and flux up to 750 g/m^{2}h. More details are available in the literature (Peng et al., 2010). Because of combining the polymeric with inorganic membranes, mixed matrix membranes has the advantage of easy fabrication. Also, cost wise, it is effective with better pervaporation performance. Hence, mixed matrix membranes are a better alternative to polymeric and inorganic membranes for separation of ethanol from aqueous solutions by pervaporation (Peng et al., 2010).

10.5.7 Pervaporation as a Green Process

Pervaporation does not need or release any toxic or hazardous chemicals; hence, it can be considered as a green process as it is environmentally friendly (Khan et al., 2013). It has many advantages and few disadvantages, which are described in the next sections (Vane, 2005).

10.5.8 Advantages of Pervaporation

- Pervaporation breaks azeotropes.
- Possible to achieve high ethanol or water flux for ethanol concentration or dehydration.
- Keep highest possible allowable permeate pressure, hence reduced vacuum requirement.
- Cost of condensation of permeate is lowest permeate condensation due to possible to operate at higher temperatures of coolant.
- High distribution and selectivity of membrane.
- Requirement of low membrane area.
- Less chance of contamination of products.
- Simple process steps.
- No requirment of heavy equipments.
- Can be operated in continuous mode.
- Less energy requirment.
- Low operating, maintenance, and capital costs as compared to other processes.
- Green operation.
- High purity of product.

10.5.9 Disadvantages of Pervaporation

- Not useful for temperature-sensitive microorganisms and compounds.
- Higher possibility to form precipitate.
- Additional cost of heater.
- Need high temperature heat source.
- Required purified feed or pre-filter is required after feed.
- Need of heat exchanger for recovery of heat from residual stream.
- Possibility of failure of material.
- Need additional properties of materials.
- Need more insulation to avoid heat losses.

10.5.10 Mode of Operation

Pervaporation can be employed in both in situ and ex situ modes for the removal of ethanol from fermentation broth. Ex situ mode of operation is preferred because it does not harm the microorganism. The simultaneous removal of ethanol from fermentation broth by pervaporation maintains a low concentration of ethanol, which avoids product inhibition. UF or MF is placed after the fermentor (Vane, 2005; Bušić et al., 2018). The presence of a solid–liquid separation device before the pervaporation system not only removes temperature-sensitive broth components, thereby allowing pervaporation operation at elevated temperatures, but it also protects the pervaporation modules from fouling by solids in the broth (Vane, 2005). The typical fermentation–pervaporation configuration for bioethanol is described in Figure 10.9.

10.5.11 Membrane Modules

Membrane modules used for the pervaporation process are available with various fabrication designs, for example, tubular, plate and frame, spiral-wound, hollow fibrous, capillary fiber, and so forth (Khan et al., 2013). In pervaporation, module liquid feed mixture is associated with the upstream side of the membrane. One component of the feed (solute) is transported preferentially through the membrane and is removed as a vapor from the downstream side. In the majority of the literature, flat-sheet membranes and hollow fiber membrane modules have been investigated for ethanol recovery from model solutions and fermentation broths (Fan et al., 2019). These investigations are performed at temperatures ranging from 25°C to 70°C with 1–8 wt% ethanol and yield fluxes of 3–5,000 g/m^{2}h with a separation factor of 2–15

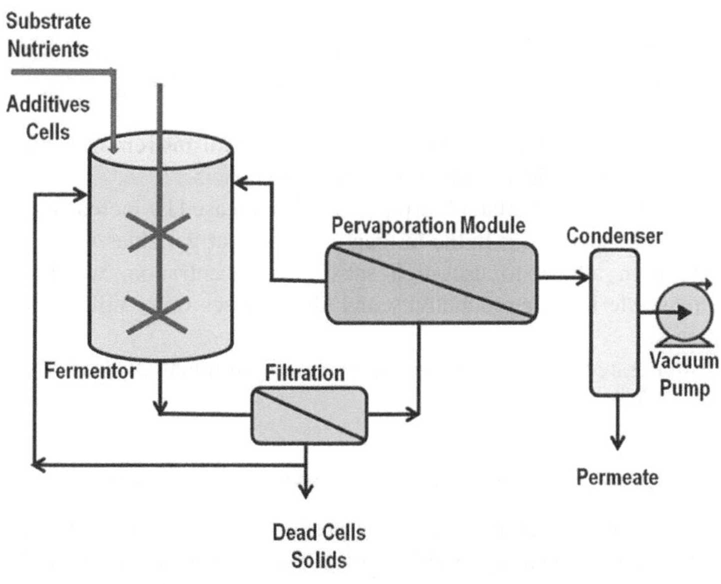

FIGURE 10.9 Typical fermentation–pervaporation configuration for bioethanol.

(Fan et al., 2019). Currently, almost all the commercial pervaporation plants employ flat-sheet and tubular membrane modules; only a few are spiral-wound modules. Compared with the flat-sheet and tubular membrane modules, the spiral-wound configuration allows for a large membrane loading density (300–1,000 m^2/m^3) and can become a less expensive alternative substitute (Liu et al., 2018).

10.6 FERMENTATION WITH PERVAPORATION FOR BIOETHANOL PRODUCTION

10.6.1 Fermentation–Pervaporation

There have been many numbers of variables in the fermentation process, such as type, strain, growth phase, and loading of microorganisms; source, pretreatment, and concentrations of nutrients and substrate; temperature and temperature history; pH and pH history; oxygen concentration; fermentor design; and level of mixing. Furthermore, the variables associated with pervaporation are also included to increase the design complexity, and these variables may be listed as follows: type of membrane, membrane characteristics, distribution and selectivity, flux, operating temperature and pressure, and so on (Vane, 2005).

In typical practice, an integrated system of fermentation and ex situ pervaporation with a UF/MF unit between them is used for ethanol production from sorghum juice, and cost evaluations observed that it was more expensive than other alternatives (Kaewkannetra et al., 2011). The hybrid composite membrane (silicalite-1/polydimethylsiloxane/polyvinylidene fluoride) is used in pervaporation with fed-batch and continuous operations (Cai et al., 2016). This integrated operation enhances separation efficiency and productivity (Bušić et al., 2018).

Based on available literature, general observations can be made on ethanol fermentation (Strathmann and Gudernatsch, 1991; Qureshi et al., 1992; Groot et al., 1992; Roca and Olsson, 2003; Vane, 2005). These are listed as:

- Ethanol fermentation productivity is increased by using removal technology at the level of product inhibition of microorganisms.
- The productivity of ethanol fermentation is increased by increasing the viable cell concentrations in the fermentor for similar variables.
- Integration allows to use high substrate concentration, which directly increases fermentor productivity and also reduces water utilization in the process.
- An integrated system is technically and environmentally sustainable for ethanol fermentation.

10.6.2 Ethanol Fermentation Coupled with Pervaporation

Ethanol fermentation is controlled by the products in the broth, and ethanol is the main product that has a strong inhibitory effect on cell growth and ethanol fermentation. During batch fermentation or the continuous stirred tank reactor (CSTR) operation process without ethanol removal, the cell growth and ethanol productivity would be gradually decreased, owing to the ethanol that had accumulated in the broth. These issues can be overcome by coupling fermentation with pervaporation. The typical schematic for ethanol fermentation coupled with pervaporation (Fan et al., 2019). A few of the other equipment/processes are not shown in the figure like filtration (UF/MF unit) between fermentation and pervaporation modules, condenser, and vacuum system. As compared to conventional ethanol fermentation configurations (fed batch or continuous), this configuration for bioethanol production has many advantages, such as enhanced productivity of ethanol, higher cell density, the ability to be operated continuously for a long time, less wastewater, and a lower energy requirement for further purification (Fan et al., 2019). Fan et al. (2019) summarized various investigations for ethanol fermentation coupled with pervaporation. Mostly, *S. cerevisiae* cells have been used for the biological transformation of glucose in ethanol at temperatures ranging from 30°C to 35°C, with a productivity of 1.6–9.6 g/(L.hr) and a production of 19–822 g/hr (Fan et al., 2019).

10.6.3 Ethanol Fermentation with Thermo-pervaporation

In standard pervaporation configuration for ethanol separation, the permeate is condensed separately in a condenser, which may require high energy. This issue is resolved in thermo-pervaporation, where permeate gases are condensed directly at the cold surface near the membrane at atmospheric pressure in the pervaporation membrane module (Golubev et al., 2018). Four different pervaporation membranes, namely poly[1-(trimethylsilyl)-1-propyne] and commercial polysiloxanes pervaporation membranes (Pervap 4060, Pervatech PDMS, PolyAn, and MDK-3), have been used for ethanol recovery from fermentation broth by thermo-pervaporation with 35% maximum permeate ethanol concentration and 10.2 as separation factor using poly[1-(trimethylsilyl)-1-propyne] pervaporation membrane and highest permeate flux (5.4 kg $m^{-2}h^{-1}$) for PolyAn membrane at 60°C (Golubev et al. 2018).

10.6.4 Pervaporation with Closed Heat Pump

In pervaporation, energy is required for evaporation and condensation. The energy requirement for the separation is directly related to membrane performance characteristics, process configurations, feed composition, and desired separation requirements. The heat that must be removed in order to condense the permeate vapor is approximately the same as the heat required for the evaporation step (Vane, 2005). Ideally, the heat released during condensation can be used to provide the heat of evaporation. Due to heat transfer resistances and the difference between the temperatures of the feed liquid and the permeate condensate necessary to maintain a permeate pressure driving force, the heat released during condensation cannot be directly used to heat the feed liquid. When pervaporation is operated at an elevated feed temperature, the temperature of the permeate condenser may also be elevated relative to ambient temperatures. Thus, the heat released during condensation can be removed with a simple forced air heat exchanger, requiring little energy input. Under these circumstances, the heat of evaporation is the dominant energy sink in the pervaporation process. Vane (2005) used a closed-cycle heat pump in pervaporation unit for maximum utilization of energy. The evaporator portion (cold side) of the heat pump is linked to the condensers of the pervaporation system, while the condenser portion (hot side) of the heat pump is linked to the feed liquid heaters of the pervaporation system.

10.6.5 Pervaporation with Dephlegmation Fractional Condenser

More efficient condensation schemes can also be used to increase the energy efficiency of pervaporation systems. Vane (2005) presented a pervaporation system with a dephlegmation fractional condenser for the effective utilization of energy in ethanol fermentation. In the dephlegmator, a rising vapor is contacted with a falling condensate to generate temperature and concentration gradients in the column. The column contains a high surface area material to enable efficient mass transfer, establishing multiple VLE stages. No reboiler is employed, and the condensate is generated either from an overhead condenser or by operation of the dephlegmator as a countercurrent heat exchanger.

10.6.6 Pervaporation for Recovery and Dehydration

When pervaporation is used for alcohol recovery from the fermentation broth, economic synergies can be realized if the dehydration is also performed by pervaporation since the permeate infrastructure is already in place (Vane, 2005). The permeate vapor from the dehydration pervaporation modules would be processed by the same dephlegmator, condenser, and vacuum pump as the permeate from the alcohol recovery pervaporation modules. The quantity of water removed by the dehydration modules is small compared to the quantity of water in the original permeate from the alcohol recovery modules. As a result, the size and cost of the permeate infrastructure are only marginally changed by the addition of permeate from the dehydration modules. Such a scheme enables the use of dehydration membranes with lower water–ethanol separation factors since the ethanol that is ‘lost’ in the permeate is recovered in the dephlegmator–condenser system. Since flux and selectivity are often inversely correlated, the lower selectivity membranes should deliver high water fluxes and require less area to produce dehydrated ethanol (Vane, 2005).

10.7 DISCUSSION

Second-generation bioethanol production has been encouraged as an alternative solution to the problem raised by the growing energy crisis and environmental insecurity (Dey et al., 2020). Based on the available literature data, it is obvious that bioethanol can be an alternative solution for the current fuel issue. There has been significant progress in renewable biomass pretreatment, cellulase production, and cofermentation of sugars (pentose and hexose), as well as bioethanol separation and purification in recent decades, but bioethanol (based on the production costs) is still not competitive (the exception being bioethanol production from sugar cane in Brazil) with fossil fuels (Bušić et al., 2018).

Conventional routes for the production of bioethanol specifically from cellulosic feed are considered feasible at the laboratory scale, but technical problems and the cost economics involved in most of these processing stages make the overall process unfit for large-scale commercial sustainability. Therefore, there has been an emergent need to develop innovative technologies that can facilitate the large-scale production of bioethanol while minimizing existing difficulties (Dey et al., 2020).

Ethanol purification is a critical process during bioethanol production. In the industry, purification is mainly done by distillation. Distillation is still an effective and favorable separation technique for the bioethanol industry due to several advantages, such as the high separation capacity of ethanol and the simplicity of its application. However, other alternative techniques have been optimized for ethanol recovery that are more energy- and cost-efficient, such as pervaporation, adsorption, gas stripping, and vacuum fermentation. Thus far, the alternative methods have been successfully used for ethanol recovery at the laboratory scale but have not yet met the same acceptance at the industrial scale. Despite the fact that these alternative techniques are less energy-consuming, their integration may imply some technical problems such as maintenance requirements, high sensitivity, and the need for qualified labor, which is not really needed for distillation. For the pervaporation technique, fouling is the most challenging problem hindering the wide application of this method (Zentou, et al., 2019).

Khan et al. (2013) have compared pervaporation with azeotropic distillation, extractive distillation, liquid–liquid extraction, and drying agent-based processes using various criteria, such as separating medium, separation driving force, separation equipment, external chemical separating agent, downstream processing as regeneration and reuse, operational pressure, operational temperature, energy consumption, space requirement, operating hazards, maintenance costs, operating costs, capital costs, environmental impact, and product purity. Based on these exhaustive investigations, pervaporation has been recommended as the best process, and this can be effectively applied for recovery of bioethanol from fermentation broth.

Membrane processes exhibit a unique potential to improve each individual process involved in second-generation bioethanol production by lowering energy requirements and increasing operational flexibility. Judicious integration of membrane processes in hybrid mode can come up with some useful strategies for the successful production and commercialization of bioethanol (Dey et al., 2020).

Pervaporation is an emerging technology with significant potential to efficiently recover alcohols and other biofuels from fermentation broths. As reviewed here, a number of studies have investigated this application, reporting on new membranes, new modules, pervaporation–fermentation integration issues, energy issues, fouling, and costs. Several issues must be addressed for pervaporation to be economically viable and enlisted for biofuel recovery (Vane, 2005):

1. Increased energy efficiency:
    a. Improved ethanol–water separation factor
    b. Heat integration/energy recovery
2. Reduction of capital cost for pervaporation systems:
    a. Reduction in the membrane/module cost per unit area
    b. Increasing membrane flux to reduce required area

3. Longer term trials with actual fermentation broths to assess membrane and module stability andfouling behavior
4. Optimized integration of pervaporation with fermentor:
   a. Filtration (MF or UF) to increase cell density in fermentor and allow higher pervaporation temperatures
   b. Removal/avoidance of inhibitors
5. Synergy of performing both alcohol recovery and dehydration by pervaporation with dephlegmation fractional condensation technology
6. Updated economic analyses of pervaporation, which provide comparisons to competing technologies on even bases at various biofuel production scales

The biggest challenge remains how to reduce the production cost of bioethanol. Therefore, the biorefinery concept is needed to utilize renewable feedstocks more comprehensively and to manufacture more value-added coproducts (e.g., bio-based materials from the lignin) that would reduce the cost of bioethanol production. This will make bioethanol more economically competitive than fossil fuels (Bušić et al., 2018).

REFERENCES

Beaumelle, D., Marin, M., Gibert, H. (1993). Pervaporation with organophilic membranes: state of the art. *Food Bioprod. Process.*, 71(C2): 77.

Bermudez Jaimes, J.H., Alvarez, M.E.T., Villarroel Rojas, J., Maciel Filho, R. (2014) Pervaporation: promissory method for the bioethanol separation of fermentation. *Chem. Eng. Trans.*, 38, 139–144. DOI: 10.3303/CET1438024.

Bowen, T.C., Noble, R.D., Falconer, J.L. (2004) Fundamentals and applications of pervaporation through zeolite membranes. *J. Membr. Sci.*, 245(1–2): 1.

Bušić, A., Marđetko, N., Kundas, S., Morzak, G., Belskaya, H., Šantek M. I., Komes D., Novak S., Šantek, B. (2018). Bioethanol production from renewable raw materials and its separation and purification: A review, *Food Technol. Biotechnol.*, 56(3): 289–311.

Cai, D., Hu, S., Chen, C., Wang, Y., Zhang, C., Miao, Q., et al. (2016). Immobilized ethanol fermentation coupled to pervaporation with silicalite-1 / polydimethyl siloxane / polyvinylidene fluoride composite membrane. *Bioresour. Technol.*, 220: 124–131. https://doi.org/10.1016/j.biortech.2016.08.036.

Choi, S., Song, C. W., Shin, J. H., Lee, S. Y. (2015). Biorefineries for the production of top building block chemicals and their derivatives. *Metab. Eng.*, 28: 223–239. https://doi.org/10.1016/j.ymben.2014.12.007.

Cysewski, G. R., Wilke, C. R. (1977). Rapid ethanol fermentations using vacuum and cell recycle. *Biotechnol. Bioeng.*, 19: 1125–1143.

Dey, A., Pal, A., Kevin, J. D., Das, D. B. (2020). Lignocellulosic bioethanol production: prospects of emerging membrane technologies to improve the process – a critical review. *Rev. Chem. Eng.*, 36(3): 333–367.

EPURE, 2019. Share of European renewable ethanol produced from each feedstock type 2018. https://www.epure.org/resources/statistics-and-infographics/Journal.

Fan, S., Liu, J., Tang, X., Wang, W., Xiao, Z., Qiu, B., Wang, Y., Jian, S., Qin, Y., Wang, Y. (2019). Process operation performance of PDMS membrane pervaporation coupled with fermentation for efficient bioethanol production. *Chin. J. Chem. Eng.*, 27: 1339–1347.

Fujita, H., Qian, Q., Fujii, T., Mochizuki, K., Sakoda, A. (2011). Isolation of ethanol from its aqueous solution by liquid phase adsorption and gas phase desorption using molecular sieving carbon. *Adsorption*, 17: 869–883.

Kollerup, F., Daugulis, A. J. (1986). Ethanol production by extractive fermentation-solvent identification and prototype development. *Can. J. Chem. Eng.*, 64: 598–606.

Golubev, G. S., Borisov, I. L., Volkov, V. V. (2018). Performance of commercial and laboratory membranes for recovering bioethanol from fermentation broth by thermopervaporation. *Russ. J. Appl. Chem.*, 91(8): 1375–1381.

Groot, W. J., Kraayenbrink, M. R., Waldram, R. H., van der Lans, R. G. J. M., Luyben, K. Ch. A. M. (1992). Ethanol production in an integrated process of fermentation and ethanol recovery by pervaporation. *Bioproc. Eng.*, 8: 99–111.

Halakoo, E. (2019). Thin film composite membranes via layer-by-layer assembly for pervaporation separation, PhD Thesis, University of Waterloo, Waterloo, Ontario, Canada.

Hilmioglu, N. D. (2009). Bioethanol recovery using the pervaporation separation technique. *Manag. Environ. Qual.: An Int. J.*, 20(2): 165–174. http://dx.doi.org/10.1108/14777830910939471.

Huang, H., Qureshi, N., Chen, M. H., Liu, W., Singh, V. (2015). Ethanol production from food waste at high solids content with vacuum recovery technology. *J. Agric. Food Chem.*, 63: 2760–2766.

IEA, 2019a. Oil Information 2019. https://www.iea.org/reports/oil-information-2019.

IRENA, 2016. Innovation Outlook Advanced Liquid Biofuels. https://www.irena.org/publications/2016/Oct/Innovation-Outlook-Advanced-Liquid-Biofuels.

Kaewkannetra, P., Chutinate, N., Moonamart, S., Kamsan, T, Chiu, T. Y. (2011). Separation of ethanol from ethanol–water mixture and fermented sweet sorghum juice using pervaporation membrane reactor. *Desalination*, 271(1–3): 88–91. https://doi.org/10.1016/j.desal.2010.12.012.

Kaewkannetra, P., Chutinate, N., Moonamart, S., Kamsan, T., Chiu, T. Y. (2012). Experimental study and cost evaluation for ethanol separation from fermentation broth using pervaporation. *Desalin. Water Treat.*, 41: 88–94.

Kaminski, W., Marszalek, J., Ciolkowska, A. (2008). Renewable energy source – dehydrated ethanol. *Chem. Eng. J.*, 135: 95–102.

Kamtsikakis, A., McBride, S., Zoppe, J. O., Weder, C. (2021). Cellulose nanofiber nanocomposite pervaporation membranes for ethanol recovery. *ACS Appl. Nano Mater.*, 4: 568–579.

Khan, S. M., Usman, M., Gull, N., Butt, M. T. Z., Jamil, T. (2013). An overview on pervaporation (an advanced separation technique). *J. Qual. Technol. Manag.*, IX(I): 155–161.

Kober, P. A. (1995). Pervaporation, perstillation and percrystallization. *J. Membr. Sci.* 100: 61–64.

Kongkaew, A., Tönjes, J., Siemer, M., Boontawan, P., Rarey, J., Boontawan, A. (2018). Extractive fermentation of ethanol from sweet sorghum using vacuum fractionation technique: Optimization and techno-economic assessment. *Int. J. Chem. React. Eng.*, 16. https://doi.org/10.1515/ijcre-2017-0160.

Krea, M., Roizard, D., Moulai-Mostefa, N., Sacco, D. (2004). New copolyimide membranes with high siloxane content designed to remove polar organics from water by pervaporation. *J. Membr. Sci.*, 241 (1): 55.

Liu, J., Li, J., Chen, Q., Li, X. (2018). Performance of a pervaporation system for the separation of an ethanol-water mixture using fractional condensation. *Water Sci. Technol.*, 77(7): 1861–1869. https://doi.org/10.2166/wst.2018.067.

Lonsdale, H. K. (1982). The growth of membrane technology. *J. Membr. Sci.*, 10: 81–181.

Mendoza-Pedroza, J. J., Segovia-Hernandez, J. G. (2018). Alternative schemes for the purification of bioethanol: A comparative study. *Recent Adv. Petrochem. Sci.*, 4(2): 24–32.

Muhammad, I. S. N., Rosentrater, A. K. (2020). Economic assessment of bioethanol recovery using membrane distillation for food waste fermentation. *Bioengineering (Basel, Switzerland)*, 7. https://doi.org/10.3390/bioengineering7010015.

Nagy, E. (2019). Chapter 16: Pervaporation. In Nagy, E. (Ed.), *Basic Equations of Mass Transport through a Membrane Layer* (Second Edition). Elsevier, pp. 429–445, https://doi.org/10.1016/B978-0-12-813722-2.00016-9.

Niemistö, J. (2014). Towards sustainable and efficient biofuels production: Use of pervaporation in product recovery and purification, PhD Thesis, University of Oulu, Finland.

O'Brien, D.J., Roth, L.H., McAloon, A.J. (2000). Ethanol production by continuous fermentation–pervaporation: a preliminary economic analysis. *J. Membr. Sci.*, 166(1): 105.

Okada, T., Matsuura, T. (1992). Predictability of transport equations for pervaporation on the basis of pore-flow mechanism. *J. Membr. Sci.*, 70: 163–175.

Pejó, E. T. (2020). Valorisation of lignocellulosic residues for lactic acid and bioethanol production in a biorefinery context, PhD Thesis, Universidad Complutense De Madrid Enrique Cubas Cano, Instituto Madrileño De Estudios Avanzados en Energía Madrid.

Peng, P., Shi, B., Lan, Y. (2010). A review of membrane materials for ethanol recovery by pervaporation. *Sep. Sci. Technol.*, 46(2): 234–246. https://doi.org/10.1080/01496395.2010.504681.

Qureshi, N., Maddox, I. S., Friedl, A. (1992). Application of continuous substrate feeding to the ABE fermentation: relief of product inhibition using extraction, perstraction, stripping and pervaporation. *Biotechnol Prog.*, 8: 382–390.

Robak, K., Balcerek, M. (2018). Review of second generation bioethanol production from residual biomass. *Food Technol. Biotechnol.*, 56: 174–187. https://doi.org/10.17113/ftb.56.02.18.5428.

Roca, C., Olsson, L. (2003). Increasing ethanol productivity during xylose fermentation by cell recycling of recombinant *Saccharomyces cerevisiae*. *Appl. Microbiol. Biotechnol.*, 60: 560–563.

Schmidt, S. L., Myers, M. D., Kelley, S. S., McMillan, J. D., Padukone, N. (1997). Evaluation of PTMSP membranes in achieving enhanced ethanol removal from fermentations by pervaporation. *Appl. Biochem. Biotechnol.*, 63–65(1): 469.

Soufiani, A. M. (2019). Immersed flat-sheet membrane bioreactors for lignocellulosic bioethanol production, PhD Thesis, Swedish Centre for Resource Recovery, University of Borås, SE-501 90 Borås, Sweden. http://urn.kb.se/resolve?urn=urn:nbn:se:hb:diva-21668.

Strathmann, H., Gudernatsch, W. (1991). Continuous removal of ethanol from bioreactor by pervaporation. In Mattiasson, B., Holst, O. (Eds.), *Extractive Bioconversions*. Marcel Dekker, New York, pp. 67–89.

Tomas-Pejo, E., Alvira, P., Ballesteros, M., Negro, M.J. (2011). Chapter 7: Pretreatment technologies for lignocellulose-to-bioethanol conversion A2- Pandey, Ashok. In Larroche, C., Ricke, S.C., Dussap, C.-G., Gnansounou, E. (Eds.), *Biofuels*. Academic Press, Amsterdam, pp. 149–176. https://doi.org/10.1016/B978-0-12-385099-7.00007-3.

Toor, M., Kumar, S.S., Malyan, S.K., Bishnoi, N.R., Mathimani, T., Rajendran, K., Pugazhendhi, A. (2020). An overview on bioethanol production from lignocellulosic feedstocks. *Chemosphere*, 242: 125080. https://doi.org/10.1016/j.chemosphere.2019.125080.

Vane, L. M. (2005). A review of pervaporation for product recovery from biomass fermentation processes, *J. Chem. Technol. Biotechnol.* 80: 603–629.

Wei, P., Cheng, L. H., Zhang, L., Xu, X. H., Chen, H., Gao, C. (2014). A review of membrane technology for bioethanol production. *Renew. Sustain. Energy Rev.*, 30: 388–400.

Zentou, H., Abidin, Z. Z., Yunus, R., Biak, D. R. A., Korelskiy, D. (2019). Overview of alternative ethanol removal techniques for enhancing bioethanol recovery from fermentation broth. *Processes*, 7: 458. https://doi.org/10.3390/pr7070458.

Zhao, L., Zhang, X., Xu, J., Ou, X., Chang, S., Wu, M. (2015). Techno-economic analysis of bioethanol production from lignocellulosic biomass in China: Dilute-acid pretreatment and enzymatic hydrolysis of corn stover. *Energies*, 8: 4096–4117.

11 Production of High-Performance/ Aviation Fuels from Lignocellulosic Biomass

Nhamo Chaukura
Sol Plaatje University

Charles Rashama and Ngonidzashe Chimwani
University of South Africa

CONTENTS

DOI: 10.1201/9781003203452-11

11.1 INTRODUCTION

Hybrid or solo electric vehicles may possibly replace conventional gasoline-propelled road vehicles. However, this might not be appropriate for the aviation industry. The use of fossil fuels causes a continuous rise in the concentration of greenhouse gases in the atmosphere, a situation that results in global warming (Wang et al., 2014; Raud et al., 2019; Wang et al., 2019; Lim et al., 2021; Neves et al., 2020; Vela-García et al., 2021). For sustainability in aeronautical transportation, the development of a substitute renewable fuel to meet the increasing demand while decreasing reliance on petroleum-derived fuels and the associated costs is urgently required (Perkins et al., 2019; Wei et al., 2019; Escalante et al., 2022). Thus, the substitution of fossil fuels with renewable sources will continue to be a significant issue in the future (Perkins et al., 2019).

Generally, biofuels can be classified according to the biomass feedstock. Specifically, first-generation biofuels, such as ethanol and biodiesel, are principally made from food crops through the use of simple technologies (Raud et al., 2019). The sustainability of first-generation biofuels is, however, problematic due to the competition with food resources. Biofuels can be produced from a range of biowastes, and among these, lignocellulosic biomass is the most suitable feedstock for the production of renewable liquid fuels because of its abundance and distinctive chemistry (Chen et al., 2020). Lignocellulosic biofuels constitute second-generation biofuels and are made from inedible biomass such as agricultural and forest residues and energy crops, which are abundant, low cost, and renewable (Han et al., 2019). Unlike first-generation biofuels, second-generation biofuels have no direct food versus fuel conflict because they use non-edible biomass (Raud et al., 2019). Biomass is an important resource that comprises all organic matter, such as food, farming and forestry residues, crops, and plants, which can serve as an energy source. It is mostly made up of plant-based material and lignocellulosic biomass, which is a key structural component of plant dry matter (Han et al., 2019). Lignocellulose is made up of lignin (10–25 wt%), hemicellulose (20–40 wt%) and cellulose (40–60 wt%) (Han et al., 2019).

Biojet fuel should be suitable for the design of the aircraft engine and the fuel distribution systems. In view of the fact that jet engine technology and the current airport infrastructure are not likely to change any time soon, it is essential to develop a drop-in fuel that can replace the fossil-based jet fuel (Tzanetis et al., 2017; Kargbo et al., 2021; Ng et al., 2021). For partial fossil-based jet fuel substitution, the physicochemical compatibility between both fuels is important due to specifications on properties, such as aromatic content, freezing point, cold flow property, energy density, viscosity, flash point, thermal stability and cloudiness phenomena (Bwapwa et al., 2017; Escalante et al., 2022; Vela-García et al., 2021). In addition, compliance must also focus on chemical kinetics, ignition and extinction features, lubricity flame speed and limits of flammability (Bwapwa et al., 2017). Besides, biojet fuel has to satisfy storage and safety standards, must be widely available and should be economically competitive against current jet fuels (Escalante et al., 2022).

Fossil fuel-derived jet fuels are made up of linear and branched alkanes, cyclic alkanes and aromatics. The densities of linear and branched alkanes are lower, resulting in comparatively poorer volumetric heating capacities, while cycloalkanes have ring structures, which result in higher combustion enthalpies (Chen et al., 2020). Despite their role in the emission of particulate matter during combustion, the presence of aromatics helps in avoiding engine leakage and confers density properties in the required range. Indeed, biojet fuels could be blended with fossil fuel-derived jet fuel in order to achieve this specification (Neves et al., 2020). The ASTM standards require that biojet fuels must be composed of hydrocarbons that do not contain heteroatoms, and the carbon chain length and physicochemical properties must have a narrow range. Oxygenated compounds have high acid values, high viscosity, a low heat of combustion and are unstable, while the presence of nitrogen or sulphur compounds in jet fuel leads to the formation of compounds such as H_2SO_4, NO_x and SO_2 during combustion (Chen et al., 2020; Neves et al., 2020). These specifications make the production of biojet fuels through existing biomass conversion strategies a challenge (Wang et al., 2019). However, biojet fuels enjoy the advantage that they use available feedstock and production technology that can be easily upscaled for industrial-scale production (Escalante et al., 2022).

Conventionally, biojet fuel is produced from biomass through the biochemical conversion of plant oil and the thermochemical conversion of biomass via the Fischer–Tropsch process (Chen et al., 2020; Kargbo et al., 2021). Nonetheless, these technologies cannot produce aromatics and naphthenes, which constitute approximately 40% of fossil fuel-derived jet fuel (Wang et al., 2014). In order to produce isoparaffin, the major constituent of jet fuels, the fuels are upgraded through cracking isomerization. Biochemical conversion produces liquid fuels as the primary product and biogas, which can also be further converted into biojet fuel (Kargbo et al., 2021).

The conversion of lignocellulosic biomass into biojet fuel has great promise in addressing sustainability, cost and environmental pollution concerns. To date, a number of companies have commenced the commercial production of biojet from biomass. In this chapter, we explore the feedstock properties and production processes used in converting lignocellulosic biomass to biojet fuel.

11.2 FEEDSTOCK

The major feedstock for biofuel production for ground transportation applications is biomass, owing to its renewable properties. Likewise, biomass-derived fuels can be potentially used in air transport, which is increasingly becoming a major form of transport. Basically, any biomass of animal or plant origin that contains carbohydrates can be a source of energy (Escalante et al., 2022). The component of interest in the biomass studied in this article is lignocellulose, which consists of cellulose (35%–50%), hemicellulose (20%–35%), lignin (15%–20%), and other minor components (15%–20%) (Kargbo et al., 2021; Vela-García et al., 2021). Lignin is an amorphous phenolic polymer with a higher energy value than cellulose and hemicellulose. The depolymerization of lignin produces a mixture of phenolic compounds, which limits their direct use as fuels (Han et al., 2019). Generally, the natural decomposition rate of biomass is sluggish, so various pretreatment techniques can be used to accelerate the process. The cost of pretreatment techniques should be taken into consideration in the selection of feedstock for biofuel production (Kargbo et al., 2021).

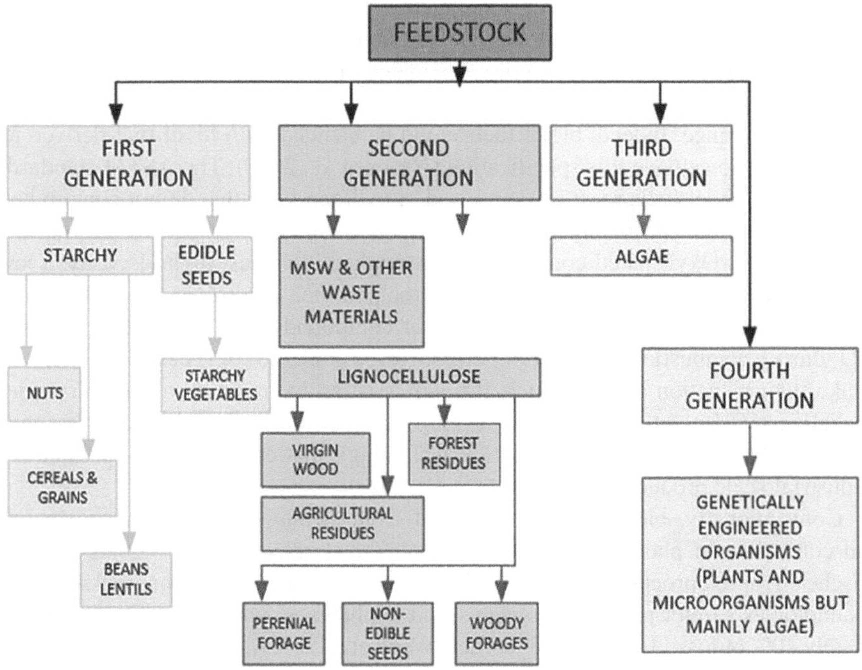

FIGURE 11.1 Biomass feedstock classifications.

In accordance with the simplicity of processing and their sustainability, biofuel feedstocks are classified as first, second, third and fourth generations (Figure 11.1) (Aron et al., 2020; Luthra et al., 2015). First-generation biofuels, mainly biodiesel and bioethanol, are produced from edible crops, which is problematic because they conflict with food supply and land use (Raud et al., 2019; Kargbo et al., 2021). To an extent, the challenges associated with first-generation biofuels can be solved through the use of non-food crops, agro-wastes and forest residues (Kargbo et al., 2021). The latter are referred to as second-generation feedstock, and they include lignocellulosic biomass, which is the focal point of this study on conversion to biojet fuel (Lim et al., 2021).

Lignocellulosic biomass, such as municipal solid waste, forestry residues and agricultural waste, is an abundant and renewable energy source and has been used to produce biojet fuel via different process routes that include gasification coupled to Fischer–Tropsch synthesis (FTS) (Raud et al., 2019; Chen et al., 2020). It mostly consists of cellulose (40–50 wt%), hemicellulose (25–35 wt%), and lignin (15–25 wt%) (Figure 11.2) (Raud et al., 2019; Vela-García et al., 2021). Biowaste is normally disposed of through landfilling or incineration after the extraction of useful components. Conversely, dedicated energy crops can be purposely cultivated for energy use. These crops are easy to cultivate, have low input requirements, produce high energy yields and are tolerant to harsh climatic conditions (Raud et al., 2019). Whereas first-generation biomass uses only seeds, second-generation biomass uses the whole plant as feedstock, and can thus be produced on smaller land (Raud et al., 2019).

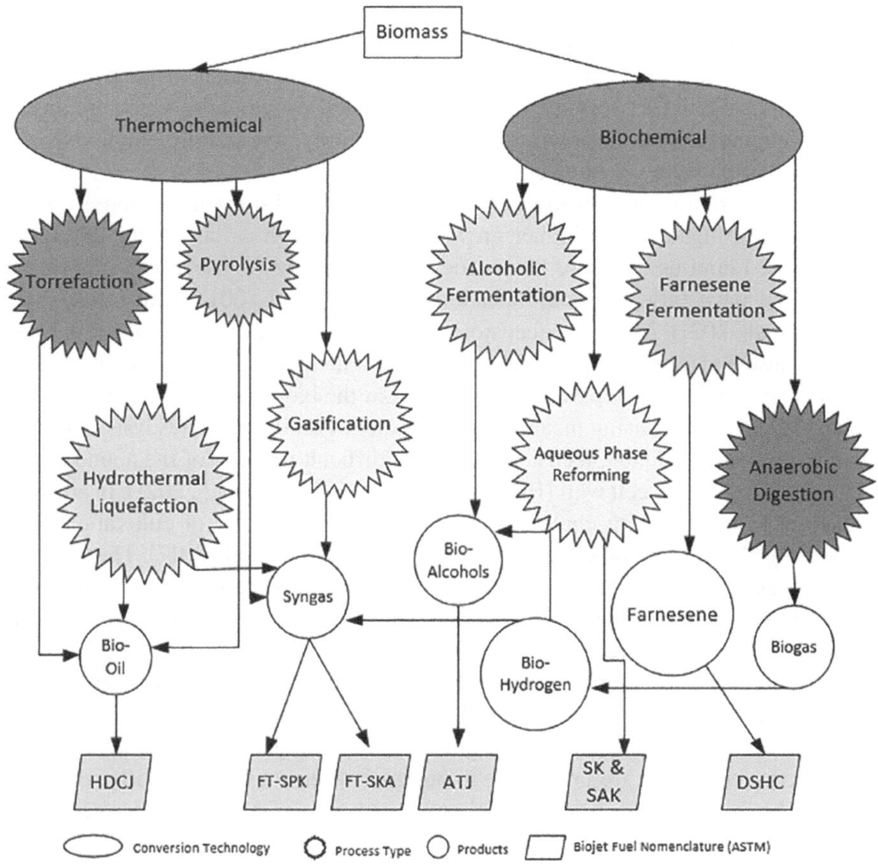

FIGURE 11.2 Lignocellulosic biomass valorization to biojet fuels with ASTM biojet nomenclature.

Furthermore, the sugar-to-jet fuel strategy can convert high sugar or starch biomass to alkane-type fuels, while pretreatment is necessary to obtain fermentable sugars in the case of lignocellulose (Lim et al., 2021). Interestingly, second-generation biofuels have the potential to reduce costs and increase production efficiencies, and are thus likely to improve sustainability and be eco-friendly to the transport sector (Kargbo et al., 2021). The advantages of lignocellulosic biomass include lower costs and wide availability, and they do not compete with food supplies (Wei et al., 2019). In addition, lignocellulose biomass can be converted into biojet fuel through a number of technologies.

Nonetheless, second-generation biofuels have some challenges arising from the complexity of biomass and difficulties linked to its production, harvesting, haulage, and pretreatment before processing into biofuel (Kargbo et al., 2021). The price of biomass is correlated to the yield, which is influenced by genetics, location, and soil fertility. Various genetic engineering techniques are currently used on potential energy feedstocks to increase the yield. However, these processes ultimately increase the costs of biojet fuel production (Kargbo et al., 2021). Besides, the directional conversion of lignocellulosic biomass into C8–C15 aromatics and cycloalkanes is still a challenge (Zhang et al., 2015). For instance, the liquid products of the pyrolysis of lignocellulosic biomass normally contain oxygenated compounds, which are unsuitable for engine applications owing to their high acidity, low stability, high viscosity, and low heating values (Zhang et al., 2015).

Third-generation biofuels are derived from algae and microalgae and are fairly new. Notably, algae have a higher growth rate, have high carbon fixing efficiency, have limited land use, require few pesticides and fertilizers, use waste as a carbon source and have little seasonal variation (Bwapwa et al., 2017; Lim et al., 2021; Kargbo et al., 2021). Algae as a feedstock is, however, limited by careful strain selection because it affects the lipid content, the requirement for meticulous control of the nutrient source and the separation of water from the biomass due to the small algal cells. Moreover, recovering meaningful quantities of biofuel requires lysis of the cell wall to discharge valuable feedstock, which is difficult because of the abundance of stable cellulose in the cell wall (Bwapwa et al., 2017; Kargbo et al., 2021). In addition, algae are associated with challenges for the maintenance cost of cultivation units, harvesting, drying, and oil extraction equipment (Bwapwa et al., 2017). These extraction processes have cost implications that result in an expensive biojet fuel (Lim et al., 2021). Therefore, the production of third-generation biofuels is mostly conceptual at present but has great potential if these limitations can be addressed.

Fourth-generation biofuel feedstocks are centred on genetically engineered microbes and plants (Doliente et al., 2020). The science around genetically modified organisms (GMOs) advocates for manipulation of organisms' genetic makeup to infer certain traits that are favourable to our goals (Lee et al., 2008; Villarreal et al., 2020). The focus of this chapter will, however, be on second-generation feedstocks, specifically lignocellulosics conversion to biojet fuel, and therefore aspects of first-generation, third-generation, and GMO-based substrates will not be pursued further in this chapter.

11.3 PRODUCTION PROCESSES

11.3.1 Overview of Lignocellulosics-to-Biojet Fuel Conversion Technologies

Lignocellulose has a complex chemical structure, which causes incomplete conversion during the production of biofuels. Accordingly, practical conversion technologies to obtain liquid biofuels from these feedstocks require a combination of specific technologies (Kargbo et al., 2021). The various lignocellulosic-to-biofuel conversion technologies include biochemical and thermochemical routes (Raud et al., 2019; Kargbo et al., 2021). These technological routes are broken down into various processes, and different products are produced from each of them. This information is depicted in Figure 11.2 and discussed in detail in the forthcoming sections.

The choice of the most appropriate conversion method is influenced by the amount and type of biomass, the preferred form of energy, end-use requirements, economic factors, environmental regulations and product specifications (Kargbo et al., 2021).

Furthermore, process efficiency depends on the particle size of the feedstock and the design of the reactor. Bio-oil, sugars (from hydrolysis of lignocellulosic before fermentation), alcohols (from fermentation or other chemical routes) and biogas are normally upgraded through different processes, including hydrotreatment and/or oligomerization, to produce biojet fuel as a final product (Gorimbo et al., 2022). Syngas from either pyrolysis or gasification is normally converted to biocrude first through the popular Fischer–Tropsch reaction before this biocrude is also refined in process operations, with processes very similar to those used in petroleum oil refining (Kunamalla et al., 2022). The various process routes for lignocellulose conversion to biojet fuel are depicted in Figure 11.2. According to the American Society for Testing and Materials standard (ASTM:D7566), biojet fuel products from each specific process route are named according to the process that produced them as follows (Shahabuddin et al., 2020):

- FT–SPK (approved in 2009): Fischer–Tropsch Synthetic Paraffinic Kerosene derived from the FT routes
- HEFA–SPK (approved in 2011): Hydroprocessed Esters and Fatty Acids Synthetic Paraffinic Kerosene derived from vegetable oils hydrotreatment
- HFS–SIP (approved in 2014): Hydroprocessed fermented sugars to synthetic Isoparaffins derived from biomass-based sugars
- FT–SPK/A (approved in 2015) : Fischer–Tropsch Synthetic Paraffinic Kerosene with Aromatics with the aromatics alkylation done based on non-petroleum-based sources
- ATJ (approved in 2016): Alcohol-to-Jet Synthetic Paraffinic Kerosene based on biomass-derived alcohol as platform chemicals
- CH–SK or CHJ (approved in 2020): Catalytic Hydrothermolysis Synthesized Kerosene which is derived from biomass hydrothermal products upgrading

Note that although HEFA is certified by ASTM:D7566 and is the most applied technology for current commercial biojet fuel companies, this technology is not applicable to the current study of lignocellulosics-to-biojet fuel production, where no esters or fatty acids are involved.

11.3.2 Biochemical Conversion

During biochemical conversion, hemicellulose is broken down enzymatically through a hydrolysis stage so that enzymes can access the cellulose. The unreacted lignin can be extracted and used as a precursor for downstream thermochemical conversion, such as combustion (Wei et al., 2019; Kargbo et al., 2021). Alternatively, the lignocellulosic biomass undergoes physicochemical pretreatment followed by hydrolysis to produce C5 and C6 sugars, which are subsequently selectively converted to liquid biofuels enzymatically or chemically (Shah et al., 2022; Kargbo et al., 2021). The choice of subprocesses to consider after the enzymatic hydrolysis of lignocellulosics to sugars is diverse. The available options include fermentation, anaerobic digestion and direct catalytic conversion. The major products of biochemical conversion are liquid biofuels (alcohol and biodiesel) and biogas. These products are later refined

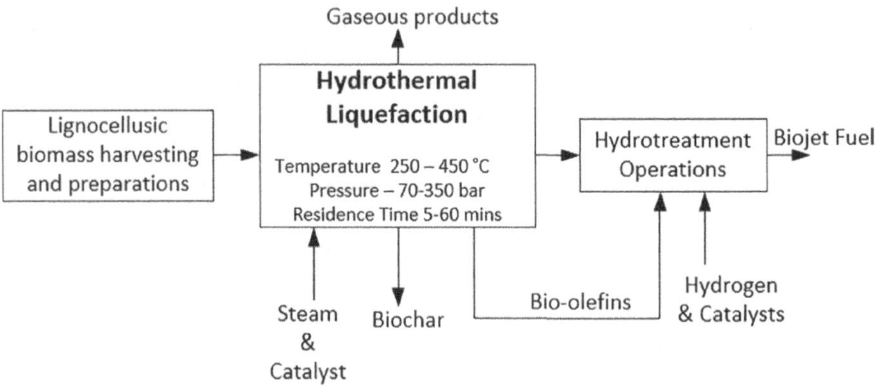

FIGURE 11.3 Hydrothermal liquefaction block flow diagram.

through various upgrading techniques such as reforming (steam or gas), oligomerization and hydrotreatment to meet drop-in biojet fuel specifications (Gorimbo et al., 2022). Figure 11.3 depicts a block flow diagram of the different biochemical pathways of converting lignocellulosics.

11.3.2.1 The Sugars-to-Alcohol Fermentation Route (ATJ)

Sugars derived from lignocellulosic hydrolysis can be fermented under anaerobic conditions to produce mainly alcohols (Ijoma et al., 2021; Rashama & Matambo, 2020). Alcohol yields from lignocellulosic-based sugars are generally low compared to those from first-generation feedstocks due to microbial activity inhibition induced by products of hemicellulose degradation during hydrolysis. These inhibitors are hydroxymethylfurfural, furfural and acetic acid (Rashama & Matambo, 2020). Genetic modification of microorganisms is a trending strategy for improving microbial tolerance to inhibitors, increasing alcohol yields and promoting the use of lignocellulosic materials in biofuel production. A wide spectrum of alcohols can be produced from different C5–C6 sugars, and fortunately most of these alcohols, which are normally dominated by butanol, ethanol and propanol, can be upgraded to biojet fuel through the dehydration, oligomerization, distillation and deoxyhydrogenation processes to be discussed later.

11.3.2.2 The Sugars-to-Biogas Anaerobic Digestion Route

The basic steps in the biological anaerobic degradation/digestion of biomass to produce biogas are hydrolysis, acidogenesis, acetogenesis and finally methanogenesis. Under appropriate conditions, sugars derived from different lignocellulosic biomass pretreatments can be processed through anaerobic degradation/digestion reactors to produce biogas (Tagne et al., 2019; Vintila et al., 2019). Once biogas has been produced, it can be upgraded in the same way as other alkanes and alkenes to biojet fuel through gas reforming or partial oxidation, which produces syngas for the FTS upgrading route (Gorimbo et al., 2022).

11.3.2.3 The Direct Sugar Fermentation Conversion Route

Direct sugar to hydrocarbon conversions, also known as direct fermentation of sugar to jet or HFS–SIP according to ASTM, employ genetically engineered microbes to convert sugars into biofuels such as farnesene (Wei et al., 2019). Alternatively, biofuel intermediates such as lipids or alkanes can also be produced for later upgrading, and this depends on the gene selections made at gene modification stages (Bauen et al., 2020). Farnesene is the popular end product, which can easily replace many hydrocarbons in biojet fuel (Gorimbo et al., 2022). Engineered microorganisms, especially *Saccharomyces cerevisaea*, have been successfully applied in this technology before by a company called Amyris that exploited a biochemical pathway called the mevalonate pathway in yeast biotechnology (Wei et al., 2019).

11.3.2.4 Sugars Conversion through Aqueous Phase Reforming and Hydrogenolysis

Sugars derived from a biochemical route (e.g., microorganism-mediated hydrolysis) of lignocellulosic feedstocks can be converted in their aqueous form to bio-olefins through a process called aqueous phase reforming (APR) (Saenz de Miera et al., 2020). This process was first popularized for the conversion of sugars and polyols into bio-hydrogen (Mounguengui-Diallo et al., 2019). If APR is performed in the presence of different catalysts and process conditions, the ultimate end product profile changes, and the process will then be known colloquially by different names, such as hydrogenolysis and aqueous phase catalytic transformation (Mounguengui-Diallo et al., 2019; Wang et al., 2015). Under the different reaction conditions of temperature, pressure and catalyst, the process has the potential to yield alkanes and syngas, which will be upgraded to biojet fuel (Wang et al., 2014, 2015). The various schemes and products possible under APR were reviewed by Alonso et al. (2010). Table 11.1 gives a comparison of the four commonly used biochemical pathways for biojet fuel productionusing lignocellulosic biomass.

TABLE 11.1
Comparison of Lignocellulosic Biochemical Pathways to Biojet Fuel

	Anaerobic Digestion	Fermentation	Farnesene Route	Aqueous Phase Reforming
Commercialization stage	Mature technology now industrially employed	Mature technology. Several industrial plants operational globally	Two companies Amyris and LS9 have commercialized two separate technologies based on this process route. Market uptake still low	Still at piloting. Few establishments running at industrial scale as yet

(*Continued*)

TABLE 11.1 (*Continued*)
Comparison of Lignocellulosic Biochemical Pathways to Biojet Fuel

	Anaerobic Digestion	Fermentation	Farnesene Route	Aqueous Phase Reforming
End products	Biomethane and carbon dioxide	Bioethanol, biobutanol	Farnesene	Syngas and alkanes. Hydrogen
Further processing required	Yes	Yes	Yes	Yes
Operating conditions	Mostly mesophilic temperatures, atmospheric pressure	Same as anaerobic digestion	Near ambient	After biomass liquor extraction biologically, chemical processes involving high temperature, high pressure and heterogeneous catalyst take place
Yields/Economics	Microbes sensitive to many variables such as temperature, pH, volatile fatty acids, retention times. Process control therefore key	Microbes inhibited in high alcohol assays and parameters stated under anaerobic digesters as well	The complexity of converting lignocellulose to sugars to fuel makes this the most expensive technologies among those studied so far	Process still to be optimized. Lab scale results affected by the wide spectrum of products possible, which reduce final biojet fuel recovered
Environmental	Also produces CO_2, which must be captured to prevent climate change effects. Digestate is a good replacement for synthetic fertilizers	Also produces CO_2, which must be captured to prevent climate change effects. High contaminated effluent generation problematic	Issues around ethical and environmental concerns about GMO use are not yet resolved in other jurisdictions	Not so much study has been directed towards this still to mature technology on safety, health and environmental aspects
Other sustainability comments	Lignocellulose as substrate can be combined with other substrates (codigestion) to resolve waste management or improve economics. Process consumes high water volumes	Requirement of distillation to concentrate the alcohols from dilute liquors is an energy-intensive process that still needs to be addressed to improve economics	Can be blended at max 10% with petrochemical based jet fuel. Beyond this ratio the biojet fuel starts to affect engine performance	Requires less energy in terms of pretreatment and upgrading of intermediate products compared to the other two processes compared here

11.3.3 Thermochemical Conversion

In thermochemical methods, heat is used to valorize lignocellulosic biomass into various products that can finally be upgraded to produce biojet fuel. Biomass handling and feeding systems represent a bottleneck in these processes because biomass properties vary, which demands complicated designs to cater to these variations. Following feed preparation stages, biomass is directed for downstream conversion processes, which may include direct combustion, hydrothermal liquefaction (HTL), gasification and pyrolysis (Kargbo et al., 2021). The biomass is subsequently heated under predetermined oxygen concentrations (Raud et al., 2019). A variety of products are thus produced through thermal decomposition and chemical rearrangement.

11.3.3.1 Pyrolysis

Through pyrolysis, biomass is thermally (at 400°C–600°C) decomposed under an inert atmosphere to produce bio-oil, syngas and biochar fractions. The moisture content of the incoming biomass is a critical factor that determines product distribution and quality, with the recommended limit of 10% being the one that gives optimal bio-oil yields. The relative amounts of pyrolytic fractions depend on the nature of the biomass and process conditions (Neves et al., 2020). Variating these process conditions, especially the operating temperature and residence time (of both the biomass and products), results in three distinguishable sub-classifications of pyrolysis, which are called slow, intermediate and fast pyrolysis (Komandur & Mohanty, 2022). To maximize the bio-oil fraction yield, fast pyrolysis is performed at high heating rates of approximately 100°C/s and short residence times of about 2 s (Kargbo et al., 2021) (Figure 11.4). Biofuels produced from fast pyrolysis have complex compositions and may include furans and oxygenated organics (35%–55% oxygen content), which confer high acid values, low heat values, high viscosities and instabilities to the biofuels (Chen et al., 2020). As a result, the potential applications of the biofuels are limited. A number of strategies can be used to circumvent this. For instance, to improve

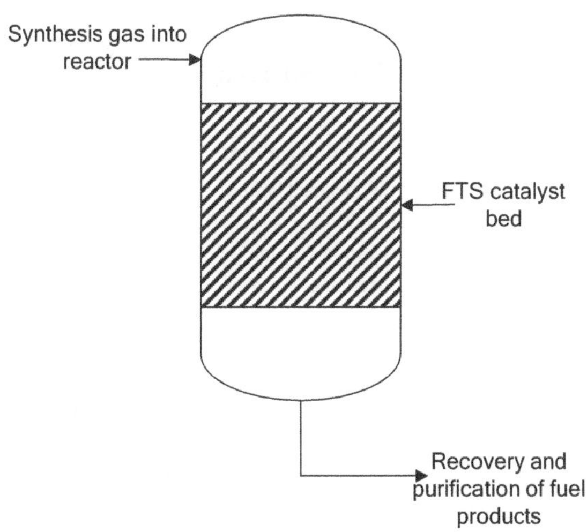

FIGURE 11.4 The production of biofuels via the Fischer–Tropsch synthesis.

the stability of the bio-oil, the pyrolysis process can be performed in a hydrogen atmosphere, a process called hydropyrolysis, in the presence of a catalyst (Perkins et al., 2019). Alternatively, catalytic cracking and other hydrotreatment processes may be instituted after the bio-oil has already been extracted from the process reactors (Figure 11.4). In fact, the physicochemical properties and oxygen and water content of bio-oil can be tuned using catalysts to yield partly deoxygenated and stabilized bio-oil (Tzanetis et al., 2017). Biomass particle sizes are recommended to be below 3 mm in pyrolytic processes to facilitate rapid heat transfer. The intermediate pyrolysis yields both bio-oil and biochar, while the slow pyrolysis will predominantly yield biochar (Komandur & Mohanty, 2022). In all the subclasses, varied amounts of syngas are also produced as a byproduct, which can also be refined through FTS to finally produce biofuels.

11.3.3.2 Gasification

The gasification method converts lignocellulosic biomass at higher temperatures (at times >2,500°C) than pyrolysis and in the presence of oxygen/steam to produce synthesis gas (syngas) (85% of the product distribution), which is a mixture of mainly carbon dioxide and hydrogen. A little bio-oil (5%) and biochar (10%) may be produced as well compared to pyrolysis, which is biased towards the bio-oil (30%–70%) and biochar (10%–35%) product fractions (Akhtar et al., 2018). For pyrolysis, the gas fraction ranges from 20% to 35%. Syngas, the main product of gasification, is subsequently upgraded towards producing drop-in biojet fuel by a combination of processes that may include FTS, cracking, decabonylation, decarboxylation, hydrotreatment and distillation technologies. The syngas can also serve as a precursor for the synthesis of ammonia, methanol and dimethyl ether (Kargbo et al., 2021). Five main classes of biomass gasification technologies can be distinguished, as captured in Table 11.2. Each technology has its own advantages and disadvantages. Extensive reviews of these technologies and applicable reactor flowsheets have been previously conducted elsewhere (Mednikov, 2018; Molino et al., 2018; Shahabuddin et al., 2020).

TABLE 11.2
Summary of Main Biomass Gasification Technologies

Technology	Key Features	Main Products	Advantages/ Disadvantages	References
Plasma	High temperature electric powered torches used to break down biomass into syngas	Hydrogen Carbon monoxide Carbon dioxide Methane	A clean syngas is produced High carbon efficiency High water content (40% moisture) biomass can also be processed Low electric efficiency	Kuo et al. (2020)

(*Continued*)

TABLE 11.2 (*Continued*)
Summary of Main Biomass Gasification Technologies

Technology	Key Features	Main Products	Advantages/ Disadvantages	References
Melting	Moving/Fixed bed technology designed to allow biomass to go through pyrolysis (800°C), then gasification with air (1,200°C) and finally advanced gasification (2,000°C) using oxygen mixed with natural gas	Syngas	Mature technology on which established plants have been built	Shahabuddin et al. (2020)
Fluidized bed	Biomass is gasified while in suspension by use of hot gas-fuel mixtures	Syngas	Fluidization improves reactant contact. This increases rates of reaction, product yields and easier process control	Shahabuddin et al. (2020)
Supercritical water	Gasification takes place at high moisture-to-biomass ratio. Milder temperatures (374°C–550°C) or high temperature (550°C–700°C) can be used with or without a catalyst	Syngas	Still at early stages of maturity. High moisture content biomass can be processed without the need for prior drying	Shahabuddin et al. (2020)
Microwave	Biomass is gasified under microwave radiation. Steam addition into the biomass is a must	Syngas	Still at laboratory and piloting stages. Uniform temperature profiles, cleaner product. Process can handle large-sized biomass thus reduce costs on milling	Arpia et al. (2022)

11.3.3.3 Torrefaction

Torrefaction is a thermochemical process that employs milder conditions than pyrolysis in terms of temperature (250°C–350°C rising at 50°C/min) and long residence times (30 min to several hours) with the aim of producing an upgraded biomass sometimes called wood charcoal (Akbari et al., 2020; Wilk & Magdziarz, 2017). The volatiles arising from this lignocellulosic material pretreatment can be condensed to give bio-oil just like in pyrolysis, though in this case the bio-oil yields are much lower than in pyrolysis. This bio-oil can also be directed towards biojet fuel production by following the same upgrading routes as those applied to bio-oil derived from pyrolysis. The solid charcoal product yield is normally around 60%, while the bio-oil fraction will be between 19% and 29%. The rest are non-condensibles (Singh et al., 2020). Wet torrefaction is a new approach that targets the use of high moisture-containing biomasses for better grindability, pelletization and other final biomass properties as opposed to dry torrefaction. Although bio-oil is a byproduct of torrefaction, the main aim is biomass upgrading; therefore, this may not be a competitive technology for biojet fuel production from lignocellulosic materials.

11.3.3.4 Hydrothermal Liquefaction

One emerging technology that overcomes some of the limitations of the other thermochemical processes described earlier is HTL. This is a process that involves simultaneous decomposition of lignocellulosic biomass and repolymerization of byproducts into biocrude, biochar, aqueous chemicals and a gas. This product mixture can be further refined to produce biojet fuel and other fuels (Dimitriadis & Bezergianni, 2017; Perkins et al., 2019). Organic material reacts under hydrothermal conditions at temperatures in the range of 250°C–450°C, pressures of 70–350 bar and residence times of 5–60 min in the presence of a catalyst (Perkins et al., 2019). This results in a biocrude with low oxygen content and high aromatic and cycloalkane content, conferring the product with a high energy density and enhanced thermal stability while reducing downstream refining costs (Tzanetis et al., 2017). Because it is performed in the presence of water, HTL does not require the drying step, saving on cost (Tzanetis et al., 2017; Chen et al., 2020). Depending on the biomass feedstock, the process can be accelerated using a variety of catalysts, ranging from alkaline solutions to iron-based catalysts. Table 11.3 compares of the four commonly used thermochemical pathways for biojet fuel production.

TABLE 11.3
Comparison of Lignocellulosic Thermochemical Pathways of Biojet Fuel Production

Parameter	Pyrolysis	Gasification	Torrefaction	Liquefaction
Drying	Necessary	May or may not be necessary	May or may not be necessary depending on targeted quality	Depend on targeted quality and yields
Pressure (MPa)	0.1–0.5	8	Atmospheric	5–20
Temperatures	350–800	Above 1,000	Below 350	250–450

(*Continued*)

TABLE 11.3 (*Continued*)
Comparison of Lignocellulosic Thermochemical Pathways of Biojet Fuel Production

Parameter	Pyrolysis	Gasification	Torrefaction	Liquefaction
Catalyst	No	Sometimes	No	Sometimes
Upgrade	Hard	Syngas is easy through FT	Hard for biocrude	Easy and many options
Oxygen content	High	Medium in syngas	High	Low
Final biojet fuel yield	Likely lesser than or equal to that in gasification due to losses in high biochar output	Operating conditions affect overall yields but numbers not so different from pyrolysis.	Very low since low quantities of bio-oil are produced	Likely to produce more biojet fuel than gasification and pyrolysis in the future. Technology still under optimization studies

Source: Adapted from Dimitriadis & Bezergianni (2017).

11.4 LIGNOCELLULOSIC VALORIZATION PRODUCTS UPGRADING TO BIOJET FUEL

11.4.1 Upgrading Pyrolytic and Gasification Products through Fischer–Tropsch Synthesis

The FTS is a well-developed method for the conversion of syngas to a mixture of hydrocarbons called syncrude, from which biojet fuel can be recovered. Syngas from gasification, pyrolysis and gas reforming unit operations can be used in the FTS process (Ng et al., 2021). The liquid products from FTS syncrude distillation comprise the five hydrocarbon homologues (paraffins, iso-paraffins, olefins, aromatics and naphthenes) along with oxygenated compounds. To produce suitable biojet fuels, the liquid fuels are further improved by hydrocracking, isomerization and distillation (Neves et al., 2020). Because the majority of the liquid products are olefins and paraffins, the content of naphthenes and aromatics in the fractions should be augmented to meet the specifications for jet fuels (Zhang et al., 2015). The FTS produces hydrocarbons of different chain lengths, including gaseous hydrocarbons (C1–C4), which are directly fed into gas turbines for heat or power generation or further developed into LPG; naphtha (C5–C10) and kerosene (C10–C16), which are blended with gasoline and biojet fuel; distillate (C14–C20), which is further processed into biodiesel; and waxes (C20+), from which biodiesel can be produced via hydrocracking (Ng et al., 2021). Notably, Fischer–Tropsch liquid products are sulphur-free and have a low aromatic content relative to fossil fuel-based diesel and gasoline, resulting in reduced environmental pollution. The resulting syngas from gasification processes should, however, be cleaned to remove tars, particulates, nitrogen- and sulphur-containing compounds, and other impurities that are likely to cause fouling (de Klerk, 2016; Chen et al., 2020). Currently, syngas cleaning is a major challenge for biomass gasification Fischer–Tropsch integrated systems, and this deserves further development to economically produce clean Fischer–Tropsch feed (Ng et al., 2021).

Although FTS is a well-established method for biofuel synthesis, its application in the production of biojet fuel has a high energy demand to gasify the biomass into syngas (Chen et al., 2020). Moreover, the main products of thermochemical conversion processes are gases and liquids, but these require additional refining to become drop-in fuels (Kargbo et al., 2021). This makes the technology costly, apart from its low efficiency (25–50%). Moreover, the resulting biofuels have low lubricity due to the absence of sulphur, although this can be addressed through blending with fossil fuels (Escalante et al., 2022). While thermochemical processes in the biomass-to-liquid fuel pathways hold great promise in biofuel production, their large-scale applications are limited by the difficulty in obtaining tar-free syngas from the gasification step. Tar can deactivate catalysts and be detrimental to engines. In addition, hazardous gases can poison catalysts and inhibit the conversion of syngas, along with blocking and fouling filters by heavier hydrocarbons (Neves et al., 2020).

11.4.1.1 Syncrude Upgrading Processes

Upgrading bio-oils or FTS products to produce biojet fuels requires extra processes collectively called upgrading techniques. The main processes are represented in Table 11.4. The main purpose of upgrading is to make sure that the final biojet fuel

TABLE 11.4
Main upgrading Processes for Biojet Fuel Production (Akhtar et al., 2018)

	Chemical Reactions[a]		
Unit Process	**Reactant/s**	**Products**	**Purpose and Description**
Catalytic cracking	$R_1CH_2CH_2$ $CH_2CH_2R_2$	$R_1CH_2CH_3$ R_2CHCH_2	Catalyst used to break long chain hydrocarbons to smaller units
Decarbonylation	RCOH	RH CO	Knocking out oxygen from carbonyl compounds by removing carbon monoxide
Decarboxylation	RCOOH	RH CO_2	Knocking out oxygen from carboxylic acids by removing carbon dioxide
Hydrocracking	$R_1CH_2CH_2R_2$ H_2	R_1CH_3 R_2CH_3	Using hydrogen to facilitate long chain hydrocarbon breaking into smaller chain molecules

(Continued)

TABLE 11.4 (*Continued*)
Main upgrading Processes for Biojet Fuel Production (Akhtar et al., 2018)

Unit Process	Chemical Reactions[a] Reactant/s	Products	Purpose and Description
Hydrodeoxygenation	ROH H_2	RH H_2O	Knocking out oxygen from organics by removing water using hydrogen as reducing agent
Hydrogenation	R_1CHCHR_2 H_2	$R_1CH_2CH_2R_2$	Saturating multiple carbon to carbon bonds by reacting unsaturated hydrocarbons with hydrogen

[a] Most of these reactions are facilitated by solid metal inorganic catalysts at optimized pressure and temperature conditions.

meets the recommended ASTM specifications by removing oxygen, regulating the amount of aromatics, and raising the level of carbon bond saturation and carbon length for aliphatics. The upgrading processes are typically integrated in a block flow diagram for bio-oil in Figure 11.5.

11.4.2 Catalysts in Lignocellulosics Conversion to Biojet Fuels

The existence of oxygenated moieties and their position influence the physicochemical properties and reactivity of biomass-derived compounds (Sakhayi et al., 2021). In order to reduce oxygenated groups, a variety of catalysts can be used. Generally, the catalysts serve to eliminate heteroatoms (sulphur, nitrogen and oxygen), improve bio-oil yield and provide easier control over the composition of bio-oil (Perkins et al., 2019). HTL commonly uses homogenous alkali catalysts owing to their reactivity with lignin and their capacity to inhibit coke formation and support small chain condensation (Perkins et al., 2019). Catalysts used in the hydrodeoxygenation process combine acidic supports with transition and noble metals (Sakhayi et al., 2021). A study found that bi-functional catalysts with hierarchization support favour hydroisomerization and keep hydrocracking to a minimum (Chen et al., 2020). Another study prepared basic Mg-La mixed oxides with adjustable interaction between the oxides and used them as catalysts for the production of biodiesel and biojet fuel precursors at low temperatures through aldol condensation (Bohre et al., 2020).

Although catalysts accelerate the reactions and improve the yield of biojet fuel or its precursors, several challenges exist. The cost of noble metals can be prohibitively high, and this can affect the economic feasibility of the whole process. Specifically, homogeneous catalysts may possibly increase the cost of the HTL process because of their corrosive properties and difficulty in recovery (Perkins et al., 2019). Other challenges include extreme reaction conditions, ambient CO_2

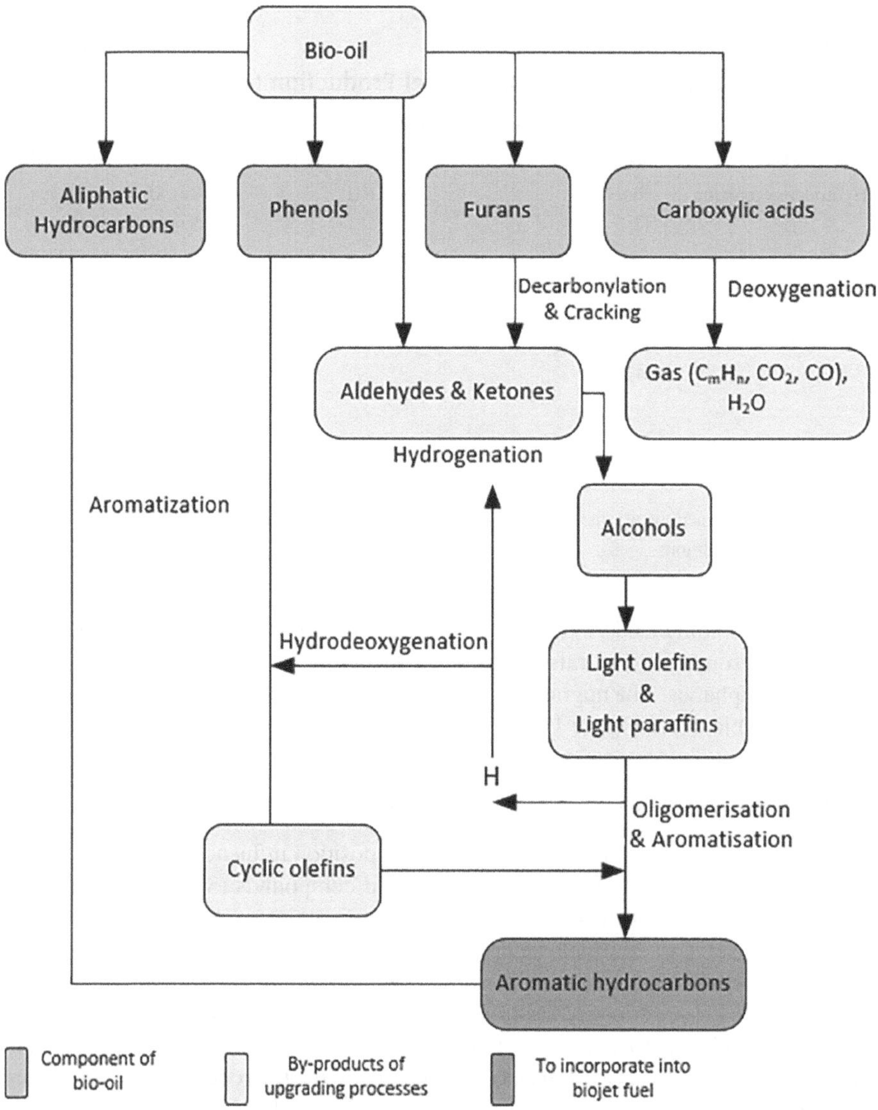

FIGURE 11.5 Block flow diagram for upgrading biofuel processes.

or the adsorption of reactive products can cause catalyst poisoning, and catalyst reusability can be limited by enhanced leaching and solubility issues (Bohre et al., 2020). In some cases, heterogeneous catalysts have been preferred since they do not react with the walls of the reactor and can thus be easily recovered from the products for reuse (Perkins et al., 2019). In this way, a sustainable catalytic HTL process can be attained. In commercial applications, the recyclability and repeated use of a catalyst are highly desirable.

11.5 STATUS AND ONGOING PROJECTS

A number of biofuel manufacturing companies and research institutions continue to explore the development of bioject fuel industry because of the merits associated with migrating from fossil to biofuels (Shahabuddin et al., 2020). Some of the technologies developed to date have already been certified for implementation by ASTM, while others are still in different stages of development (Bauen et al., 2020; Doliente et al., 2020; Shahabuddin et al., 2020). So far, the only known operational (since 2016) and commercial-scale biojet fuel production facility globally that is producing 10 kt per year of biojet fuel is AltAir Fuels (IRENA, 2020). The main obstacles to fast adoption of these biojet fuel technologies are unstable regulatory policies, uncertainties around technology support from developers and a cost of production that is still double that of their fossil counterparts (IRENA, 2020). A review and assessment of technology readiness versus resource availability has been performed before (Bauen et al., 2020; Doliente et al., 2020). The results are projected in Figure 11.6 to give an overall picture of the current and future mix of biojet fuel technologies on the market.

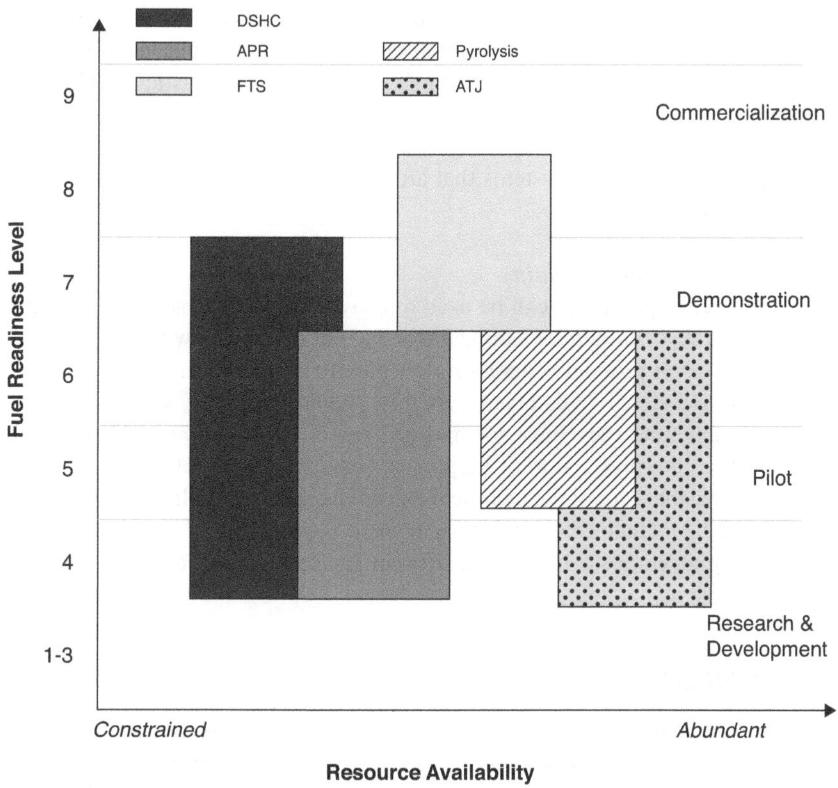

FIGURE 11.6 The current state of technology readiness for lignocellulosic biojet fuel.

11.6 GAPS AND FUTURE PERSPECTIVES

Biojet fuel derived from lignocellulosic biomass is increasingly attracting interest and has the potential to have global impacts. Because they seek to replace petroleum-based jet fuels, biojet fuels can reduce the effects of climate change and global warming (Escalante et al., 2022). As a result, the operating costs in the aviation sector will be reduced (Bwapwa et al., 2017). In addition, the technologies for producing biojet fuels are fairly established. In order to facilitate the uptake of the technology on a larger scale, a number of issues, however, have to be addressed. These include:

1. *Dedicated energy crops*
   It is important that, unlike in the case of first-generation fuels, biomass feedstock does not compete with the food supply. For this reason, biojet production from lignocellulose is sustainable because it uses non-edible parts of plants as feedstock. Further research on increasing the scope of energy crops and genetically modifying them to increase yield will be useful.

2. *Biomass handling and feeding systems*
   The handling and feeding systems for many conversion processes create a bottleneck. For smooth downstream operations, appropriate handling and feeding are important, considering the variable characteristics of biomass (Neves et al., 2020). Parameters such as foreign matter, particle size and moisture might require adjusting at this stage. There is scope to optimize handling and feeding systems that are efficient without causing challenges to downstream processes.

3. *Techno-economic feasibility*
   A number of processes can be used to convert lignocellulosic biomass into biojet fuel. Commercial applications favour processes with low capital costs, simpler product separation, low energy demand and low waste generation. Besides, if a catalyst is used, it should be highly selective, easily recoverable and recyclable for multiple reuses (Bohre et al., 2020). Apart from this, the production of biojet fuel requires a comprehensive analysis of the possible economic, technical and environmental challenges (Bwapwa et al., 2017). Once this is addressed, there is potential to make biojet fuel from lignocellulosic biomass a drop-in fuel that complies with aviation standards.

11.7 CONCLUSION

The use of petroleum-based fuels is unsustainable because they release greenhouse gases that cause serious environmental impacts. The aviation industry largely depends on non-renewable jet fuel, and the demand continues to grow. There is thus a need to develop alternative sources of jet fuel. In this regard, biojet, which is derived from lignocellulosic biomass, is gaining prominence owing to its renewability, advantageous physicochemical characteristics and lower emissions.

Lignocellulosic biomass can be converted into biojet fuel through various forms of thermochemical processes, including pyrolysis and HTL. Although FTS is a well-established method for biofuel synthesis, its application in the production of biojet fuel has a high energy demand to gasify the biomass into syngas. As such, further process development is required in order to economically produce biojet fuel at a commercial scale. Other process improvements include removing bottlenecks in biomass handling and feeding systems. This should take into account the variable properties of lignocellulosic biomass and not adversely affect downstream processing operations.

The substitution of conventional jet fuel with biojet fuel should easily fit into existing engine designs and fuel distribution systems. Overall, biojet fuel represents a long-term solution for the aviation sector, which can potentially reduce operating costs and the environmental impact.

REFERENCES

Akbari, M., Oyedun, A. O., & Kumar, A. (2020). Techno-ceonomic assesment of wet and dry torrefaction of biomass feedstock. *Energy*, 118287. https://doi.org/10.1016/j.energy.2020.118287.

Akhtar, A., Krepl, V., & Ivanova, T. (2018). A combined overview of combustion, pyrolysis and gasification of biomass. *Energy & Fuels*, 32, 7294–7318. https://doi.org/10.1021/acs.energyfuels.8b01678.

Alonso, D. M., Bond, J. Q., & Dumesis, J. A. (2010). Catalytic conversion of biomass to biofuels. *Green Chemistry*, 12, 1493–1513. https://doi.org/10.1039/c004654j.

Aron, N. S. M., Khoo, K. S., Chew, K. W., Show, P. L., Chen, W.-H., & Nguyen, T. H. P. (2020). Sustainability of the four generations of biofuels - A review. *International Journal of Energy Research*, 44, 9266–9282. https://doi.org/10.1002/er.5557.

Bauen, A., Bitossi, N., German, L., Harris, A., & Leow, K. (2020). Sustainable aviation fuels. *Johnson Matthey Review*, 64(3), 263–278. https://doi.org/10.1595/205651320X15816756012040.

Bohre, A., Alam, M. I., Avasthi, K., Ruiz-Zepeda, F., & Likozar, B. (2020). Low temperature transformation of lignocellulose derived bioinspired molecules to aviation fuel precursor over magnesium–lanthanum mixed oxide catalyst. *Applied Catalysis B: Environmental*, 276, 119069. https://doi.org/10.1016/j.apcatb.2020.119069.

Bwapwa, J. K., Anandraj, A., & Trois, C. (2017). Possibilities for conversion of microalgae oil into aviation fuel: A review. *Renewable and Sustainable Energy Reviews*, 80, 1345–1354. http://dx.doi.org/10.1016/j.rser.2017.05.224.

Chen, Y., Lin, C., & Wang, W. (2020). The conversion of biomass into renewable jet fuel. *Energy*, 201, 117655. https://doi.org/10.1016/j.energy.2020.117655.

de Klerk, A. (2016). Aviation turbine fuels through the Fischer–Tropsch process. In *Biofuels for Aviation* (pp. 241–259). Elsevier. https://doi.org/10.1016/B978-0-12-804568-8.00010-X.

Dimitriadis, A., & Bezergianni, S. (2017). Hydrothermal liquefaction of various biomass and waste feedstocks for biocrude production: A state of the art review. *Renewable and Sustainable Energy Reviews*, 68, 113–125. https://doi.org/10.1016/j.rser.2016.09.120.

Doliente, S. S., Narayan, A., Tapia, J. F. D., Samsatli, N. J., Zhao, Y., & Samsatli, S. (2020). Bio-aviation fuel: A comprehensive review and analysis of the supply chain components. *Frontiers in Energy Research*, 8(110), 1–38. https://doi.org/10.3389/fenrg.2020.00110.

Escalante, E. S. R., Ramos, L. S., Coronado, C. J. R., & de Carvalho Júnior, J. A. (2022). Evaluation of the potential feedstock for biojet fuel production: Focus in the Brazilian context. *Renewable and Sustainable Energy Reviews*, 153, 111716. https://doi.org/10.1016/j.rser.2021.111716.

Gorimbo, J., Moyo, M., & Xinying, L. (2022). Oligomerisation of bioolefins to biojet. In S. K. Maity, K. Gayen, & T. K. Bhowmick (Eds.), *Hydrocarbon Biorefinery, Sustainable Processing of Biomass to Biofuels* (pp. 271–291). Elsevier. https://doi.org/101016/B978-0-12-823306-1.00010-8.

Han, X., Guo, Y., Liu, X., Xia, Q., & Wang, Y. (2019). Catalytic conversion of lignocellulosic biomass into hydrocarbons: A mini review. *Catalysis Today*, 319, 2–13. https://doi.org/10.1016/j.cattod.2018.05.013.

Ijoma, G. N., Adegbenro, G., Rashama, C., & Matambo, T. S. (2021). Perculiar response in the co-culture fermentation of Leuconostoc mesenteroides and Lactobacillus plantarum for the production of ABE solvents. *Fermentation*, 7(212), 1–13. https://doi.org/10.3390/fermentation7040212.

IRENA. (2020). *Reaching Zero with Renewables*. International Renewable Energy Agency, Abu Dhabi.

Kargbo, H., Harris, J. S., & Phan, A. N. (2021). "Drop-in" fuel production from biomass: Critical review on techno-economic feasibility and sustainability. *Renewable and Sustainable Energy Reviews*, 135, 110168. https://doi.org/10.1016/j.rser.2020.110168.

Komandur, J., & Mohanty, K. (2022). Fast pyrolysis of biomass and deoxygenation of bio oil for sustainable production of hydrocarbon biofuels. In S. K. Maity, K. Gayen, & T. K. Bhowmick (Eds.), *Hydrocarbon Biorefinery, Sustainable Processing of Biomass to Biofuels* (pp. 47–73). Elsevier. https://doi.org/10.1016/B978-0-12-823306-1-00003-0.

Kunamalla, A., Mailaram, S., Shrirame, B. S., Kumar, P., & Maity, S. K. (2022). Hydrocarbon Biorefinery: A sustainable approach. In S. K. Maity, K. Gayen, & K. T. Bhowmick (Eds.), *Hydrocarbon Biorefinery, Sustainable Processing of Biomass to Biofuels* (pp. 2–32). Elsevier. https://doi.org/10.1016/B978-0-12-823306-1.0004-2.

Lee, D., Chen, A., & Nair, R. (2008). Genetically engineered crops for biofuel production: Regulatory perspectives. *Biotechnology and Genetic Engineering Reviews*, 25(1), 331–362. https://doi.org/10.5661/bger-25-331.

Lim, J. H. K., Gan, Y. Y., Ong, H. C., Lau, B. F., Chen, W., Chong, C. T., Ling, T. C., & Kleme, J. J. (2021). Utilization of microalgae for bio-jet fuel production in the aviation sector: Challenges and perspective. *Renewable and Sustainable Energy Reviews*, 149, 111396. https://doi.org/10.1016/j.rser.2021.111396.

Luthra, S., Kumar, S., Garg, D., & Haleem, A. (2015). Barriers to renewable/sustainable energy technologies adoption: Indian perspective. *Renewable and Sustainable Energy Reviews*, 41, 762–776.

Mednikov, A. S. (2018). A review of technologies for multistage wood biomass gasification. *Thermal Engineering*, 65(8), 531–546.

Molino, A., Larocca, V., Chianese, S., & Musmarra, D. (2018). Biofuels production by biomass gasification: A Review. *Energies*, 11(811), 1–31.

Mounguengui-Diallo, M., Sadier, A., Noly, E., Perez, D. D. S., Pinel, C., Perret, N., & Besson, M. (2019). C-O bond hydrogenolysis of aqueous mixtures of sugar polyols and sugars over ReOx-Rh/ZrO2 catalys: Application to an hemicellulose extracted liquor. *Catalysts*, 9(740), 1–22.

Neves, R. C., Klein, B. C., da Silva, R. J., Rezende, M. C. A. F., Funke, A., Olivarez-Gomez, E., Bonomi, A., & Maciel-Filho, R. (2020). A vision on biomass-to-liquids (BTL) thermochemical routes in integrated sugarcane biorefineries for biojet fuel production. *Renewable and Sustainable Energy Reviews*, 119, 109607. https://doi.org/10.1016/j.rser.2019.109607.

Ng, K. S., Farooq, D., & Yang, A. (2021). Global biorenewable development strategies for sustainable aviation fuel production. *Renewable and Sustainable Energy Reviews*, 150, 111502. https://doi.org/10.1016/j.rser.2021.111502.

Perkins, G., Batalha, N., Kumar, A., Bhaskar, T., & Konarova, M. (2019). Recent advances in liquefaction technologies for production of liquid hydrocarbon fuels from biomass and carbonaceous wastes. *Renewable and Sustainable Energy Reviews*, 115, 109400. https://doi.org/10.1016/j.rser.2019.109400.

Rashama, C., & Matambo, T. S. (2020). Investigating the fermentability of synthetic medium density fiberboard wastewater to alcohol and effects of cofermenting wastewater with sucrose. *Biofuels*, 11(2), 135–140. https://doi.org/10.1080/17597269.2018.1475712.

Raud, M., Kikasa, T., Sippula, O., & Shurpali, N. J. (2019). Potentials and challenges in lignocellulosic biofuel production technology. *Renewable and Sustainable Energy Reviews*, 111, 44–56. https://doi.org/10.1016/j.rser.2019.05.020.

Saenz de Miera, B., Oliveira, A., Baeza, J., Calvo, L., Rodriguez, J. J., & Gilarranz, M. (2020). Treatment aqeous phase reforming: Effect of pH, organic load and salinity. *Journal of Cleaner Production*, 252(119849), 9. https://doi.org/10.1016/j.jclepro.2019.119849.

Sakhayi, A., Bakhtyari, A., & Rahimpour, M. R. (2021). Cleaner production of liquid fuels and chemicals by synthesis-gas-assisted hydroprocessing of lignin compounds as biomass-derivates. *Journal of Cleaner Production*, 316, 128331. https://doi.org/10.1016/j.jclepro.2021.128331.

Shah, A. A., Seehar, T. H., Sharma, K., & Toor, S. S. (2022). Biomass pretreatment technologies. In S. K. Maity, K. Gayen, & K. Bhowmick (Eds.), *Hydrocarbon Biorefinery, Sustainable Processing of Biomass to Biofuels* (pp. 203–223). Elsevier. https://doi.org/10.1016/B978-0-12-823306-1.00014-5.

Shahabuddin, M., Alam, M. T., Krishna, B. B., Bhaskar, T., & Perkins, G. (2020). A review on the production of renewable aviation fuels from the gasification of biomass and residual wastes. *Bioresource Technology*, 312, 123596. https://doi.org/10.1016/j.biortech.2020.123596.

Singh, R. K., Jena, K., Chakraborty, J. P., & Sarkar, A. (2020). Energy and exergy analysis for torrefaction of pigeon pea stalk (cajanus cajan) and eucalyptus (eucalyptus tereticornis). *International Journal of Hydrogen and Energy*, 45, 18922–18936. https://doi.org/10.1016/j.ijhydene.2020.05.045.

Tagne, R. F. T., Anagho, S. G., Ionel, I., Matiuti, A. C., & Ungureanu, C. I. (2019). Experimental biogas production from Cameroon lignocellulosic waste biomass. *Journal of Environmental Protection and Ecology*, 20(3), 1335–1344.

Tzanetis, K. F., Posada, J. A., & Ramirez, A. (2017). Analysis of biomass hydrothermal liquefaction and biocrude-oil upgrading for renewable jet fuel production: The impact of reaction conditions on production costs and GHG emissions performance. *Renewable Energy*, 113, 1388–1398. http://dx.doi.org/10.1016/j.renene.2017.06.104.

Vela-García, N., Bolonio, D., García-Martínez, M., Ortega, M. F., Streitwieser, D. A., & Canoira, L. (2021). Biojet fuel production from oleaginous crop residues: Thermoeconomic, life cycle and flight performance analysis. *Energy Conversion and Management*, 244, 114534. https://doi.org/10.1016/j.enconman.2021.114534.

Villarreal, J. V., Burgues, C., & Rosch, C. (2020). Acceptability of genetically engineered algae biofuels in Europe: Opinions of experts and stakeholders. *Biotechnology for Biofuels*, 13(92), 1–21. https://doi.org/10.1186/s13068-020-01730-y.

Vintila, T., Ionel, I., Rufis Fregue, T. T., Wächter, A. R., Julean, C., & Gabche, A. S. (2019). Residual biomass from food processing industry in Cameroon as feedstock for second-generation biofuels. *BioResources*, 14(2), 3731–3745. https://doi.org/10.15376/biores.14.2.3731-3745.

Wang, M., Dewil, R., Maniatis, K., Wheeldon, J., Tan, T., Baeyens, J., & Fang, Y. (2019). Biomass-derived aviation fuels: Challenges and perspective. *Progress in Energy and Combustion Science*, 74, 31–49. https://doi.org/10.1016/j.pecs.2019.04.004.

Wang, T., Li, K., Liu, Q., Zhang, Q., Qiu, S., Long, J., Chen, L., Ma, L., & Zhang, Q. (2014). Aviation fuel synthesis by catalytic conversation of biomass hydrolysate in aqueous phase. *Applied Energy*, 136, 775–780. https://doi.org/10.1016/j.apenergy.2014.06.035.

Wang, T., Tan, J., Qiu, S., Zhang, Q., Long, J., Chen, L., Ma, L., Liu, K. L., & Zhang, Q. (2014). Liquid fuel production by aqueous phase catalytic transformation of biomass for aviation. *Energy Procedia*, 61, 432–435. http://creativecommons.org/licenses/by-nc-nd/3.0/.

Wang, T., Weng, Y., Qiu, S., Chen, L., Liu, Q., Long, J., Tan, J., Zhang, Q., Zhang, Q., & Ma, L. (2015). Liquefied fuel production by aqueous phase catalytic transformation of biomass for aviation. *Applied Energy*, 160, 329–335. https://doi.org/10.1016/j.apenergy.2015.08.116.

Wei, H., Liu, W., Chen, X., Yang, Q., Li, J., & Chen, H. (2019). Renewable bio-jet fuel production for aviation: A review. *Fuel*, 254, 115599. https://doi.org/10.1016/j.fuel.2019.06.007.

Wilk, M., & Magdziarz, A. (2017). Hydrothermal carbonization, torrefaction and slow pyrolysis of Miscanthus giganteus. 140, 1292–1304. https://doi.org/10.1016/j.energy.2017.03.031.

Xia, Q., Chen, Z., Shao, Y., Gong, X., Wang, H., Liu, X., Parker, S. F., Han, X., Yang, S., & Wang, Y. (2016). Direct hydrodeoxygenation of raw woody biomass into liquid alkanes. *Nature Communications*, 7, 11162.

Zhang, Y., Bi, P., Wang, J., Jiang, P., Wu, X., Xue, H., Liu, J., Zhou, X., & Li, Q. (2015). Production of jet and diesel biofuels from renewable lignocellulosic biomass. *Applied Energy*, 150, 128–137. http://dx.doi.org/10.1016/j.apenergy.2015.04.023.

12 Role of Thermophilic Microorganisms and Thermostable Enzymes in 2G Biofuel Production

Govindarajan Ramadoss and Saravanan Ramiah Shanmugam
SASTRA Deemed University

Ramachandran Sivaramakrishnan
Chulalongkorn University

CONTENTS

12.1 THERMOSTABLE ENZYMES IN LIGNOCELLULOSE HYDROLYSIS

12.1.1 Cellulose in Enzymatic Hydrolysis

In general, the cellulose in plants is in crystalline form, and it is linked with hemicellulose by covalent bonding, which is surrounded by lignin molecules. For the utilization of cellulose in various processes, it is necessary to remove the cellulose from lignin and hemicellulose. Physical and mechanical pretreatment processes are widely considered to remove the lignin molecules and solubilize the hemicellulose. Physical and mechanical pretreatment also increases the pore volume of the cells and

DOI: 10.1201/9781003203452-12

decreases their crystallinity, which allows for higher hydrolysis efficiency. For the hydrolysis of cellulose, it is necessary to reduce the complex structure of the cells. After the pretreatment, cellulolytic enzymes are used to hydrolyze the cellulose. The cellulolytic enzymes are classified into three types, namely endo-β-1,4-glucanases, cellobiohydrolysases and β-glucosidases (Cantarel et al. 2009). The cellulolytic enzymes come under the family of glycosyl hydrolase due to its similarities in the sequence and structural properties. *Trichoderma reesei* is a mesophilic fungus that produces powerful extracellular cellulase and is widely considered as the source of the cellulase enzymes. So far, from the genome of *T. reesei*, eight important cellulase genes have been identified; they are Cel7A and Cel6A (cellobiohydrolases) and Cel7B, Cel5a, Cel12A, Cel45A, Cel61A, Cel74A (endoglucanases) (Foreman et al. 2003). Among these enzymes, Cel7A is a major cellulase and comprises 60% of total cellulase in *T. reesei* (Nidetzky and Claeyssens 1994). The mode of action and diversity of xylanases and cellulases share similar features, which is an important advantage in terms of commercial scale. Thermostable cellulases and xylanases play an important role in industrial applications; these enzymes were used as biocatalysts in the paper and pulp, chemical, pharmaceutical and food industries (Haki and Rakshit 2003). In the initial days, the researchers mainly focused on *T. reesei* and the development of its process, genetics and biochemistry. In general, the thermal stability of both fungal and bacterial cellulases is higher than that of other enzymes. However, when compared to bacteria, fungal cellulases are mainly focused due to their ability to produce large amount of enzymes extracellularly. The production of cellulases can be increased by implying recombinant DNA technology. The thermal stability of cellulases can be increased by the cloning of different fungal enzymes. Among the various cellulolytic enzymes, cellobiohydrolase, β-glucosidase and endoglucanase were categorized as prominent cellulases. All three enzymes can be cloned and produce a mixture of thermostable cellulases in *T. reesei* (Haki and Rakshit 2003).

12.1.2 Thermostable Cellulases

Thermostable enzymes have much attention in biotechnology and biofuel production due to their high stability in industrial process conditions. The thermal stability of thermophilic enzymes is much higher when compared to mesophilic enzymes, and they can withstand higher temperatures for prolonged periods of time. The thermal stability can be measured by calculating the half-life of enzyme activity at different temperatures. Although mesophilic organisms produce thermostable enzymes, those produced by thermophilic organisms showed higher stability over prolonged periods. The important advantages of thermostable cellulolytic enzymes over mesophilic cellulolytic enzymes are their high specificity, flexibility, stability and suitability for various process conditions. The characteristics of specificity and stability increase the overall enzyme performance. These characteristics allowed for the limited usage of enzymes, which can reduce hydrolysis costs. At higher temperatures, the process performance would be higher and the reaction time would be shorter, which would also decrease the hydrolysis costs. At a higher temperature, the rate of the reaction is higher and reduces the mass transfer resistance, which allows the enzymes to easily access the substrates inside the cells (Sivaramakrishnan and Muthukumar 2012).

A study reported that the thermostable xylanase from *Myceliopthora thermophila* showed higher efficiency in the paper pulp industry than the thermostable xylanase from *T. reesei* (Kulkarni and Rao 1996). A good source of thermostable xylanases for saccharification of lignocellulosic biomass (LB) is thermophilic fungi (Kaur and Satyanarayana 2004).

Numerous thermostable cellulolytic enzymes have been isolated from several thermophilic bacteria (Bok, Yernool, and Eveleigh 1998; Bronnenmeier et al. 1995). The important thermophilic strains that produce thermostable cellulolytic enzymes are *Rhodothermus* strain (34) and *Anaerocellum thermophilum*. An anaerobic strain, *Thermotoga*, can also produce thermostable cellulolytic enzymes (Hreggvidsson et al. 1996). Enormous efforts have been made to study the thermostable cellulolytic enzymes in the Clostridia family of enzymes (Demain, Newcomb, and Wu 2005). Another study reported that the co-culture of clostridial organisms can produce ethanol from lignocellulose by the direct conversion method (33). Another important fungal family organism called ascomycete has been studied and characterized for its ability to produce thermostable cellulases (Ferrari, Gautier, and Silar 2021; Jang and Chen 2003). In addition, some mesophilic fungal strains and partial thermophilic fungal strains can also produce thermostable cellulases. This kind of enzyme shows more stability than the optimal growth temperature of its organisms (Sukharnikov et al. 2011). *Talaromyces emersonii* is the filamentous fungi producing cellulases with higher thermal stability and relative cellulase activity that can be retained at higher temperatures (Wang et al. 2014). Another study reported that the different forms of cellulases were produced by *Humicola grisea*, which produced family 7 cellobiohydrolases in four different forms (Li 2011).

Other filamentous fungi that can produce thermostable cellulases are *Chaetomium thermophilum* (Millner 1977), *Myceliophthora thermophila* (Dahiya, Kumar, and Singh 2020), *Thermoascus aurantiacus* (Ping et al. 2018), *Corynascus thermophilus* and *Thielavia terrestris* (Maheshwari, Bharadwaj, and Bhat 2000). Thermostable β-glucosidases can be produced from *Chaetomium thermophila* (Jiang et al. 2010), *Aureobasidium* sp. (Gautério et al. 2021), *Thermoascus aurantiacus* (Ping et al. 2018), *Talaromyces emersonii* (Wang et al. 2014) and *Thermomyces lanuginosa* (Gomes et al. 1993). Most of these thermostable cellulases are stable at around 70°C, and the characteristics of these thermostable enzymes are assessed in Tables 12.1 and 12.2.

TABLE 12.1
Thermostable Cellulases from Different Bacteria

Source	Enzyme	Optimal Temperature (°C)	Temperature	References
Staphyothermus hellenicus	EGSh	85		Wang et al. (2014)
Pyrococcus abyssi	CelB like	96		Thomas, Ram, and Singh (2017)

(Continued)

TABLE 12.1 (*Continued*)
Thermostable Cellulases from Different Bacteria

Source	Enzyme	Optimal Temperature (°C)	Temperature	References
Pyrococcus furiosus	CelB	102–105	50% at 100°C, 85 h	Kengen et al. (1993)
Pyrococcus horikoshii	EGPh	>97	80% at 97°C, 3 h	Ando et al. (2002)
Desulfurococcus fermentans	EBI-244	82	50% at 100°C	Graham (2011)
Sulfolobus solfataricus	SSO1949	80	50% at 80°C, 8 h	Huang et al. (2005)
Thermococcus sp.	β-Glucosidase	78	50% at 78°C, 860 min	Sinha and Datta (2016)
Rhodothermus marinus	Cel12A	>90	80% at 90°C, 16 h	Crennell, Hreggvidsson, and Nordberg Karlsson (2002)
Thermobifida fusca	Cel5A	50–55	50% at 40°C, 30 h	Gomez del Pulgar and Saadeddin (2014)
Caldicellulosiruptor bescii	CelA	75		Brunecky et al. (2017)
Cellulosimicrobium funkei	CelL	50	50% at 50°C, 12 min	Kim et al. (2016)
Thermotoga maritima	TmCel12B	85	50% at 90°C, 9 h	Liebl et al. (1996)
Bacillus licheniformis	EG	65	72% at 55°C, 42 h	Bischoff et al. (2006)
Cellulosimicrobium funkei	CelL	50	50% at 50°C, 12 h	Kim et al. (2016)
Thermus brockianus	TbGH2	90	50% at 70°C, 12 h	Bischoff et al. (2006)
Clostridium thermocellum	CenC	70	50% at 74°C, 24 min	Kim et al. (2016)
Clostridium thermocellum	Cel48S	70	-	Olson et al. (2010)
Thermotoga sp. strain FjSS3-B.1	Exo-1,4-β-cellobiohydrolase	105	50% at 108°C, 70 min	Haq et al. (2015)
Caldicellulosiruptor saccharolyticus	BglA	85	50% at 70°C, 38 h	Klippel and Antranikian (2011)
Thermus nonproteolytics HG102	β-Cellobiohydrolase	90	50% at 90°C, 2.5 h	Wang et al. (2003)
Thermotoga petrophila	Egl	95	100% at 80°C, 8 h	Muneer et al. (2015)
Thermotoga napththophila	TnBglB	85	100% at 85°C, 9 h	Akram, Haq, and Mukhtar (2018)

TABLE 12.2
Thermostable Cellulases from Different Fungi

Source	Enzyme	Optimal Temperature (°C)	Temperature	References
Myceliophthora thermophila	β-Cellobiohydrolase	60	91.8% at 60°C, 2 h	Zhao et al. (2015)
Aspergillus terreus RWY	Endoglucanase	45–55	99% at 50°C, 150 min	Narra et al. (2014)
Neurospora crassa	Cellobiohydrolase	>45	70% at 90°C, 1 h	Yang et al. (2019)
Talaromyces emersonii	Cellobiohydrolase	68	50% at 80°C, 1 h	Grassick et al. (2004)
Humicola grisea	β-Glucosidase	60	75% at 80°C, 10 min	Takashima et al. (1999)
Fusarium oxysporum	β-Glucosidase	70	83% at 70°C, 180 min	Olajuyigbe, Nlekerem, and Ogunyewo (2016)
Phanerochaete chrysosporium	Endoglucanase	60	NR	Westereng et al. (2011)
Chaetomium thermophilum	Cellobiohydrolase	60	50% at 70°C, 45 min	Li et al. (2009)
Magnaporthe oryzae	Cellobiohydrolase	50	NR	Takahashi et al. (2010)
Thermoascus aurantiacus	Cellobiohydrolase	70	70% at 60°C, 45 min	Voutilainen et al. (2008)
Aspergillus fumigatus	Endoglucanase	60–80	50% at 70°C, 866 min	Saqib et al. (2012)
Trichoderma harzianum	Endo-β-1,4-D-glucanase	60	80% at 50°C, 3 h	Bagewadi, Mulla, and Ninnekar (2016)

12.2 THERMOSTABLE CELLULASES IN ETHANOL PRODUCTION

The pretreated raw material is used for the enzymatic hydrolysis, and the hydrolyzed sugars are used for fermentation. The hydrolysis and fermentation can be performed separately by separate hydrolysis and fermentation (SHF) or simultaneously by simultaneous saccharification and fermentation (SSF). According to the economic point of view, SSF is widely considered for ethanol production (Guo, Chang, and Lee 2018). During SHF hydrolysis, the process can be inhibited by the cellobiose and glucose end products (Philippidis, Smith, and Wyman 1993). However, hydrolysis by SSF show limited inhibition due to the removal of produced glucose being utilized simultaneously by the fermentative organism for fermentation. On the other hand, inhibition caused in SHF was decreased by the addition of β-glucosidase in high amount. Hence, due to the inhibition factors, the enzymes used in the SSF are low when compared to the SHF. The presence of ethanol and a low concentration of glucose protect the contaminations from contaminating microorganisms. In addition, the contamination in SSF is low when compared to that in SHF, which reduces the reactor cost when it is performed in combined reactors. The drawback of the SSF is that maintaining suitable and optimal conditions for the combined process, especially pH and temperature, is difficult for both processes since the hydrolysis process requires higher temperatures than the fermentation process. It can be overcome using thermostable enzymes, because thermostable cellulolytic and fermentative enzymes can tolerate high temperatures. In the conventional SSF method, temperatures above 35°C are difficult to set. However, in SHF, the temperature can be set at 45°C–50°C for conventional enzymes and 60°C or above for thermostable enzymes, and the suitable temperature can be used for the fermentation. Partial pre-hydrolysis can also be performed using thermostable cellulases. Using thermostable enzymes for prehydrolysis allows for higher temperatures, which improve the rheological properties of the medium and the contact between substrate and enzymes, resulting in higher hydrolysis efficiency (Öhgren et al. 2007). Partial pre-hydrolysis requires only little quantity of enzymes, which is sufficient to reduce the viscosity of the substrate mixture and ensure high sugar formation (Öhgren et al. 2007). It is necessary to decrease the high viscosity of the initial substrate to achieve a high hydrolysis rate and ethanol production. Partial pre-hydrolysis causes a quick decrease in the viscosity of the substrate, which is necessary to overcome the obstacles related to its high viscosity. After the partial pre-hydrolysis, the performance of the thermostable cellulolytic enzymes shows very high hydrolysis efficiency in both SHF and SSF. The suitable enzyme concentration is chosen according to the nature of the substrate. Thermostable enzymes can also be supplemented with anaerobic and ethanol-producing bacterial strains in order to achieve the high conversion of cellulolytic substrates into sugars. Hence, novel thermostable enzymes can be used for the flexible process.

12.3 GENETIC ENGINEERING FOR THE THERMOSTABLE CELLULOLYTIC AND XYLANOLYTIC ENZYMES

In the economic aspect, it is necessary to improve the efficiency of thermostable cellulases to degrade the LB. The efficiency can be improved by using genetic engineering methods. Site-saturation and site-directed mutagenesis can improve the thermal stability of the cellulases (Amore, Giacobbe, and Faraco 2013). Another study reported that site-directed mutagenesis improved the thermal stability of xylanase from *Caldocellum saccharolyticum* (Lüthi et al. 1990). The cellulose-encoding genes such as Asn173Asp and Arg156Glu overexpressed in *Thielavia terrestris* NRRL 8126 produced a two-fold increase in thermal stability at 60°C (Amore, Giacobbe, and Faraco 2013). Mutations in the disulphide bridges of xylanase genes in *Bacillus circulans* increased thermal stability by 15°C (Wakarchuk et al. 1994). Gene overexpression and knockout strategies in *Thermoanaerobacterium saccharolyticum* improved the ethanol yield when compared to the wild strains (Shaw et al. 2008). A gene from *Sporotrichum thermophile* encoding the endoglucanase gene StCel5A was expressed in *Aspergillus niger* and its efficiency toward carboxymethyl cellulose degradation was 5–6 times higher than the *T. reesei* endoglucanase (Tambor et al. 2012). Other important ways to improve the thermal stability of the enzymes involve modifying the 3D structure of enzymes by altering the amino acid sequence (Amore, Giacobbe, and Faraco 2013).

12.4 MOLECULAR MECHANISMS OF INTERACTIONS BETWEEN ENZYME AND LIGNOCELLULOSIC BIOMASS

Hydrophobic interactions, electrostatic interactions and hydrogen bonding are the three major types of interactions involved during the adsorption of enzymes to LB (Ying et al. 2018; Tokunaga et al. 2019; Rahikainen et al. 2013; Liu et al. 2016; Yarbrough et al. 2015; Liu et al. 2016). Studies conducted by Yarbrough et al. (2015) and Liu et al. (2016) indicated that cellulases from fungi were adsorbed onto the lignin present in biomass due to hydrophobic interactions. During acid hydrolysis of LB, deprotonation of carboxylic groups present in lignin takes place, thus conferring a negative charge on the surface of lignin. The cellulase enzymes present in fungi are positively charged, resulting in better adsorption. At an alkaline pH range, the repulsive force between the enzymes and lignin increases, which in turn results in poor sorption of enzymes to the LB (Lou et al. 2013). Physical soaking of biomass in aqueous solution exposes the substrate fibers of the LB, which in turn aids in enzyme adsorption (Baig 2020). Enzymes involved in the pretreatment of LB adhere more to the lignin component of the biomass compared to the non-lignin portion, which is counterproductive in nature. Hence, disintegration of lignin fibers via physico-chemical pretreatment such as the steam explosion process must take place before subjecting the biomass to enzyme pretreatment. Upon disintegration, the enzymes adsorb to the crystalline regions of cellulose, resulting in successful hydrolysis and yielding fermentable sugars (Baig 2016; Podgorbunskikh, Bychkov, and Lomovsky 2019). There are basically 12 types of charged amino acid residues present in the cellulase enzyme, which interact with C1 and C4 of the glucose unit of the cellulose in LB. The strength of adsorption is increased by hydrogen bonding between cellulose and enzyme via hydroxyl groups.

12.5 MECHANISM OF ENZYME ADSORPTION

Physical forces such as Brownian movement, van der Waals attraction forces and gravitational forces govern the rate of enzymatic adsorption to the surface of a lignocellulosic substrate (Baig 2020). The effects of surface charge and hydrophobic interactions will also play a huge role in the adsorption of enzymes to the LB. The presence of chemical factors such as sugars, amino acids and oligopeptides regulates the LB adsorption components and aids in interactions between enzyme cell surfaces and enzyme surfaces (Hoarau, Badieyan, and Marsh 2017). Bulk transport of enzyme from the solution to the enzyme–surface interface occurs via long-range attractive interactions depending on the free energy available and the distance between them. If the distance is greater than 50 nm, long-range interactions play a significant role in enzyme–LB interaction (Katsikogianni and Missirli 2004; Qin and Buehler 2014). The van der Waals force will come into play as the distance decreases to 10–20 nm (Hermann, DiStasio, and Tkatchenko 2017). As the distance between the enzyme and LB surface is less than 5 nm, short-range interface actions will come into play (Al-Haddad et al. 2013). As the enzyme reaches the cell surface, molecular or cellular adhesion takes place. Following this, molecular adsorption of enzymes to the LB mainly depends on the surface structure of the enzyme and substrate. In simple terms, chemical interactions occur via attractive or repulsive forces during the adsorption process (Figure 12.1).

Adsorption of enzymes is governed by a number of factors such as fluid rheology (shear stress), temperature of the fluid, enzyme contact time, concentration of substrate and enzymes, and concentration of contaminants. Rheology of the fluids strongly influences adsorption by governing the number of enzymes attached to the LB substrate (Lippok et al. 2016; Bekard et al. 2011). Shear stress is created as a result of agitation or shaking. Hence, selection of these conditions will play a significant role in the hydrolysis of LB, as we know that enzymes that are effectively adsorbed will only yield a higher hydrolysis. There is a general misconception that low agitation speeds will result in higher adsorption. However, as the shear rates increase from 50 to 300 s^{-1}, binding interactions become ineffective (Baig 2020). The reasons for this could be due to the following hypotheses: enzymes attach to LB in layers, which hinder the access of enzymes to the substrate; modification of enzyme configuration might have occurred; and high shear rates give rise to disengagement forces that decrease the number of adsorbed enzymes.

Temperature affects the adsorption of enzymes onto LB. Some of the studies from the literature indicated that it is an exothermic process controlled by enthalpy (Medve, Stahlberg, and F. Tjerneld 1994; Kim and Hong 2000; Ooshima, Sakata, and Y. Harano 1983). On the other hand, some other researchers have proposed that the adsorption of enzymes is an endothermic process controlled by entropy (Hoshino et al. 1992; Creagh et al. 1996). To prove these contradictory claims, Tu, Pan, and Saddler (2009) performed experiments at different temperatures on softwood biomass, including lodgepole pine. The results of his study proved that the activity of enzymes increased with increasing temperature, implying that adsorption

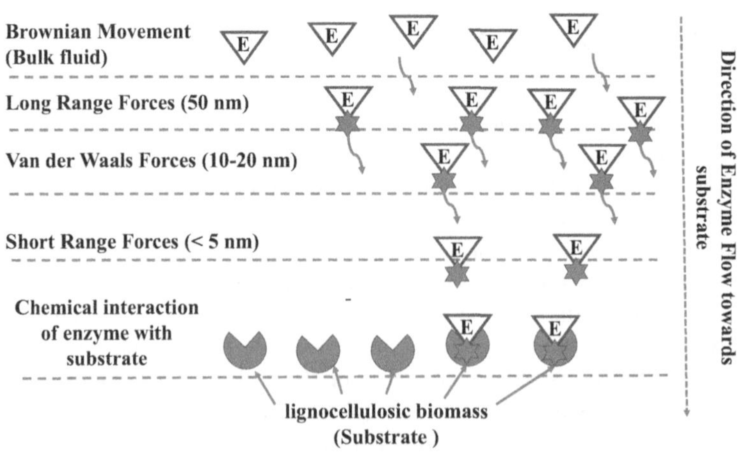

FIGURE 12.1 Mechanism of enzyme adsorption onto lignocellulosic biomass.

is an endothermic process that is indeed entropy-driven (Tu, Pan, and Saddler 2009; Zheng et al. 2013).

Another important factor affecting the enzyme–substrate interaction is the contact time, which depends on the environmental condition of the system. The pulp fraction obtained following LB pretreatment using steam explosion has shown high enzyme adsorption compared to microcrystalline cellulose (Avicel) (Baig 2020). Most of the studies conducted in the literature have shown equilibrium adsorption values within 30 min (Singh, Kumar, and Schugerl 1991; Jager et al. 2010; Steiner, Sattler, and Esterbauer 1998). While Pareek et al. (2013) found that the time required for 60% cellulose adsorption on spruce lignin and black cottonwood lignin was different. In the case of spruce lignin, most of the adsorption occurs in 30 min, whereas in the case of black cotton wood lignin, it takes around 2 h to reach a similar level of adsorption. To conclude, the adsorption process depends on the types of enzymes used in the LB pretreatment, the nature of biomass and the treatment conditions.

12.6 APPLICATION OF THERMOSTABLE CELLULOLYTIC AND XYLANOLYTIC ENZYMES

LB is considered as the second-generation biofuels and produces various value-added products for industrial applications. The most prominent applications of hemicellulases and cellulases are the production of biofuels, in textiles, paper and pulp industries, and as prebiotics (Kuhad et al. 2010). Different cellulases and xylanases are available for the various applications. However, most of the processes are performed at higher temperatures and different pH. Hence, thermotolerant and pH-stable enzymes are necessary for the bioprocesses. Thermophilic organisms can produce thermostable enzymes to fulfill the needs of industrial applications (Gusakov 2013). However, cellulases and xylanases are predominantly preferred for biofuel production. The improvement and production of thermostable cellulase and its bioethanol production were presented in Figure 12.2. The market value of the enzyme industry was $5.0 billion in 2015, which is expected to reach $7.0 million in 2022 (www.bccresearch.com).

Thermophilic microorganisms such as fungi and bacteria are considered as the excellent sources of thermostable cellulases and xylanases. Those enzymes have a high impact in biotechnological applications and protect the environment from harmful chemicals. A huge amount of readily available lignocellulosic substrates is available for the thermostable cellulases and hemicellulases from thermophilic organisms. Thermostable cellulases and xylanases show high specificity toward lignocellulosic materials when compared to mesophilic enzymes. Mostly, these enzymes are used in biofuel production and related value-added products. However, the current thermostable cellulase production is very low when compared to the mesophilic cellulases. Therefore, it is necessary to improve the production of thermostable cellulases by implying various biotechnological tools such as genetic engineering and overexpression. In addition, bioinformatic and genetic assessments may improve our understanding of the enzyme structure and its improvements. The improved production of thermostable enzymes is leading to an increase in various industrial opportunities around the globe.

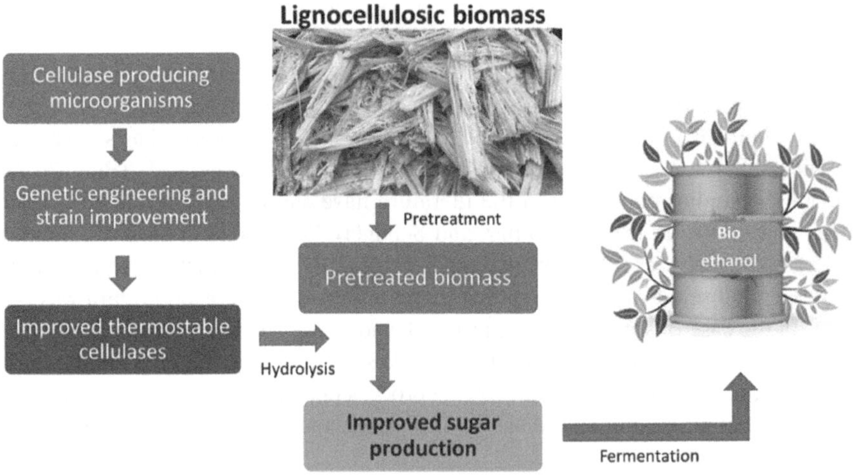

FIGURE 12.2 Production and improvement strategies of thermostable cellulases for bioethanol production.

REFERENCES

Akram, F., I. ul Haq, and H. Mukhtar. 2018. Gene cloning, characterization and thermodynamic analysis of a novel multidomain hyperthermophilic GH family 3 β-glucosidase (TnBglB) from Thermotoga naphthophila RKU-10T. *Process Biochemistry* 66: 70–81.

Al-Haddad, M., A. Al-Jumaily, J. Brooks, and J. Bartley. 2013. Biophysical effects on chronic rhinosinusitis bacterial biofilms, respiratory disease and infection–a new insight. In: B. H. Mahboub, Intech Open. https://doi.org/10.5772/53860. https://www.intechopen.com/books/respiratory-disease-and-infection-a-new-insight/biophysical-effects-on-chronic-rhinosinusitis-bacterial-biofilms.

Amore, A., S. Giacobbe, and V. Faraco. 2013. Regulation of cellulase and hemicellulase gene expression in fungi. *Current Genomics* 14 (4): 230–249.

Ando, S., H. Ishida, Y. Kosugi, and K. Ishikawa. 2002. Hyperthermostable endoglucanase from Pyrococcus horikoshii. *Applied and Environmental Microbiology* 68 (1): 430–433.

Bagewadi, Z. K., S. I. Mulla, and H. Z. Ninnekar. 2016. Purification and characterization of endo β-1,4-d-glucanase from Trichoderma harzianum strain HZN11 and its application in production of bioethanol from sweet sorghum bagasse. *3 Biotech* 6 (1): 101.

Baig, K. S. 2016. Strategic adsorption/desorption of cellulases NS 50013 onto/from Avicel PH 101 and protobind 1000. Doctoral Dissertation, presented to School of Graduate Studies at Ryerson University, Toronto, Canada.

Baig, K. S. 2020. Interaction of enzymes with lignocellulosic materials: causes, mechanism and influencing factors. *Bioresources and Bioprocessing*, 7, 21. https://doi.org/10.1186/s40643-020-00310-0.

Bekard, I. B., P. Peter Asimakis, J. Bertolini, and D. E. Dunstan. 2011. The effects of shear flow on protein structure and function. *Biopolymers* 95(11): 733–745. https://doi.org/10.1002/bip.21646.

Bischoff, K. M., A. P. Rooney, X-L. Li, S. Liu, and S. R. Hughes. 2006. Purification and characterization of a family 5 endoglucanase from a moderately thermophilic strain of Bacillus licheniformis. *Biotechnology Letters* 28 (21): 1761–1765.

Bok, J. D., D. A. Yernool, and D. E. Eveleigh. 1998. Purification, characterization, and molecular analysis of thermostable cellulases CelA and CelB from Thermotoga neapolitana. *Applied and Environmental Microbiology* 64 (12): 4774–4781.

Bronnenmeier, K., A. Kern, W. Liebl, and W. L. Staudenbauer. 1995. Purification of Thermotoga maritima enzymes for the degradation of cellulosic materials. *Applied and Environmental Microbiology* 61 (4): 1399–1407.

Brunecky, R., B. S. Donohoe, J. M. Yarbrough, et al. 2017. The multi domain Caldicellulosiruptor bescii CelA cellulase excels at the hydrolysis of crystalline cellulose. *Scientific Reports* 7 (1):9622.

Cantarel, B. L., P. M. Coutinho, C. Rancurel, T. Bernard, V. Lombard, and B. Henrissat. 2009. The Carbohydrate-Active EnZymes database (CAZy): an expert resource for Glycogenomics. *Nucleic acids Research* 37 (Database issue): D233–D238.

Creagh, A. L., E. Ong, E. Jervis, D. G. Kilburn, and C. A. Haynes. 1996. Binding of the cellulose-binding domain of exoglucanases Cex from *Cellulomonas fimi* to insoluble microcrystalline cellulose is entropically driven. *Proceedings of the National Academy of Sciences of the United States of America* 93: 12229–12234.

Crennell, S. J., G. O. Hreggvidsson, and E. Nordberg Karlsson. 2002. The structure of Rhodothermus marinus Cel12A, a highly thermostable family 12 endoglucanase, at 1.8Å resolution. *Journal of Molecular Biology* 320 (4): 883–897.

Dahiya, S., A. Kumar, and B. Singh. 2020. Enhanced endoxylanase production by Myceliophthora thermophila using rice straw and its synergism with phytase in improving nutrition. *Process Biochemistry* 94: 235–242.

Demain, A. L., M. Newcomb, and J. H. D. Wu. 2005. Cellulase, clostridia, and ethanol. *Microbiology and Molecular Biology Reviews: MMBR* 69 (1): 124–154.

Ferrari, R., V. Gautier, and P. Silar. 2021. Chapter Three - Lignin degradation by ascomycetes. In M. Morel-Rouhier, and R. Sormani (Eds.) *Advances in Botanical Research*, New York: Academic Press, vol. 99, pp. 77–113.

Foreman, P. K., D. Brown, L. Dankmeyer, et al. 2003. Transcriptional regulation of biomass-degrading enzymes in the filamentous fungus Trichoderma reesei. *Journal of Biological Chemistry* 278 (34): 31988–31997.

Gautério, G. V., L. G. Garcia da Silva, T. Hübner, T. da Rosa Ribeiro, and S. J. Kalil. 2021. Xylooligosaccharides production by crude and partially purified xylanase from Aureobasidium pullulans: biochemical and thermodynamic properties of the enzymes and their application in xylan hydrolysis. *Process Biochemistry* 104: 161–170.

Gomes, J., I. Gomes, W. Kreiner, H. Esterbauer, M. Sinner, and W. Steiner. 1993. Production of high level of cellulase-free and thermostable xylanase by a wild strain of Thermomyces lanuginosus using beechwood xylan. *Journal of Biotechnology* 30 (3): 283–297.

Gomez del Pulgar, E. M., and A. Saadeddin. 2014. The cellulolytic system of Thermobifida fusca. *Critical Reviews in Microbiology* 40 (3): 236–247.

Graham, R. S. 2011. Molecular modelling of flow-induced crystallisation in polymers. *Journal of Engineering Mathematics* 71 (3): 237–251.

Grassick, A., P. G. Murray, R. Thompson, et al. 2004. Three-dimensional structure of a thermostable native cellobiohydrolase, CBH IB, and molecular characterization of the cel7 gene from the filamentous fungus, Talaromyces emersonii. *European Journal of Biochemistry* 271 (22): 4495–4506.

Guo, H., Y. Chang, and D.-J. Lee. 2018. Enzymatic saccharification of lignocellulosic biorefinery: research focuses. *Bioresource Technology* 252: 198–215.

Gusakov, A. V. 2013. Cellulases and hemicellulases in the 21st century race for cellulosic ethanol. *Biofuels* 4(6): 567–569.

Haki, G. D., and S. K. Rakshit. 2003. Developments in industrially important thermostable enzymes: a review. *Bioresource Technology* 89 (1): 17–34.

Hermann, J., R. A. DiStasio, and A. Tkatchenko. 2017. First-principles models for van der Waals interactions in molecules and materials: concepts, theory, and applications. *Chemical Reviews* 117(6): 4714–4758.

Hoarau, M., S. Badieyan, and E. N. G. Marsh. 2017. Immobilized enzymes: understanding enzyme-surface interactions at the molecular level. *Organic & Biomolecular Chemistry* 15: 9539–9551. https://doi.org/10.1039/C7OB01880K.

Hoshino, E., T. Kanda, Y. Sasaki, and K. Nisizawa. 1992. Adsorption mode of exo-cellulases and endocellulases from *Irpex lacteus* (Polyporus tulipiferae) on cellulose with different crystallinities. *Journal of Biochemistry* 111: 600–605.

Hreggvidsson, G. O., E. Kaiste, O. Holst, G. Eggertsson, A. Palsdottir, and J. K. Kristjansson. 1996. An extremely thermostable cellulase from the thermophilic eubacterium Rhodothermus marinus. *Applied and Environmental Microbiology* 62 (8): 3047–3049.

Huang, Y., G. Krauss, S. Cottaz, H. Driguez, and G. Lipps. 2005. A highly acid-stable and thermostable endo-β-glucanase from the thermoacidophilic archaeon Sulfolobus solfataricus. *Biochemical Journal* 385 (2): 581–588.

ul Haq, I., F. Akram, M. A. Khan, et al. 2015. CenC, a multidomain thermostable GH9 processive endoglucanase from Clostridium thermocellum: cloning, characterization and saccharification studies. *World Journal of Microbiology and Biotechnology* 31 (11): 1699–1710.

Jager, G., Z. Wu, K. Garscchammer, P. Engel, T. Klement, A. C. Rinaldi Spiess, and J. Buchs. 2010. Practical screening of purified cellobiohydrolases and endoglucanases with α cellulose and specification of hydrodynamics. *Biotechnology for Biofuels* 3(18): 1–12. https://doi.org/10.1186/1754-6834-3-18.

Jang, H-D., and K-S. Chen. 2003. Production and characterization of thermostable cellulases from Streptomyces transformant T3-1. *World Journal of Microbiology and Biotechnology* 19 (3): 263–268.

Jiang, Z., Q. Cong, Q. Yan, N. Kumar, and X. Du. 2010. Characterisation of a thermostable xylanase from Chaetomium sp. and its application in Chinese steamed bread. *Food Chemistry* 120 (2): 457–462.

Katsikogianni, M., and Y. F. Missirli. 2004. Concise review of mechanism of bacterial adhesion to biomaterials and of techniques used in estimating bacteria-material interaction. *European Cells & Materials* 8: 37–57. https://doi.org/10.22203/eCM.v008a05.

Kaur, G., and T. Satyanarayana. 2004. Production of extracellular pectinolytic, cellulolytic and xylanoytic enzymes by thermophilic mould Sporotrichum thermophile Apinis in solid state fermentation. *Indian Journal of Biotechnology* 3 (4):552–557.

Kengen, S.W. M., E. J. Luesink, A. J. M. Stams, and A. J. B. Zehnder. 1993. Purification and characterization of an extremely thermostable β-glucosidase from the hyperthermophilic archaeon Pyrococcus furiosus. *European Journal of Biochemistry* 213 (1):305–312.

Kim, D. W., and Hong, Y. G. 2000. Ionic strength effect on adsorption of cellobiohydrolase I and II on microcrystalline cellulose. *Biotechnology Letters* 22: 1337–1342.

Kim, D. Y., M. J. Lee, H-Y. Cho, et al. 2016. Genetic and functional characterization of an extracellular modular GH6 endo-β-1,4-glucanase from an earthworm symbiont, Cellulosimicrobium funkei HY-13. *Antonie van Leeuwenhoek* 109 (1): 1–12.

Klippel, B., and G. Antranikian. 2011. Lignocellulose converting enzymes from thermophiles. In K. Horikoshi (Ed.) *Extremophiles Handbook*, Springer, Tokyo, pp. 443–474.

Kuhad, R. C., R. Gupta, Y. Pal Khasa, and A. Singh. 2010. Bioethanol production from Lantana camara (red sage): pretreatment, saccharification and fermentation. *Bioresource Technology* 101 (21): 8348–8354.

Kulkarni, N., and M. Rao. 1996. Application of Xylanase from Alkaliphilic Thermophilic Bacillus sp. NCIM 59 in biobleaching of bagasse pulp. *Journal of Biotechnology* 51 (2): 167–173.

Li, D. 2011. Discussion on the problems about low carbon logistics in express companies. *Modern Business Trade Industry* 5: 9–10.

Li, Y. L., H. Li, A. N. Li, and D. C. Li. 2009. Cloning of a gene encoding thermostable cellobiohydrolase from the thermophilic fungus Chaetomium thermophilum and its expression in Pichia pastoris. *Journal of Applied Microbiology* 106 (6): 1867–1875.

Liebl, W., P. Ruile, K. Bronnenmeier, K. Riedel, F. Lottspeich, and I. Greif. 1996. Analysis of a Thermotoga maritima DNA fragment encoding two similar thermostable cellulases, CelA and CelB, and characterization of the recombinant enzymes. *Microbiology* 142 (9): 2533–2542.

Lippok, S., M. Radtke, T. Obser, L. Kleemeier, R. Schneppenheim, U. Budde, and J. O. Rädler. 2016. Shear-induced unfolding and enzymatic cleavage of full-length VWF multimers. *Biophysics Journal* 110(3): 545–554. https://doi.org/10.1016/j.bpj.2015.12.023

Liu, H, J. Sun, S. Leu, and S. Chen. 2016. Toward a fundamental understanding of cellulase-lignin interactions in the whole slurry enzymatic saccharification process. *Biofuels, Bioproducts and Biorefining* 10: 648–663.

Lou, H., J. Y. Zhu, T. Q. Lan, H. Lai, and X. Qiu. 2013. pH-induced lignin surface modification to reduce nonspecific cellulase binding and enhance enzymatic saccharification of lignocelluloses. *Chem Suspens Chem* 6: 919–927.

Lüthi, E, D R Love, J McAnulty, et al. 1990. Cloning, sequence analysis, and expression of genes encoding xylan-degrading enzymes from the thermophile "Caldocellum saccharolyticum". *Applied and Environmental Microbiology* 56 (4):1017–1024.

Maheshwari, R., G. Bharadwaj, and M. K. Bhat. 2000. Thermophilic fungi: their physiology and enzymes. *Microbiology and Molecular Biology Reviews: MMBR* 64 (3): 461–488.

Medve, J., J. Stahlberg, and F. Tjerneld. 1994. Adsorption and synergism of cellobiohydrolase I and II of Trichoderma reesei during hydrolysis of microcrystalline cellulose. *Biotechnology and Bioengineering* 44(9): 1064–1073. https://doi.org/10.1002/bit.260440907.

Millner, P. D. 1977. Radial growth responses to temperature by 58 chaetomium species, and some taxonomic relationships. *Mycologia* 69 (3): 492–502.

Muneer, B., Z. Hussain, M. A. Khan, et al. 2015. Thermodynamic and saccharification analysis of cloned GH12 endo-1, 4-β-glucanase from Thermotoga petrophila in a mesophilic host. *Protein and Peptide Letters* 22 (9): 785–794.

Narra, M., G. Dixit, J. Divecha, K. Kumar, D. Madamwar, and A. R. Shah. 2014. Production, purification and characterization of a novel GH 12 family endoglucanase from Aspergillus terreus and its application in enzymatic degradation of delignified rice straw. *International Biodeterioration & Biodegradation* 88: 150–161.

Nidetzky, B., and M. Claeyssens. 1994. Specific quantification of trichoderma reesei cellulases in reconstituted mixtures and its application to cellulase-cellulose binding studies. *Biotechnology and Bioengineering* 44 (8): 961–966.

Öhgren, K., J. Vehmaanperä, M. Siik-Aho, M. Galbe, L. Viikari, and G. Zacchi. 2007. High temperature enzymatic prehydrolysis prior to simultaneous saccharification and fermentation of steam pretreated corn stover for ethanol production. *Enzyme and Microbial Technology* 40: 607–613.

Olajuyigbe, F. M, C. M. Nlekerem, and O. A. Ogunyewo. 2016. Production and characterization of highly thermostable β-glucosidase during the biodegradation of methyl cellulose by Fusarium oxysporum. *Biochemistry Research International* 2016: 1–6.

Olson, D. G., S. A. Tripathi, R. J. Giannone, et al. 2010. Deletion of the Cel48S cellulase from Clostridium thermocellum. *Proceedings of the National Academy of Sciences* 107 (41): 17727.

Ooshima, H., M. Sakata, and Y. Harano. 1983. Adsorption of cellulase from Trichoderma viride on cellulose. *Biotechnology and Bioengineering* 25: 3103–3114.

Pareek, N., T. Gillgren, and L.J. Jönsson. 2013. Adsorption of proteins involved in hydrolysis of lignocellulose on lignins and hemicelluloses. *Bioresource Technology* 148: 70–77.

Philippidis, G. P., T. K. Smith, and C. E. Wyman. 1993. Study of the enzymatic hydrolysis of cellulose for production of fuel ethanol by the simultaneous saccharification and fermentation process. *Biotechnology and Bioengineering* 41 (9): 846–853.

Ping, L., M. Wang, X. Yuan, et al. 2018. Production and characterization of a novel acidophilic and thermostable xylanase from Thermoascus aurantiacu. *International Journal of Biological Macromolecules* 109: 1270–1279.

Podgorbunskikh, E. M., A. L. Bychkov, and O. I. Lomovsky. 2019. Determination of surface accessibility of the cellulose substrate according to enzyme sorption. *Polymers* 11: 1201. https://doi.org/10.3390/polym11071201.

Qin, Z., and M. J. Buehler. 2014. Molecular mechanics of mussel adhesion proteins. *Journal of the Mechanics and Physics of Solids* 62: 19–30

Rahikainen, J. L., J. D. Evans, S. Mikander, A. Kalliola, T. Puranen, T. Tamminen, K. Marjamaa, and K. Kruus. 2013. Cellulase–lignin interactions—the role of carbohydrate-binding module and pH in non-productive binding. *Enzyme and Microbial Technology* 53: 315–321

Saqib, A. A. N., A. Farooq, M. Iqbal, J. U. Hassan, U. Hayat, and S. Baig. 2012. A thermostable crude endoglucanase produced by aspergillus fumigatus in a novel solid state fermentation process using isolated free water. *Enzyme Research* 2012, Article ID 196853, 1–6.

Shaw, A. J., K. K. Podkaminer, S. G. Desai, et al. 2008. Metabolic engineering of a thermophilic bacterium to produce ethanol at high yield. *Proceedings of the National Academy of Sciences* 105 (37): 13769–13774.

Singh, A., P. K. R. Kumar, and K. Schugerl. 1991. Adsorption and reuse of cellulases during saccharification of cellulosic materials. *Journal of Biotechnology* 18: 205–212.

Sinha, S. K., and S. Datta. 2016. β-Glucosidase from the hyperthermophilic archaeon Thermococcus sp. is a salt-tolerant enzyme that is stabilized by its reaction product glucose. *Applied Microbiology and Biotechnology* 100 (19): 8399–8409.

Sivaramakrishnan, R., and K. Muthukumar. 2012. Isolation of thermo-stable and solvent-tolerant Bacillus sp. lipase for the production of biodiesel. *Applied Biochemistry and Biotechnology* 166 (4): 1095–1111.

Steiner, W., W. Sattler, and H. Esterbauer. 1988. Adsorption of Trichoderma reesei cellulase on cellulose: experimental data and their analysis by different equations. *Biotechnology and Bioengineering* 32: 853–865.

Sukharnikov, L. O., B. J. Cantwell, M. Podar, and I. B. Zhulin. 2011. Cellulases: ambiguous nonhomologous enzymes in a genomic perspective. *Trends in Biotechnology* 29 (10): 473–479.

Takahashi, M., H. Takahashi, Y. Nakano, T. Konishi, R. Terauchi, and T. Takeda. 2010. Characterization of a cellobiohydrolase (MoCel6A) produced by Magnaporthe oryzae. *Applied and Environmental Microbiology* 76 (19): 6583–6590.

Takashima, S., A. Nakamura, M. Hidaka, H. Masaki, and T. Uozumi. 1999. Molecular cloning and expression of the novel fungal β-Glucosidase genes from Humicola grisea and Trichoderma reesei1. *The Journal of Biochemistry* 125 (4): 728–736.

Tambor, J. H., H. Ren, S. Ushinsky, et al. 2012. Recombinant expression, activity screening and functional characterization identifies three novel endo-1,4-β-glucanases that efficiently hydrolyse cellulosic substrates. *Applied Microbiology and Biotechnology* 93 (1): 203–214.

Thomas, L., H. Ram, and V. Pal Singh. 2017. Evolutionary relationships and taxa-specific conserved signature indels among cellulases of Archaea, Bacteria, and Eukarya. *Journal of Computational Biology* 24 (10): 1029–1042.

Tokunaga, Y., T. Nagata, T. Suetomi, S. Oshiro, K. Kondo, M. Katahira, and T. Watanabe. 2019. NMR analysis on molecular interaction of lignin with amino acid residues of carbohydrate-binding module from Trichoderma reesei Cel7A. *Scientific Reports* 9: 1977.

Tu, M., X. Pan, and J. N. Saddler. 2009. Adsorption of cellulase on cellulolytic enzyme lignin from Lodgepole pine. *Journal of Agricultural and Food Chemistry* 57: 7771–7778.

Voutilainen, S. P., T. Puranen, M. Siika-Aho, et al. 2008. Cloning, expression, and characterization of novel thermostable family 7 cellobiohydrolases. *Biotechnology and Bioengineering* 101 (3): 515–528.

Wakarchuk, W. W., W. L. Sung, R. L. Campbell, A. Cunningham, D. C. Watson, and M. Yaguchi. 1994. Thermostabilization of the Bacillus circulansxylanase by the introduction of disulfide bonds. *Protein Engineering, Design and Selection* 7 (11): 1379–1386.

Wang, K., H. Luo, P. Shi, H. Huang, Y. Bai, and B. Yao. 2014. A highly-active endo-1,3-1,4-β-glucanase from thermophilic Talaromyces emersonii CBS394.64 with application potential in the brewing and feed industries. *Process Biochemistry* 49 (9): 1448–1456.

Wang, X., X. He, S. Yang, X. An, W. Chang, and D. Liang. 2003. Structural basis for thermostability of β-glycosidase from the thermophilic eubacterium *Thermus nonproteolyticus* HG102. *Journal of Bacteriology* 185 (14): 4248–4255.

Westereng, B., T. Ishida, G. Vaaje-Kolstad, et al. 2011. The putative endoglucanase PcGH61D from Phanerochaete chrysosporium is a metal-dependent oxidative enzyme that cleaves cellulose. *PloS One* 6 (11): e27807.

Yang, J., L. Deng, C. Zhao, and H. Fang. 2019. Heterologous expression of Neurospora crassa cbh1 gene in Pichia pastoris resulted in production of a neutral cellobiohydrolase I. *Biotechnology Progress* 35 (3): e2795.

Yarbrough, J. M., A. Mittal, E. Mansfield, L. E. Taylor, S. E. Hobdey, D. W. Sammond, Y. J. Bomble, M. F. Crowley, S. R. Decker, M. E. Himmel, and T. B. Vinzant. 2015. New perspective on glycoside hydrolase binding to lignin from pretreated corn stover. *Biotechnology for Biofuels* 8:124. https://doi.org/10.1186/s13068-015-0397-6.

Ying, W., Z. Shi, H. Yang, G. Xu, Z. Zheng, J. Yang. 2018. Effect of alkaline lignin modification on cellulase–lignin interactions and enzymatic saccharification yield. *Biotechnology for Biofuels*. https://doi.org/10.1186/s13068-018-1217-6.

Zhao, J., C. Guo, C. Tian, and Y. Ma. 2015. Heterologous expression and characterization of a GH3 β-glucosidase from thermophilic fungi Myceliophthora thermophila in Pichia pastoris. *Applied Biochemistry and Biotechnology* 177 (2): 511–527.

Zheng, Y., S. Zhang, S. Miao, Z. Su, and P. Wang. 2013. Temperature sensitivity of cellulase adsorption on lignin and its impact on enzymatic hydrolysis of lignocellulosic biomass. *Journal of Biotechnology* 166(3): 135–145. https://doi.org/10.1016/j.jbiotec.2013.04.018.

Index

Note: **Bold** page numbers refer to tables, *italic* page numbers refer to figures.

For Product Safety Concerns and Information please contact our EU representative GPSR@taylorandfrancis.com Taylor & Francis Verlag GmbH, Kaufingerstraße 24, 80331 München, Germany